SUDAN ARCHAEOLOGICAL RESEARCH SOCIETY
PUBLICATION NUMBER 12

# ROAD ARCHAEOLOGY IN THE MIDDLE NILE

## Volume 2

## Excavations from Meroe to Atbara 1994

by

Michael Mallinson
and
Laurence Smith

with contributions by

Fathi Abdel Hamid Salih Khidr
Joyce Filer
Dorian Q Fuller
Salima Ikram
John MacGinnis
Jane Sanford Gaastra
Chris Stevens
David Thickett
Francis Thornton
Willemina Wendrich
Rebecca Whiting
Barbara Wills

**SARS**
**LONDON**
**2017**

Archaeopress Publishing Ltd
Gordon House
276 Banbury Road
Oxford OX2 7ED

www.archaeopress.com

Sudan Archaeological Research Society
Publication Number 12
General Editor: W. V. Davies
Editor of this volume: D. A. Welsby

ISBN 978 1 78491 646 6
ISBN 978 1 78491 647 3 (e-Pdf)

© The Sudan Archaeological Research Society and Archaeopress 2017

Front cover: Painted potstand T5/92C
Back cover: C17/29 Burial 159.2/T5, detail of pottery vessels in situ

All rights reserved. No part of this book may be reproduced, in any form or
by any means, electronic, mechanical, photocopying or otherwise,
without the prior written permission of the copyright owners.

Printed and bound in Great Britain by Marston Book Services Ltd, Oxfordshire

This book is obtainable from The Sudan Archaeological Research Society
c/o Department of Ancient Egypt and Sudan, The British Museum, London WC1B 3DG
or from the Society's website http://www.sudarchrs.org.uk

Also available direct from Archaeopress or from our website www.archaeopress.com

# ROAD ARCHAEOLOGY IN THE MIDDLE NILE

## Volume 2

## Excavations from Meroe to Atbara 1994

# Contents

| | |
|---|---:|
| **List of Tables** | vi |
| **List of Figures** | vii-viii |
| **List of Plates** | ix-x |
| | |
| **Preface**<br>*Michael Mallinson* | xi |
| **Site List** | xii |
| | |
| **1. Introduction and Summary of the Test Excavations**<br>*Michael Mallinson* | 1-11 |
| **2. Site Descriptions**<br>*Fathi Abdel Hamid Salih Khidr, Salima Ikram, John MacGinnis, Joyce Filer and Francis Thornton* | 13-43 |
| **3. Small Finds from the Surface Survey and from Test Excavations excluding Gabati Cemetery**<br>*Laurence Smith* | 45-58 |
| **4. Report on the Pottery from the Test Excavations: Sites 101.4, 112.3, 112.4, 118FS2, 153.8, 155.4, 165BM, 166.2 and 170.1**<br>*Laurence Smith* | 59-97 |
| **5. Gabati Basketry and Cordage**<br>*Willemina Wendrich* | 99-100 |
| **6. The Leather Samples from Test Excavations at Gabati (site 159.2)**<br>*Barbara Wills* | 101-105 |
| **7. Analysis of tanning agents from the excavated leather samples**<br>*David Thickett* | 107-108 |
| **8. Gabati grave goods from the Test Excavations and the consideration of funerary practices**<br>*Laurence Smith* | 109-120 |
| **9. Bioarchaeological Report from the Excavations from Meroe to Atbara 1994**<br>*Rebecca Whiting* | 121-140 |
| **10. The Animal Remains**<br>*Jane Sanford Gaastra* (with a contribution by *Salima Ikram*) | 141-148 |
| **11. Environmental Material from the Begrawiya-Atbara Survey 1994**<br>*Chris Stevens and Dorian Q Fuller* | 149-150 |
| **Bibliography** | 151-155 |
| | |
| **Arabic summary** | |

# List of Tables

**Small Finds**

| | | |
|---|---|---|
| 3.1 | Beads from site 112.4. | 53 |

**Pottery Report**

| | | |
|---|---|---|
| 4.1 | Groups present in the ceramics from Test Excavations and the period or periods to which they are dated. | 60 |
| 4.2 | Terms used in fabric descriptions for abundance and size of inclusions. | 61 |

**Basketry Report**

| | | |
|---|---|---|
| 5.1 | Catalogue of Gabati Basketry and Cordage. | 99 |

**Leather Report**

| | | |
|---|---|---|
| 6.1 | Leather Sample Descriptions. | 103-105 |

**Analysis of Tanning Agents**

| | | |
|---|---|---|
| 7.1 | Results of tannin tests with ammonium ferric sulphate and aliminon. | 108 |

**Bioarchaeological Report from the Excavations from Meroe to Atbara 1994**

| | | |
|---|---|---|
| 9.1. | Scores for pubic symphysis according to Todd and Suchey-Brooks as set out in Buikstra and Ubelaker (1994) and scores for auricular surface as set out in Lovejoy (1985). | 123 |
| 9.2. | Cranial non-metric traits (after Buikstra and Ubelaker 1994, 85-94). | 134 |
| 9.3. | Post-cranial non-metric traits (after Buikstra and Ubelaker 1994, 85-94). | 135 |
| 9.4. | Osteometric measurements (after Buikstra and Ubelaker 1994, 69-84). | 136 |
| 9.5. | Cranial Osteometric measurements (after Buikstra and Ubelaker 1994, 69-84). | 136-137 |
| 9.6. | Post-Cranial Osteometric measurements (after Buikstra and Ubelaker 1994. 69-84). | 137-138 |
| 9.7. | Occurrences of osteoarthritis, by skeleton. | 139-140 |

**The Animal Remains**

| | | |
|---|---|---|
| 10.1 | Osteological finds by context, given in Number of Identified Specimens (NISP) | 142 |
| 10.2 | Osteological finds by context, given in minimum number of individuals (MNI) | 142 |
| 10.3 | Non-osteological finds | 142 |
| 10.4 | Summary of the animal bone data | 144-148 |

# List of Figures

|  | Site Distribution Map. | xiii |
|---|---|---|

**Introduction. Summary of the Test Excavations**

| 1.1 | Site 118.FS.4, rock inscriptions. | 2 |
|---|---|---|
| 1.2 | Site 118. FS.4.1-3, hut circles. | 3 |
| 1.3 | Site 155.4, House S1. | 4 |
| 1.4 | Site 159.2, Tumulus 1, Unit 124. | 5 |
| 1.5 | Site 159.2, Tumulus 2, Unit 118. | 5 |
| 1.6 | Site 159.2, Tumulus 4, Unit 136. | 6 |
| 1.7 | Site 159.2, Tumulus 5, Unit 150. | 7 |
| 1.8 | Site 159.2, Tumulus 6, Unit 151 before final clearance. | 8 |
| 1.9 | Site 159.2, Tumulus 11, lower excavated levels. | 9 |
| 1.10 | Site 159.2, Gabati, site plan. | 11 |

**Site Descriptions**

| 2.1 | Site 101.4, Tumulus 1. | 13 |
|---|---|---|
| 2.2 | Site 101.4, Tumulus 1.3. | 14 |
| 2.3 | Site 101.4, Tumulus 2. | 15 |
| 2.4 | Site 101.4, Tumulus 2, north-south section, west face. | 15 |
| 2.5 | Site 101.4, Tumulus 3, section. | 15 |
| 2.6 | Site 102.1, Tumulus 1, skeleton in relation to the tumulus. | 15 |
| 2.7 | Site 112.4, Tumulus 1. | 16 |
| 2.8 | Site 112.4, Tumulus 2. | 16 |
| 2.9 | Site 112.4, Tumulus 3. | 17 |
| 2.10 | Site 112.4, Tumulus 4, plan. | 17 |
| 2.11 | Site 112.4, Tumulus 5, section and plan. | 18 |
| 2.12 | Site 112.4, Tumulus 6. | 18 |
| 2.13 | Site 112.3/X1, plan. | 19 |
| 2.14 | Site 118.FS.2.1 S3, plan of stone structure in House 3. | 19 |
| 2.15 | Site 154.5, Tumulus 1, section. | 20 |
| 2.16 | Site 155.4/S1. | 20 |
| 2.17 | Site 155.4/S1/FP1, plan of fire pit south of House 155.4. | 21 |
| 2.18 | Site 170.1, general plan. | 21 |
| 2.19 | Site 170.1, Tumulus 1 and Tumulus 2. | 22 |
| 2.20 | Site 159.2, Tumulus 1, plan. | 24 |
| 2.21 | Site 159.2, Tumulus 3. | 24 |
| 2.22 | Site 159.2, Tumulus 1. | 26 |
| 2.23 | Site 159.2, Tumulus 1. | 27 |
| 2.24 | Site 159.2, Tumulus 1. | 27 |
| 2.25 | Site 159.2, Tumulus 2, east-west section. | 28 |
| 2.26 | Site 159.2, Tumulus 2, blocking stones. | 28 |
| 2.27 | Site 159.2, Tumulus 3. | 28 |
| 2.28 | Site 159.2, Tumulus 4. | 29 |
| 2.29 | Site 159.2, Tumulus 5, burial chamber blockings, levels 2, 3 and 4. | 29 |
| 2.30 | Site 159.2, Tumulus 5, section. | 31 |
| 2.31 | Site 159.2, Tumulus 5, plan. | 32 |
| 2.32 | Site 159.2, Tumulus 5. | 33 |
| 2.33 | Site 159.2, Tumulus 5, section along east-west long axis. | 33 |
| 2.34 | Site 159.2, Tumulus 5, plan. | 34 |
| 2.35 | Site 159.2, Tumulus 6, plan. | 34 |
| 2.36 | Site 159.2, Tumulus 6. | 34 |
| 2.37 | Site 159.2, Tumulus 7, north-south section. | 36 |
| 2.38 | Site 159.2, T11, plan. | 37 |
| 2.39 | Site 159.2, T11, reconstruction of the pyramid and offering chapel. | 38 |

| | | |
|---|---|---|
| 2.40 | Site 159.2, T11, north-south section of pyramid structure. | 38 |
| 2.41 | Site 159.2, T11, section of pit. | 39 |
| 2.42 | Site 159.2, T11, section through the descendary and burial chamber. | 39 |
| 2.43 | Site 159.2, T11, complete and reconstructed jars recovered from test excavations. | 40 |
| 2.44 | Site 159.2. Pot marks and Meroitic graffito from T11. | 41 |
| 2.45 | Site 159.2, T11, complete and reconstructed jars recovered from test excavations. | 42 |
| 2.47 | Site 159.2, Tumulus 13, plan of the burial pit. | 42 |
| 2.48 | Site map of the Gabati area (sheet BM 159-161,93). | 43 |

**Small Finds (excluding Gabati)**

| | | |
|---|---|---|
| 3.1 | Examples of Lithic Artefacts. | 47 |
| 3.2 | Grindstones, Type 1-5. | 50 |
| 3.3 | Grindstones, Type 7-11. | 52 |
| 3.4 | Grindstones, Type 12-20 and other small finds. | 54 |

**Pottery**

| | | |
|---|---|---|
| 4.1 | Forms 1:1.1 - 2:3.3. | 70 |
| 4.2 | Forms 2:5.1 - 10:2.1. | 71 |
| 4.3 | Decoration D1.1 - D13.7. | 73 |
| 4.4 | Decoration D13.8 - B11. | 74 |
| 4.5 | Percentages of featured sherds calculated over all sites considered in the present report. | 94 |
| 4.6 | Percentages of the sherds calculated for Site 101.4. | 94 |
| 4.7 | Percentages of the sherds calculated for Site 112.3. | 95 |
| 4.8 | Percentages of the sherds calculated for Site 112.4. | 95 |
| 4.9 | Percentages of the sherds calculated for Site 118FS.2. | 96 |
| 4.10 | Percentages of the sherds calculated for Site 153.8. | 96 |
| 4.11 | Percentages of the sherds calculated for Site 155.4. | 97 |

**Basketry**

| | | |
|---|---|---|
| 5.1 | All string found at Gabati was made of three z-spun yarns which were S-plied (zS3 string). | 100 |
| 5.2 | Schematic drawing of an *angarib* bed with string at the head for maintaining the tension of the webbing. | 100 |
| 5.3 | Fragment of coiled basket. | 100 |
| 5.4 | Fragment of coiled basket. | 100 |
| 5.5 | Fragment of twill plaited mat. | 100 |

**Leather Samples**

| | | |
|---|---|---|
| 6.1 | Examples of Leather Samples noted in Table 6.1. | 102 |

**Gabati Grave Goods**

| | | |
|---|---|---|
| 8.1 | Items from the tomb chamber of Tumulus 1. | 110 |
| 8.2 | Items from Tumulus 2. | 111 |
| 8.3 | Items from the tomb chamber of Tumulus 4. | 112 |
| 8.4 | Items from Tumulus 5 and earrings from Tumulus 13. | 113 |
| 8.5 | *Kohl* pot and mortar from Tumulus 5; carved wooden piece of bier from Tumulus 6. | 116 |

**Bioarchaeological Report from the Excavations from Meroe to Atbara 1994**

| | | |
|---|---|---|
| 9.1 | Approximate percentage of each individual skeleton present upon analysis, in 5% increments. | 122 |
| 9.2 | Bar chart showing the distribution of individuals in each age category across the collection. | 123 |
| 9.3 | Distribution of individuals by sex category. Unknown refers to skeletons for which sex could not be determined due to poor preservation. | 123 |

# List of Plates

**Site Descriptions**

2.1 Tumulus 101.4 T1 with T2 and T3.
2.2 Tomb 307 from Garstang's excavations at Meroe.
2.3 Burial 101.4/T1.3.
2.4 Burial 101.4/T1.4.
2.5 Burial 101.4/T.2.1.
2.6 Burial 101.4/ T.3.
2.7 Burial 102.1/T1.1.
2.8 Tumulus 112.4/T1.
2.9 Burial 112.4/T3.1.
2.10 Site 118/FS Rock engraving.
2.11 Site 118/FS.3 Rock engravings.
2.12 Site 118/FS.2 Rock inscription.
2.13 Site 118/FS.1.
2.14 Tumulus 153.4/T1 excavation.
2.15 Burial 154.5/T1.1.
2.16 Burial 154.5/T1.2.
2.17 Burial 170.1/T5, showing fragmentary state of skeletal material.
2.18 C4/17 Gabati site before excavation.
2.19 Burial 102.1/T1.1.
2.20 C18B/23 House 155.4 /S1 as cleared showing Phase I from the east.
2.21 C18B/22 House 155.4/S1 as cleared – Phase I.
2.22 C8/20 House 155.4 – Phase II.
2.23 JF8/20 118/FS.4 Rock engravings.
2.24 C17/7 Burial 159.2/T4 showing detail of skeleton.
2.25 C9/6 159.2/T1 general view of burial.
2.26 C10/11 Burial 159.2/T4.
2.27 C10/4 159.2/T4 Second layer of stones.
2.28 C12/14 Burial 159.2/T7 Christian burial.
2.29 C17/19 Burial 159.2/T5.
2.30 C12/35 Burial 159.2/T5. Detail of ivory *kohl* pot as found.
2.31 Post-Meroitic potstand from Garstang's Meroe excavation (Garstang *et al.* 1911, pl. XLI, no. 9, fig. 8.4).
2.32 C17/28 Burial 159.2/T5, Detail of skeleton.
2.33 C17/29 Burial 159.2/T5. Detail of pottery vessels *in situ*.
2.34 C10/25 T5 Shaft Final Blocking with T1 in the background.
2.35 C19/6 159.2/T6 Tomb chamber blocking. Compare with Plate 2.34.
2.36 C18/31 'Tumulus' 159.2/T11 and T13. Jars at bottom of grave shaft.
2.37 'Tumulus' 159.2/T11, general view of the superstructure.
2.38 C10/2 Bed leg detail T1.
2.39 C12/12 159.2/T11/102C Detail of incised 'owner's mark' and Meroitic inscription.
2.40 Burial 170.1/T2.1/3or.
2.41 Burial 170.1/T1.
2.42 Aerial photograph of the Gabati/Aliab area.

**Pottery**

4.1 Globular jars from T11.
4.2 Globular jars from T11.
4.3 Large bowl (Type 2:5.1).
4.4 Large bowl (Type 2:5.1).

**Gabati Grave Goods**

8.1 Broad open bowl T1/41C.
8.2 Bowl with upright sides T1/42C.
8.3 Leaf-shaped arrowhead T1/18.

8.4 Knife blade or spear point T1/24.
8.5 Imported oil bottle T2/31C.
8.6 Examples of beads T2/33.
8.7 Worked cowrie shell.
8.8 Painted imported oil bottle T4/49C.
8.9 Spatula or *kohl* spoon T4/12.
8.10 Examples of beads from Tumulus 4.
8.11 Examples of beads from Tumulus 4.
8.12 Examples of beads from Tumulus 4.
8.13 Painted potstand T5/92C.
8.14 Painted bowl T5/96C.
8.15 Ovoid bowl with lug T5/95C.
8.16 Spouted jar T5/93C.
8.17 Table amphora T5/94C.
8.18 Circular mirror T5/6.
8.19 Spatula and nail T5/3.
8.20 Stone bowl or mortar T5/97S.
8.21 Ivory *kohl* pot T5/5.
8.22 Wooden comb T5/21.
8.23 Examples of beads from Tumulus 5.
8.24 Part of necklace T5/9.
8.25 Examples of beads from Tumulus 5.
8.26 Necklace of beads T5/2.
8.27 Beads from Christian burial, T13.

### Bioarchaeological Report from the Excavations from Meroe to Atbara 1994

9.1 Begrawiya skeleton 101.4/T1.2/38*or* during excavation.
9.2 Skeleton 101.4/T1.2/38*or*, both clavicles showing false joints.
9.3 Skeleton 101.4/T1.2/38*or*, showing fusion of $5^{th}$ and $6^{th}$ cervical vertebrae.
9.4 Skeleton 101.4/T1.2/38*or*, showing Schmorl's nodes and osteophytic growths.
9.5 Skeleton 101.4/T1.6/51*or*, during excavation.

### Environmental Material from the Begrawiya-Atbara Survey 1994

11.1. Desiccated stone of *Balanites aegyptiaca* from 112.4/T.1.1/1.
11.2. Date stone (*Phoenix dactylifera*) (dorsal and ventral views) from 159.2/T4.1/89Or.
11.3. Possible bird of prey/carrion pellet from 159.2/T4.1/89Or.
11.4. Possible bird of prey/carrion pellet from 159.2/T4.1/89Or. Reverse and front view.
11.5. Fragment of possible bird of prey/carrion pellet from 159.2/T4.1/89Or.
11.6. Close-up of hairs within a possible bird of prey/carrion pellet from 159.2/T4.1/89Or.
11.7. Two shells of *Limicolaria* sp. from 159.2/T1/27 Or, Unit 128.
11.8. Two fragments of freshwater mussel and one of cf. *Limicolaria* sp. (left side). All from 153.8/S1/28Or.
11.9. Fragment of river mussel shell (Unionidae/Iridinidae) from 112.3/9Or.
11.10. Fragment of freshwater oyster shell, cf. *Etheria elliptica* from 155.4/51/5Or (dorsal and ventral views).
11.11. Cowrie shell *Monetaria moneta/annulus* from 101.4/T2/23 Or, Unit 72.

# Preface

*Michael Mallinson*

The first season of survey work in 1993 was undertaken in advance of the construction of the North Challenge Road initially between Geili and Atbara. This work was carried out in the SARS concession area from BM98, opposite the Pyramids of Meroe, to Atbara. A total of 170 sites was recorded and this was published in the first volume of Road Archaeology in the Sudan (Mallinson *et al*. 96). In addition, a report was prepared advising the Sudan National Committee for Roads and Bridges of areas which were likely to be damaged by the road construction. The following year it was indicated that due to the advanced development of the road design no rerouting would be possible.

In response to this a rescue season was proposed to excavate the sites clearly at risk in the remaining few months before construction and grading began. A limited amount of funds was provided by the Haycock Fund and within this resource a project was assembled with SARS directed by Laurence Smith and Michael Mallinson. As a total of eight sites with 30 archaeological structures was clearly directly on the road line a methodology was needed that would permit these to be properly excavated and recorded in the available time of three weeks that the funds would accommodate.

As a number of burials was expected and a minimum of two teams would be required, we concentrated on finding suitable specialists in this area. Francis Thornton of the University of Bradford and Joyce Filer of the British Museum, who had worked previously with the British Institute in Eastern Africa at Soba East, agreed to join the project. Dr Salima Ikram returned as a site supervisor and to follow up on her animal bone studies in the first season. Dr John MacGinnis also joined as a site supervisor. The project was conceived as a team effort and the publication involved all team members. Fathi Abdel Hamid Salih Khidr, who was our NCAM inspector for the excavations, carried out site supervision during the season and has been a great assistance since with the publication.

The two teams worked 3-7km apart during the three weeks moving site every three to four days as the work was completed. The excavation was assisted by a small local work force. As the sites were due to be destroyed they were excavated as completely as possible, and only a minimal back filling was done to make the sites safe for passing camels. Full area excavation, as was carried out in the subsequent seasons at Gabati, was not attempted due to the shortness of time and number of sites needed to be cleared. Individual features were gridded out and excavated in quadrants. This permitted whatever local stratigraphy existed to be identified; it obviously did not permit relation of stratigraphy from neighbouring sites. In most cases this was not appropriate as the surrounding areas, even between closely situated sites, were mostly virgin desert.

The stratigraphic evidence was most useful in the cemetery sites to identify from which layer of gravel the burial pit was dug from, and afterwards at which period the burials were first plundered. In the Meroitic and Post-Meroitic sites this burial layer was a white gravel that left a characteristic surface covering after the burial was completed. In the later Christian/Medieval period pits were cut to a shallower yellow gravel level and the bottom then filled with a yellow/orange sand before the inhumation was carried out. The earliest remains which were excavated, and which were associated with some second millennium BC pottery found nearby in the first season, were very shallow burials of several children and one adult/adolescent. No carbon dating was possible but the absence of the characteristic Post-Meroitic/Christian mounds, the layer of yellow sand, and the very close proximity to the surface which suggested a long period of surface erosion indicates that these burials, at BM170, could be associated with the early period of settlement in this part of the Nile Valley.

The work of the two teams produced a large amount of material which is briefly discussed below and in detail in the following chapters. The interest aroused by the discovery of the multi-period site at Gabati permitted funding for a special season in the following year. This was developed from our experience in the first two seasons and is a tribute to the effort put in by the first season's teams. The material recovered from that special season was published first (Edwards 1998) as many of the specialists went on to work on the Gabati project and were thus delayed in completing the earlier season's research. The results from the first excavation season have confirmed the estimated dating of many of the sites recorded in the surface survey season.

The human skeletal remains from all the work at Gabati has recently been published by Margaret Judd (Judd 2012).

# SITE LIST

## 1. MEROE DIG HOUSE 24.3.1994 - 31.3.1994

SITE 101.1 3 Tumuli, 1 Mixed Period, 2 Christian. Team A

SITE 102.1 1 Tumuli Christian. Team A

SITE 112.4 5 Tumuli, 2 Empty, 3 Post-Meroitic/Christian. Team B

SITE 112.5 1 Structure Late Christian. Team B

SITE 118 FS2.1 3 Christian Structures, 1 Empty Tumulus, Inscription. Team B

**Team A:**
Francis Thornton - Bone Analysis 101.1/T1.1, 101.1/T2.1,102.1/T1.1,
Joyce Filer - Bone Analysis 101.1/T1.2 - 6,101.1/T2.1, 101.1/T3.1
Salima Ikram - Site Supervisor 101.1/T1 - T3,
Fathi Abdel Hamid Salih Khidr - Site Supervisor 102.1/T1

| | |
|---|---|
| Michael Mallinson | - Survey, Photography |
| Fedullah | - Excavator |
| Mahmoud | - Excavator |
| El Amin Ali | - Excavator |
| Abdel Minam Mohamed | - Excavator |
| Abdullah Suliman | - Excavator |
| Alker Akhmed | - Excavator |
| Mohamed Abdullah | - Excavator |

**Team B:**
Joyce Filer - Bone Analysis 112.4/T3.1,112.4/T4.1,112.4/T5.1
John MacGinnis - Site Supervisor 112.4/T1,112.4/T3 -5, 118 FS2.1/S1- 3
Fathi Abdel Hamid Salih Khidr - Site Supervisor 112.4/T2, 112.5/S1, 118, FS2.1/T1
Michael Mallinson - Survey Drawings, Photography

| | |
|---|---|
| Akhmed Mohamed Ali | - Excavator |
| Bereid Mohamed Ali | - Excavator |
| Abu Sid | - Excavator |
| Majoub Ali | - Excavator |
| Anwar Mohamed Ali | - Excavator |
| Fatah er-Rakman Akhmed | - Excavator |

House:
| | |
|---|---|
| Laurence Smith | - Small Finds/Pottery |
| Akhmed Mostafa | - Cook |
| Hamid Abdullah | - House Assistant |

## 2. GABATI DIG HOUSE 1.4.1994 - 18.4.1994

SITE 153.8 1 Structure Late Christian. Team B

SITE 154.5 2 Tumuli, 1 Empty, 1 Christian. Team B

SITE 155.4 1 Structure Late Christian. Team B

SITE 159.2 9 Tumuli, 5 Meroitic, 4 Christian, 1 Tomb, Meroitic. Team A

SITE 170.1 5 Tumuli Neolithic or Post-Meroitic/Christian. Team B

**Team A:**
Joyce Filer - Bone Analysis, 159.2/T1.1,159.2/T2.1,159.2/T3.1, 159.2/T4.1 159.2/T5.1,159.2/T6.1,159.2/T7.1,159.2/T11.1,159.2/T13.1.
Salima Ikram - Site Supervisor 159.2/T1,159.2/T3,159.2/T5,159.2/T7
Fathi Abdel Hamid Salih Khidr - Site Supervisor 159.2/T2, 159.2/T4, 159.2/T6,159.2/T13
John MacGinnis - Site Supervisor 159.2/T11

| | |
|---|---|
| Michael Mallinson | - Survey, Photography |
| Majoub Ali | - Excavator |
| Khullah Fullah | - Excavator |
| Osama | - Excavator |
| Zein Bellah | - Excavator |
| Akhmed el-Mahi | - Excavator |
| Bellal Taib | - Excavator |
| Bohadein Sheif | - Excavator |
| Abdel Rahman | - Excavator |
| Soliman Adaroub | - Excavator |
| Shamsadeen Mirgani | - Excavator |
| Deffullah Dreis | - Excavator and *ghaffir* |
| Mohamed Ali Hamid | - Excavator and night *ghaffir* |

**Team B:**
Francis Thornton - Bone Analysis, 154.5/T1.1, 154.5/T2.1, 170.1/T1.1 - 5.1 - Site Supervisor 154.5/T1.1, 154.5/T2.1, 153.8/S1.
John MacGinnis - Site155.4/S1, 170.1/T1.1 - 5.1
Michael Mallinson - Survey Drawing, Photography

| | |
|---|---|
| Fedullah | - Excavator |
| Osam Said Akhmed | - Excavator |
| Abdel Majid | - Excavator |
| Taib Ali | - Excavator |
| Abdullah Taif Samii | - Excavator |
| Emad Said Akhmed | - Excavator |
| Bellal Abdel Farid | - Excavator |
| Mustafa Ali | - Excavator |
| Alloa esh-Sharif Hussein | - Excavator |
| Sheikh Khalid | - Night *ghaffir* |

# SITE DISTRIBUTION MAP (scale 1:500,000).

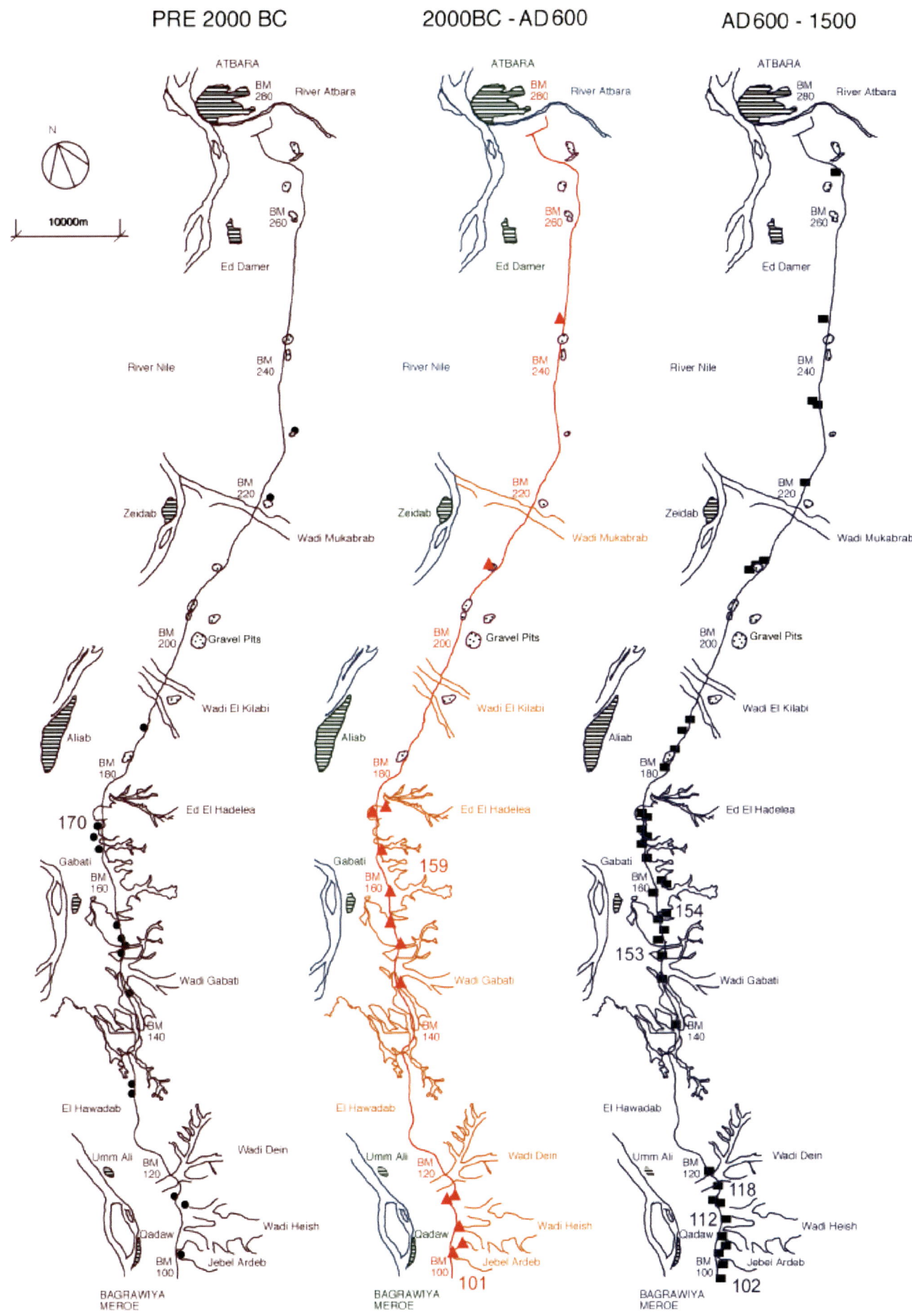

# 1. Introduction
# Summary of the Test Excavations

*Michael Mallinson*

## Meroe

The expedition began work based at the Sudanese Antiquities Service Rest House next to the Pyramids at Meroe. From here three sites were excavated at the south end of the SARS concession area (BM99-274). The first consisted of four tumuli directly on the proposed road line at Jebel Ardeb (BM101-2). The second site was at Jebel Abayud (BM112) and was a series of small tumuli and a hut circle located at the bottom of a *wadi* which the road had to pass through. At the third a camp site was excavated, of possible military origin, on the other side of Wadi Dein (BM118).

At Jebel Ardeb (BM101-2) the tumuli proved to have an origin in the late Meroitic period with the largest, 101.4/T1, (Plates 2.1, 2.3 and 2.4) having a shallow grave surrounded by large stones. The burial was accompanied by pottery, a few beads and a metal band and was laid on a mat. In this tumulus the primary burial was followed by later, probably Post-Meroitic, burials, two of which were in a semi-flexed position aligned roughly north-south with their heads to the south, seemingly disturbed but apparently unaccompanied by grave goods. The last burial was of a small child placed just to the north of the first burial. All four of these burials were on the south of the mound. In the centre of the tumulus was found a Christian burial with a child buried at the feet. The body was extended east-west with the head to the east without grave goods. The grave may have been intrusive; it appears to have been dug through the tumulus at a later period. The alternative sequence is that the grave had been dug beside the earlier burials and then the original tumulus was either moved over the new burial, or constructed over both.

The other two burials at this site to the south were both Christian, one a young adult male, 101.4/T2.1 (Plate 2.5) and one a female, 101.4/T3.1 (Plate 2.6), sealed under a smaller mound. The single burial excavated at 102.1/T1 (Plates 2.7 and 2.19) also proved to be a female of the Christian period under a small mound. The Christian burials were characterised by the remarkable preservation of the bones. The mounds were built of ferricrete sandstone from the *jebels*, mounded over the fill excavated from the grave cuts.

The burials at Jebel Abayud (112.4/T1-T5) (Plates 2.8 and 2.9) were very shallow, almost on the surface, and covered by a mound of ferricrete sandstone, and were either robbed or eroded. The bodies were aligned east-west, although insufficient remained to establish where the head was located or whether the bodies were contracted or extended. The presence of beads in two of the burials suggests a possible post-Meroitic rather than Christian date, although such grave goods can occur in Christian burials (see Gabati, below). The excavated hut circle (112.5/S1) contained Group 2 (Christian/Early Islamic) pottery, on the surface, and a hearth. Some beautiful decorated Classic Christian Period sherds were present on the site surface.

The camp at Wadi Dein (BM118 FS2/S1-S3) was surveyed in 1993 and in 1994 three hut circles were excavated (Figure 1.2a-c, Plate 2.13). The huts had compacted floors, with Group 3 and 13 (post-Meroitic-Christian) pottery, and a well preserved hearth and depressions probably made to support pottery vessels. No post-holes were apparent among the stones at the perimeter of the hut circle, which suggests that the stones may have been used to weigh down the edge of a tented structure. On the nearby *jebel* were found carved pilgrims' feet similar to those found at pilgrimage sites in Egypt, and throughout Sudan (Figure 1.1, Plates 2.10-2.12). The carvings were distinctive due to the details of hob-nails, and thong patterns, and they included an Islamic inscription of two girls' names which were also recorded. One tumulus at the site was tested but the grave was empty. Excavation of a tumulus here in the previous season had revealed a single Christian burial (see pg. 131).

## Gabati

On April 1st the expedition moved to a new dig house at the village of Gabati 30km to the north. The house was hired from the local sheikh, Hashim. Gabati is a small village without services but it was located close to a number of sites which our previous season had shown to be seriously threatened by the road works.

From the new base the area marked as Gabati on the 1:250,000 series maps (from 17° 10' - 17° 14' North, and 33° 44' - 33° 45' East) was worked on between BM 153-170. A total of five sites was studied. The first three to the south of the area were located in the Khor Shangarite. Here we excavated a Late Christian hut, two tumuli, and a fortified Christian house. The Khor Shangarite, when surveyed the previous year proved an extremely rich area, full of structures, several hundred tumuli, and with a diversity of pottery and other surface finds. The hut circle 153.8/S1, which was at the head of the valley, contained a large quantity of Group 1 pottery, a hearth and a wind break, the position of which indicated that the door was on the eastern side. Of the two tumuli excavated one (154.5/T2.1) was empty apart from three fragmentary bones, the other (154.5/T1.1), had an east-west aligned burial of a male buried face down with head to west in an extended position, without grave goods (Plate 2.15), and a second skeleton (154.5/T1.2) of which only part remained, lying east-west on its back, facing north (Plate 2.16). No other tumuli were directly on the road line, so the date of the large cemeteries either side remains uncertain, although they are probably of the Christian period.

The final excavation in the area was of a substantial

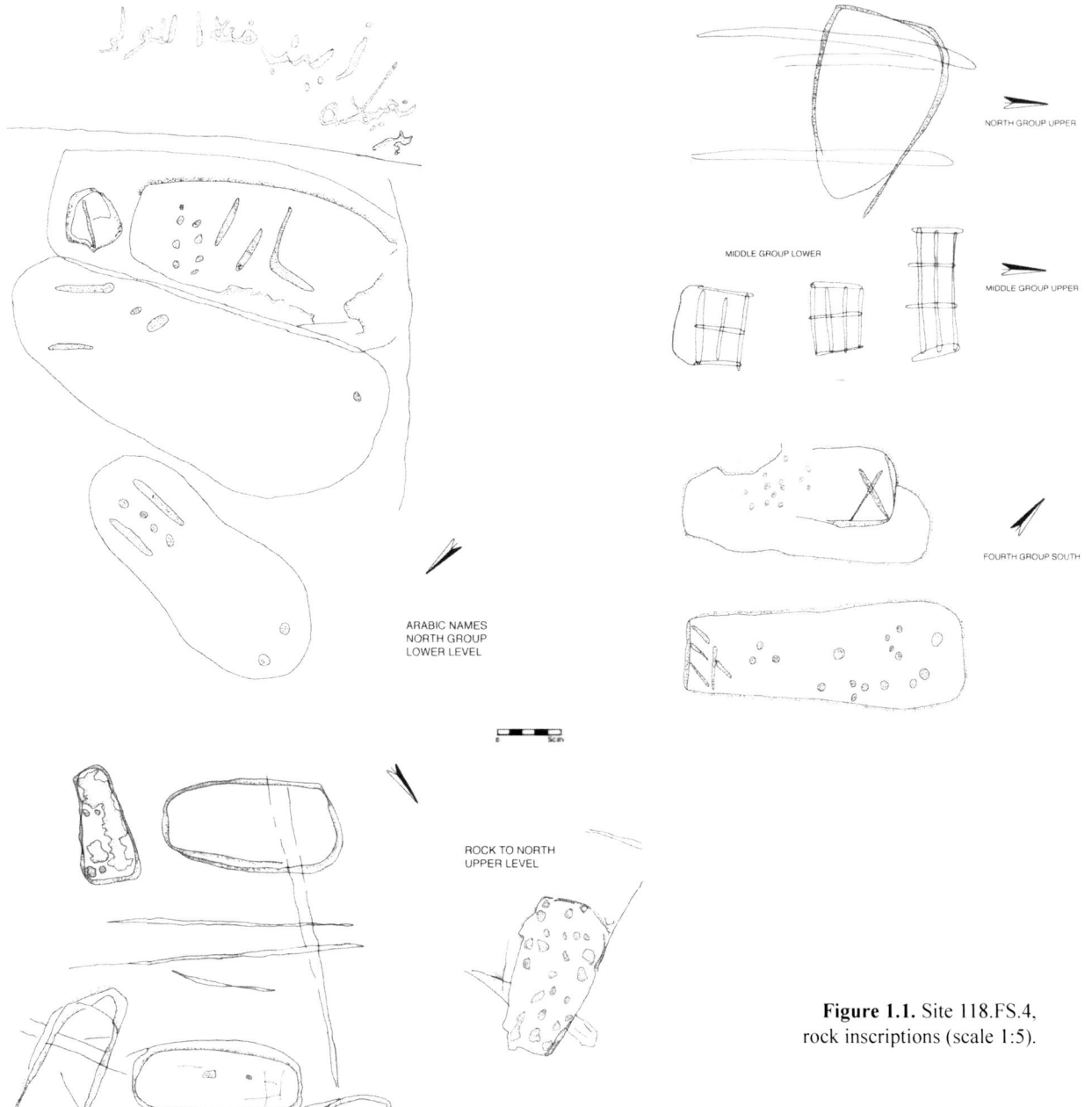

**Figure 1.1.** Site 118.FS.4, rock inscriptions (scale 1:5).

house (155.4/S1). This was cleared and planned, and two phases of construction established (Figure 1.3, Plates 2.20-2.22). The first phase was a substantial wall built in dry-stone construction similar to that used at Meroe, but which also continued in use well into the Medieval period. This enclosed an area 10 x 15m with a doorway in the south east. The north wall was convex. In the interior of the house, immediately adjacent to the north and south walls, were post-holes and pits, the functions of which are unclear. The later stage involved the construction of light-weight stone walls to form two rooms on the east side, one of which contained a hearth and the other, two substantial pits, presumably for the storage of grain. The post-holes indicate that the structure had a sheltered courtyard, as the central area was much worn, and lacked pits. Outside the house to the south was a large hearth surrounded by stones.

The remote position of the house was due probably either to its function to control a trade route through the desert, as the Khor Shangarite provides a short cut between the Wadi Gabati and the plain to the north, or possibly, judging by the other structures surrounding it, to it being within a small village related to a wet season utilisation of the *wadi*. A final possibility is some activity related to the surrounding cemeteries and Jebel Abu Sheifa. The site is close to the base of the *jebel*, which is covered in tumuli.

The penultimate site studied was a group of five tumuli 170.1/T1-T5, (Plates 2.17, 2.40 and 2.41) which proved to be associated with relatively young burials. The bodies were all fully contracted, heads to east or north; the sex was indeterminate. The graves were shallow, and the mounds insubstantial heaps of the locally available ferricrete sandstone. Neolithic pottery has been uncovered in the area, and one burial contained a single sherd. It is possible, but will need substantiation with $C^{14}$ dating, that these are, therefore, extremely ancient burials. The condition of the bones was very poor, and certainly a Meroitic or post-Meroitic date is probable. Two other sites in the

**Figure 1.2a.** Site 118. FS.4.1-3, hut circle 1 (scale 1:200).

area were checked, at BM 164 and BM 175, but it was felt that, although close to the road, they might avoid damage due to their proximity to houses. They were both marked by signs. At BM 164 further possible Neolithic pottery was found but no associated graves or structures were apparent in the area, which evidently floods regularly. The most northern site studied from Gabati was adjacent to BM 168, next to the village of el-Adalea.

The final site studied was at BM 159, east of the village of en-Natalia. The site, 159.2 (Plates 2.18 and 2.42), is on a low mound in the middle of the plain between the tumuli-covered foothills and the cultivation. Adjacent on the *jebel* is a quarry, possibly modern, or a modern working of an ancient mine. The village of Gabati lies between the cemetery and the river. Local tradition has it

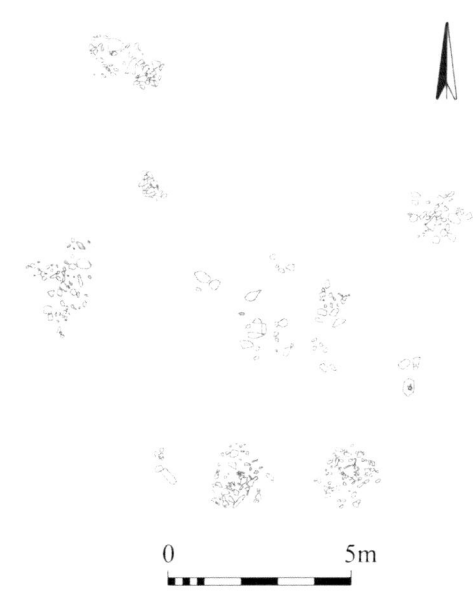

**Figure 1.2b.** Site 118. FS.4.1-3, hut circle 2 (scale 1:200).

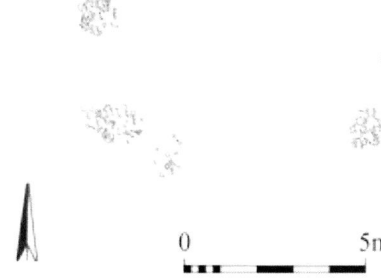

**Figure 1.2c.** Site 118. FS.4.1-3, hut circle 3 (scale 1:200).

that the village was founded anciently by Queen Koptoi (or Kaptin).

The new road line was expected to pass directly through the middle of the site and would destroy at least 50% of the 80 visible mounds, and render inaccessible the remainder, as banking at this point would cover the remaining area. After excavation of nine of the mounds it was possible to determine the approximate extent and nature of the cemetery (See the layout of the site in Figure 1.10). About 22 mounds are of probable Christian date, and the other 58 of Late Meroitic and post-Meroitic date. A clear area in the centre of the cemetery possibly contains a number of early Meroitic date structures, one of which was excavated in the 1994 season. The importance and uniqueness of the site will be discussed below.

The nine tumuli excavated were chosen due to their proximity to the eastern edge of the cemetery, which made them the most vulnerable ones even if the road had been moved eastwards as discussed with the road engineers the previous year. In fact after the season's work it was discovered that it had been decided already, following a meeting in January between the National Corporation for Antiquities and Museums of Sudan and the Director General of Road Building and Public Works, that the cost of moving the road was too great, and that the road line was to remain unaltered. This increased the urgency of excavating this cemetery particularly due to its likely contents as indicated by the test excavations, described below, and the starting of the road works later that year.

## *Site 159.2 - Gabati*

**Tumulus 1**
A large gravel tumulus with a post-Meroitic tomb structure beneath. The tomb consists of a rectangular shaft, its long axis aligned at right angles to the river, here approximately east-west. It was cut down 2m through orange gravel to the white gravel underlayer. At the west end a chamber had been hollowed out of the white gravel (Figure 1.4, Plate 2.25). It is similar to the form of those tombs uncovered by John Garstang in 1909-10, and published in *Meroe The City of the Ethiopians* (Garstang *et al*. 1911) (see Plate 2.2).

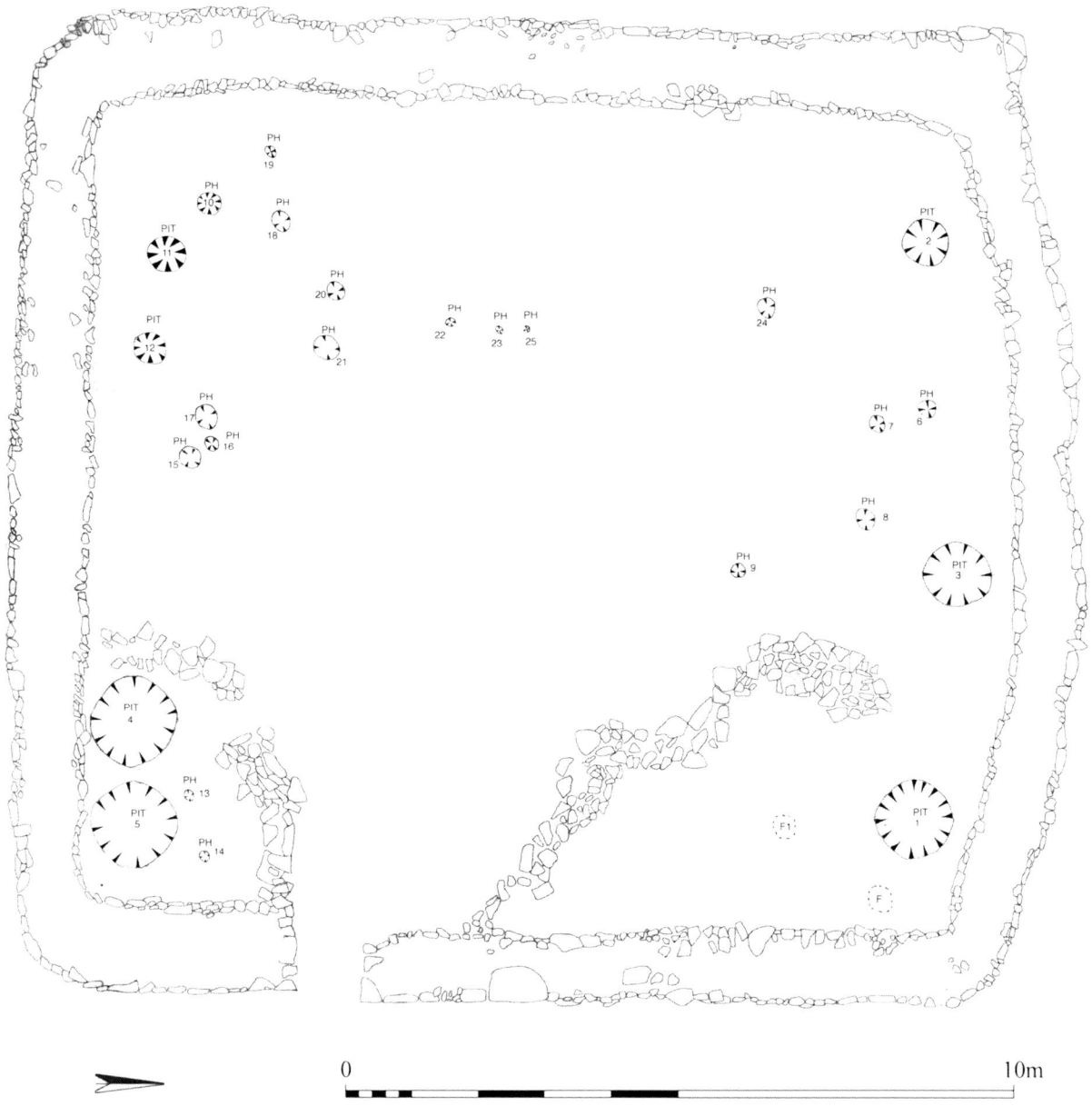

**Figure 1.3.** Site 155.4, House S1 (scale 1:100).

The deceased was a man laid head to south in a semi-flexed position facing east. The body had been placed originally on a bed (*angareeb*), but this had subsequently collapsed, and remains of the bed surrounded the burial. Accompanying the burial at the head was a bowl, and a wide dish, similar to that of Type H as defined by Garstang (1911, 37-49), with incised zig-zag paint similar to Garstang's Type ZW but decorated differently. The dish lay in the centre of the grave beside the bed.

Both these types at Meroe are from the tombs numbered 302-307, and like these tombs the burials at Gabati were on beds with carved legs around 300mm high (Plates 2.24 and 2.38), although the beds are lighter. At Gabati there is a notable absence of inscriptions in contrast with those at Meroe. The only other burial of this size at Gabati, T5, also had pottery similar to these Meroe tombs. Garstang dates these tombs to be the Classic Meroitic Period (150 BC - AD 150), although they are now considered to be Late Meroitic to Post-Meroitic (AD 150-350). It is pos-

sible that these larger tombs at Gabati could be of this date; the absence of any inscription may have been due to the distance from a temple or city where a literate scribe might have been available or may be due to the apparent lack, or drastic decline, of literacy at this period.

*Summary of Objects Registered: 159.2/T1*
Unit 124 (See Figure 8.1, Plates 8.1-8.4)

159.2/T1.1/41C - Open black bowl
159.2/T1.1/42C - Small bowl with upright sides, brown and red.

Item No.
1. Rust coloured textile fabric
2. Rope fragments (a-c)
3. Wood sample
4. Leather work (a-b)
5. Rust coloured textile fabric
6. Bed frame south
7. Bed frame west
8. Bed leg
9. Bed leg

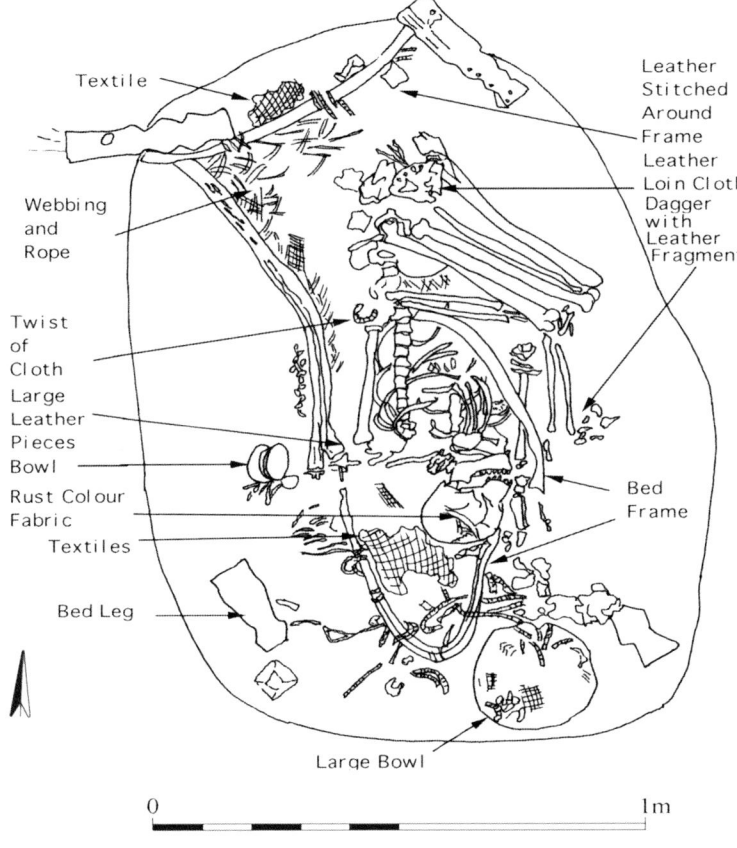

Figure 1.4. Site 159.2, Tumulus 1, unit 124 (scale 1:15).

on a bed which had short legs 200mm high. The bed had collapsed, or was disassembled, and grave goods buried with her lay at head and foot. They consisted of a red wheel-made bottle, considered to be an oil bottle, with a narrow flared top, at her head, and a small black burnished bowl with obliquely-incised vertically applied cordons on two sides and incised cross-shaped motifs, at her feet. The bottle which is similar to Form G32 dated to the later X-Group by W. Y. Adams (1986, 538-542) is a type rarely seen south of Lower Nubia. The burnished bowl, Type ZT in Garstang's classification, is of a common shape, although the decoration is unusual. The bottle may be for burial libations, or food for the dead, as the other bowl contained black seeds, possibly grain. The burial was remarkable in the variety of the textiles and the different types of bead decoration worn.

*Summary of Objects Registered 159.2/T2 Unit 118 (See Figure 8.2, Plates 8.5-8.7)*

159.2/T2.1/31C Red oil bottle
159.2/T2.1/32C Black bowl
Item No.
 1. Cloth
 2. Textile, leather fragments and beads
 3. Leather and textile
 4. Textile and tanned skin

10. Bed leg
11. Striped fabric
12. Leather work
13. Fabric and leather
14. Leather (a-c)
15. Stitched leather
16. Stitched leather
17. Cloth
18. Arrowhead, iron
19. Stitched leather (a-b)
20. Rope
21. Plaited leather.
22. Leather/Textile sample
23. Leather thongs
24. Spear tip or dagger blade, iron
25. Leather work.
26. Plaited rope and leather

## Tumulus 2

This was the first grave of the post-Meroitic period excavated during the season (Figure 1.5). This time the tumulus was much shallower, and a ring of black stones marked the perimeter, in a manner similar to that recorded by Garstang at the 300-399 Cemetery at Meroe. The shaft descended 1.5m to the bottom of the pit, and large stones in four layers blocked the tomb entrance. The burial was of a woman, fully clothed with her head to the north facing east. She had a baby buried under her legs in a bag, and had been laid

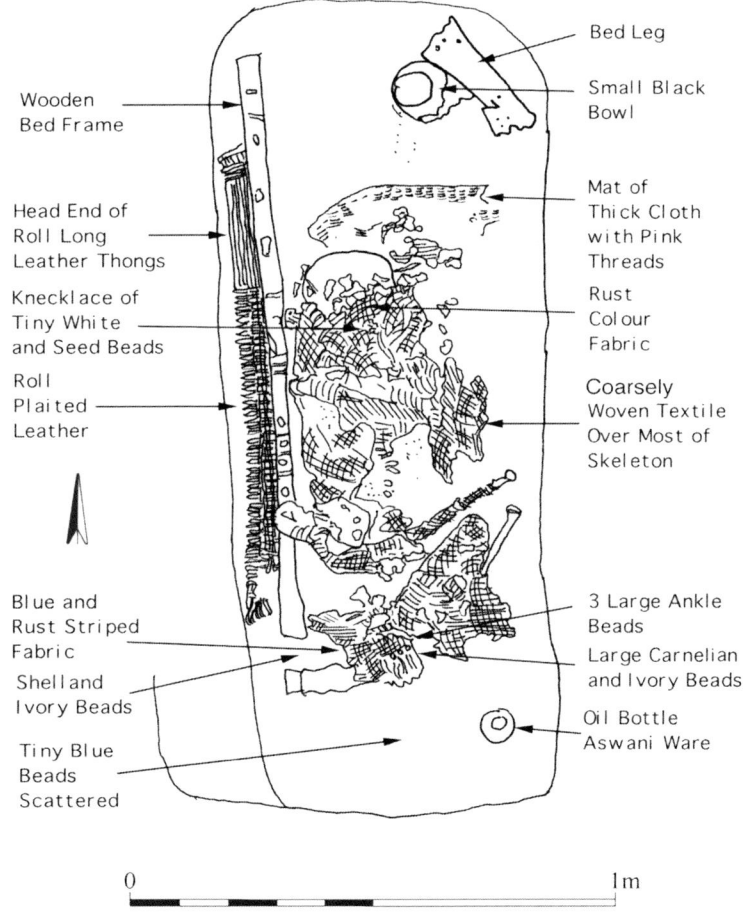

Figure 1.5. Site 159.2, Tumulus 2 unit 118 (scale 1:15).

5. Pelvis sample
6. Leather work
7. Baby skeleton
8. Textile fabric
9. Worked wood and leather
10. Worked wood, leather and textile
11. Skin/leather/fabric
12. Skin/leather fabric
13. Cloth
14. Pink cloth threads
15. Scalp, hair and textiles
16. Textiles
17. Skin/ear/leather
18. Scalp, hair and cloth
19. Cloth
20. Cloth and leather
21. Cloth and leather and woven matting
22. Cloth and leather
23. Cloth and leather
24. Textile
25. Ginger, grey and buff textile
26. Cowrie shell and small beads
27. Seed beads and white beads
28. Baby bones
29. Striped textiles
30. Bed leg
31. Plaited leather from bed stringing
32. Bed frame sample
33. Beads

**Tumulus 3**

This was a Post-Meroitic/Christian burial under a small ferricrete sandstone mound. The grave cut was shallow and there were no grave goods. Some small fragments of cloth remained. The body was laid with its head to the south, facing east, and the skeleton was in a good state of preservation.

**Tumulus 4**

This was a grave very similar to that beneath Tumulus 2. It was selected for excavation to see if these small tumuli were related to the large burial at Tumulus 1. The tomb contained an adolescent male(?), head to the south, facing east laid on an *angareeb* (Figure 1.6, Plates 2.24, 2.26 and 2.27). The grave goods included a wheel-made oil bottle similar to that in T2 but damaged, and a gourd. Two dom-palm nuts lay beside the body, and a copper-alloy fibulae/spatula lay near by. A similar object, of a type described by Woolley in the burials from Karanog as a kohl stick (Woolley 1910, pl. 36), also occurred in a bent form in one other burial at Gabati (Tumulus 5).

The burial lacked substantial quantities of fabric over it, but was still adorned with a ring, necklaces, and an anklet with large beads. The skeleton was in excellent condition, and the unfused nature of the bones was clear to see. The burial pots may reflect the status or youth of the deceased. After excavation was finished here, and prior to closing up, it was observed that the tomb had cut through the underside of an earlier burial laid in a shallow grave exposing the skeleton embedded in gravel. The grave cut was not apparent from above, and as it was not possible to excavate it, it was left for a future season.

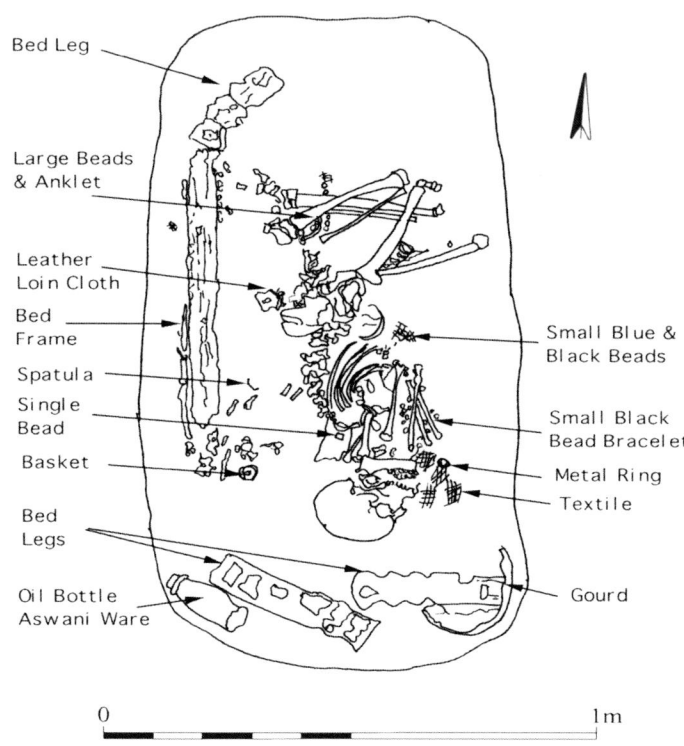

**Figure 1.6.** Site 159.2, Tumulus 4, unit 136 (scale 1:15).

*Summary of Objects Registered 159.2/T4*
Unit 136 (See Figure 8.3, Plates 8.8-8.12)

Red oil bottle 159.2/T4/49C
Item No.
1. Anklet and large beads
2. Leather work
3. Small beads
4. Textile
5. Ring
6. Gourd
7. Bed leg
8. Bed legs
9. Bed leg
10. Bed frame
11. Small blue and variously-coloured beads
12. Copper-alloy spatula or *kohl* stick
13. Sample from small baskets
14. Beads
15. Large bead

**Tumulus 5**

This was a substantial gravel mound with a tomb similar in structure to that beneath Tumulus 1. The tomb, however (Figure 1.7, Plates 2.29-2.30, 2.32-2.34), was larger, and the burial the most impressive uncovered. One of the stones of the tomb blocking had been cut into an oval shape, and inscribed with an axe head or cross-shaped motif. The skeleton was of a female buried head to the south, facing east, and the bed had massive carved legs 350mm long and 100mm square. The grave goods were placed at the foot, although it was hard to tell whether these had originally been on the bed or not.

There were five pots, again similar to some of those of tombs 300-399 at Meroe. One was a pot-stand of Type P, (illustrated in Garstang's *Meroe* as No. 26 pl. XLV, found in Tomb 307), with white painted decoration. A small red bowl was decorated to match. A small red-burnished bowl beside this is similar to Garstang's Type J, found in Tomb 304, and also a red jar with a handle or 'table amphora' similar to R2 Type as defined by Adams, of Nubian production (W. Adams 1986, 470, Form G39) similar to the Aswani Ware R30, and closest in form to J13, in Adams' classification (W. Adams 1986, 534). This was in almost perfect condition, apart from one missing handle. Beside this was a red-burnished spouted vessel with motifs painted on small white backgrounds. The form is similar to Garstang's Type ZK, found in Tomb 15. This tomb was thought to have been of the early Meroitic period by Garstang but these spouted vessels are normally associated with a much later, even Christian, period. This, combined with the absence of inscriptions, indicates a post-Meroitic to 'Transitional' period for these burials.

On the other hand amongst the grave goods was found, a turned ivory object similar to that found in tomb Beg. W.34 in the West Cemetery (Dunham 1963, 220, fig. 155, no. 13) and this was dated to Generation 24-37, a 1st century BC date. Grave goods, particularly high quality products, can date from much earlier than the burial in which they are found.

Other grave goods include a wooden comb, a bent copper-alloy fibulae or spatula, and a copper-alloy mirror, possibly silvered originally, in two parts. This again would point to a Roman import and a later date, or an item either from the later Meroitic period or continuing Meroitic traditions. The importance of this grave is shown by the scale of the burial, and concentration of quality grave goods. Although the Meroe cemeteries may have had richer grave goods prior to plundering none of them survived undisturbed, and the intact nature of the cemetery at Gabati offers the opportunity to understand what those burials may have been like. It also indicates the importance of the Gabati region in this period and the level of trade that may have taken place. The tomb was presumably of an individual of quite high status with access to expensive goods which suggests that she may have been a local ruler or noble. In the Christian period the top of the mound was pitted and a large quantity of Christian pottery was found in the fill. These may have been failed robbing pits, but David Edwards (pers. comm.) suggested that this could reflect the use of cemeteries as a safe place to store food and grain in times of trouble, as holes dug in cemeteries are less conspicuous, and perhaps even thought to have been protected by the presence of the burials.

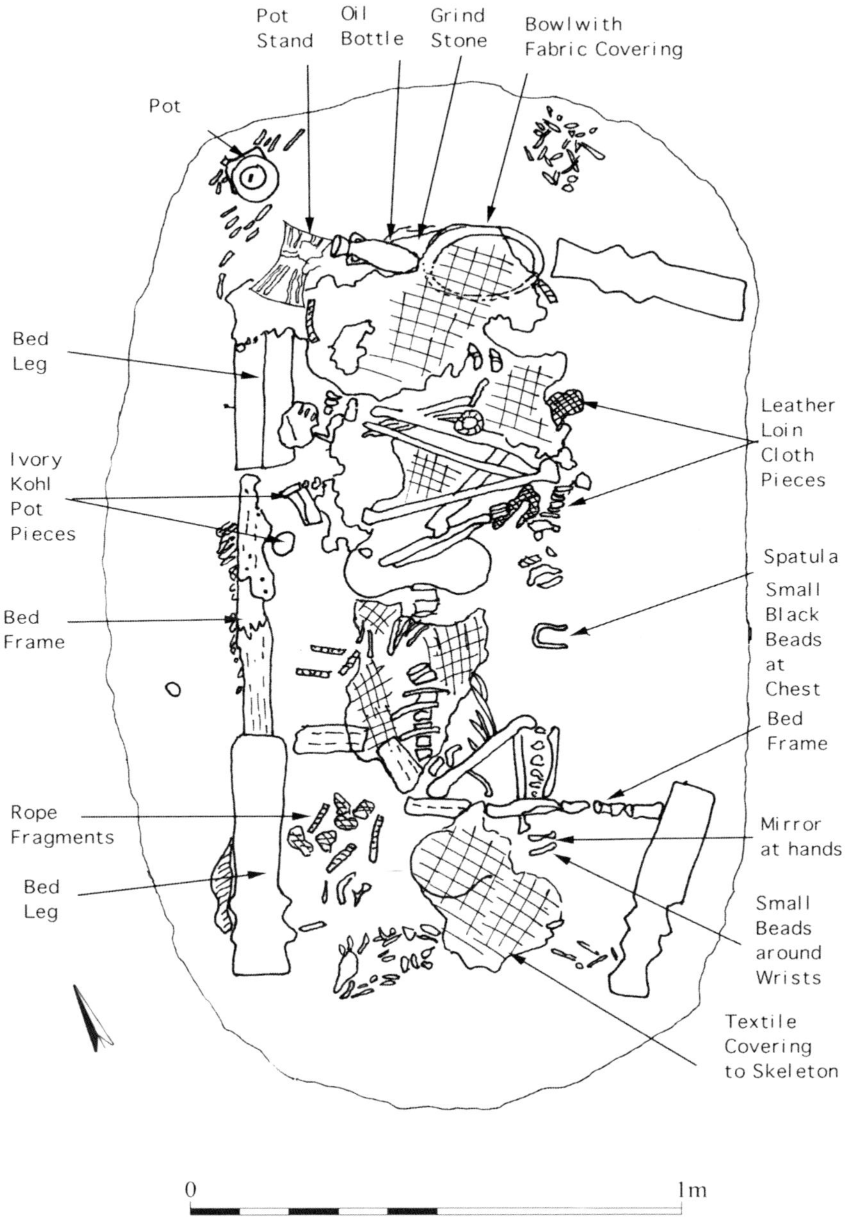

*Summary of Objects Registered 159.2/T5*
Unit 150 (See Figures 8.4 and 8.5, Plates 8.13-8.26)

159.2/T5.1/92C Painted red pot-stand
159.2/T5.1/93C Spouted jar
159.2/T5.1/94C Red jar or 'table amphora' with handles
159.2/T5.1/95C Red bowl with lug
159.2/T5.1/96C Red-burnished painted bowl
159.2/T5.1/97S Stone bowl or mortar with attenuated lug corners
1. Large beads
2. Small variously-coloured and white beads
3. Bent copper-alloy fibulae or spatula, and nail
4. Textile
5. Ivory scent jar or *kohl* pot
6. Mirror, copper alloy

**Figure 1.7.** Site 159.2, Tumulus 5, unit 150 (scale 1:15).

7. Small beads
8. Leather
9. Small blue-green and white beads
10. Variously-coloured beads
11. Leather bag or apron (a-k)
12. Textile
13. Textile and threads
14. Bed leg
15. Bed leg
16. Bed leg
17. Bed leg
18. Bed frame
20. Leather pieces
21. Wooden comb
22. Beads
22a. White bead
23. Rope and leather

**Tumulus 6**
This tumulus was similar to Nos 2 and 4 in structure, except that the grave pit was shallower, and in plan less square and more kite-shaped (Figure 1.8, Plate 2.35). In this respect it more closely resembled the post-Meroitic graves found by Francis Geus and Patrice Lenoble at el-Kadada. (Geus and Lenoble 1985, 67-92). Tumulus 6, though, was lacking grave goods, and the burial was the only one not on an *angareeb*. The body of a elderly man lay head to the south facing east. The body was covered in a thick shroud, and its head lay on a blanket rolled into a pillow. The only other content was a leather object, possibly an apron, which lay over the feet.

In all five of the tomb burials, leather fragments were found associated with the legs, and it is supposed that this was either part of the offerings or an item of clothing. Leather was also found as part of the structure of the bed frame. Curiously, in the upper part of the fill of this burial the remains of wood with leather ties were found. It can only be supposed that this was the remains of the bier on which the deceased was carried. The absence of grave goods in this tomb would tend to confirm that, if this was a late representative of the type of burials discussed above, it dates to immediately prior to the conversion of the region to Christianity. Alternatively, it may just reflect the extreme poverty of the individual concerned.

*Summary of Objects Registered 159.2/T6.1*

1-7 Lower half of shroud, blanket interlayered
8-20 Shroud, blanket interlayered
21-22 Shroud, blanket interlayered
23-24 Shroud, blanket interlayered
25-26 Shroud, blanket interlayered
27-28 Biological samples
29 Sample from leather object
55S Samples of carved wood (Figure 8.5)

**Tumulus 7**
This tumulus covered a conventional Christian burial, aligned east-west with the head to the east. The skeleton was in a fine state of preservation, and the tumulus was distinguished by the grave cut being in a silt layer not found elsewhere on the site (Plate 2.28). The upper part of this tumulus, to a depth of half a metre, was made of local ferricrete sandstone cobbles.

**Tumulus 8 - 10 and 12**
These were not excavated as it was clear that they were intact and similar to Tumuli 1- 6, and would require more recording than time permitted.

**'Tumulus' 11**
Tumulus 11 was a small Christian burial which (Figure 1.9, Plates 2.36 and 2.37) was intended to be the last excavated. On uncovering the tumulus top, a mud-brick structure (bricks 370 x 170 x 100mm) was revealed which at first was thought to be the top of a Christian tomb as it was aligned true east-west rather than at 100° from north as were the Meroitic tombs. Further clearance revealed a structure approximately 4m square filled with large black stones and a quantity of mud bricks. The top of the fill was full of fine quality Meroitic pottery sherds. It was therefore decided to clear the structure, as it seemed to be Christian in date.

Immediately a tomb shaft was uncovered which went down through mud layers to a depth of 2m, whereupon the remains of a tomb chamber were found, much disturbed but still containing seven intact large globular vessels, and five broken ones. The vessels were clearly Meroitic, and it was decided to record this last Meroitic tumulus

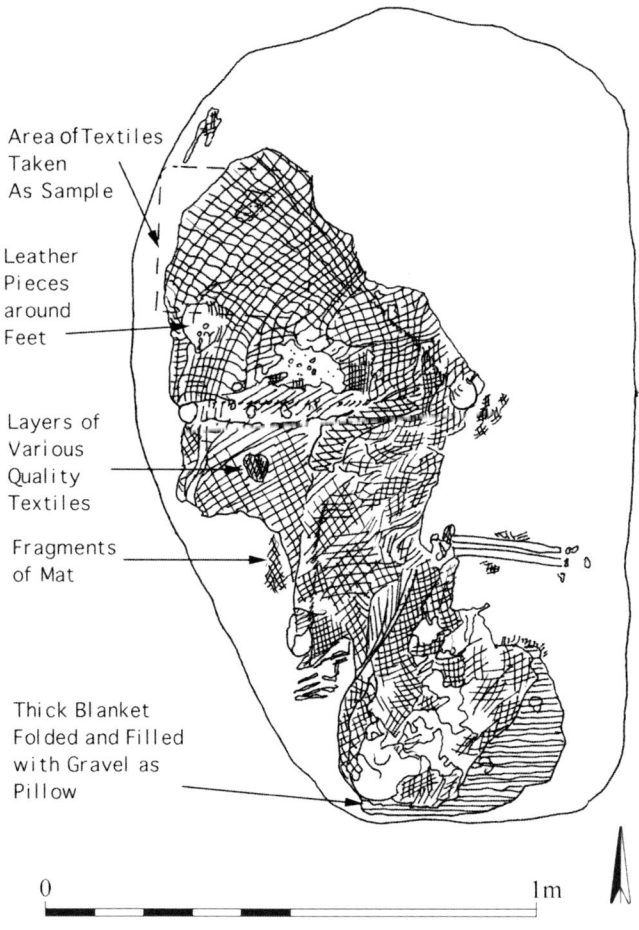

**Figure 1.8.** Site 159.2, Tumulus 6, unit 151 before final clearance (scale 1:15).

to the level of excavation possible in the remaining time.

The final excavation revealed a structure 8.4 x 4.1m divided into a western section exactly 4.1m square, and a attached chapel 4.1 x 4.3m. The walls were 600mm thick of 1½-brick construction, and had been rendered and painted white on the outside. At the east there was a doorway into the building which originally led straight into the tomb shaft, 1.2m wide, that sloped down to 2m below foundation level. The tomb was cut into the white gravel layer under the square western section of the structure. Its entrance had been blocked with mud bricks and then back-filled with the original fill. At some time in the past, the tomb had been disturbed, a shaft had been cut into the gravel fill and two attempts at robbing had been made, as two pits were found in the gravel, and the blocking had been removed. The large stones found in the front of the chapel may have been part of this blocking, or may have been part of the superstructure. The robbers, after the disturbance, had thrown back into the grave the larger vessels which they presumably considered to be of no value. The tomb seems to have been not at right angles to the shaft but a continuation of it, and the vessels would have been at the foot of the deceased by the entrance. The tomb was completely full of mud-brick rubble, and could not be excavated down to the floor in the time available. Outside the original mud-brick blocking wall there had been four pots including a painted white ware vessel with a serpentine motif, and floral decoration. Inside the doorway were a further eight vessels.

The globular vessels (Plates 4.1 and 4.2) were of four kinds. The first type, of which there were four examples, were black burnished and decorated with rocker marks, and impressions around the neck and rim. The vessels in form are very similar to that found at Shendi (Geus *et al.* 1986, 12, SHQL4), and also by Reisner in the Western Cemetery at Meroe tomb Beg.W.29 (Dunham 1963, 220, fig. 155, no. 9); the date of both these examples is thought to be 1st century BC.

The second type is a red-burnished vessel, three of which were found. These are similar in form to that illustrated by Garstang (1911, pl. XLI, no. 8) but unclassified. These are from the northern group of tombs (no. 400) at Meroe. Of particular interest on one of these pots was an inscription in Meroitic. This is the first example recorded in Sudan, but at Karanog in Egypt, Woolley noted inscriptions on a few of the vessels in Meroitic (Woolley 1910, pl. 107, no. 26 G187. Fxii). The third type, of which two were found, are painted onto a white-slipped vessel, the decoration is serpentine with flowers, and has parallels only much further north, where until now they were thought to have been produced.

Two examples are published by Bonnet (1978, fig. 16, p.125; 1990, nos 357 and 360, also colour pg. 14); they are from Tombs 10 and 3B in the Meroitic Cemetery excavated at Kerma, and date from 100 BC – AD 100. The final type of vessel was a very small globular jar with neck and foot ring. These again have been found further north, and were published by Adams as oil jars (*lekythoi*), examples of which were also published in the Kerma Catalogue by Bonnet (1990, 246, nos 385). Five of the group of vessels were marked with a symbol which

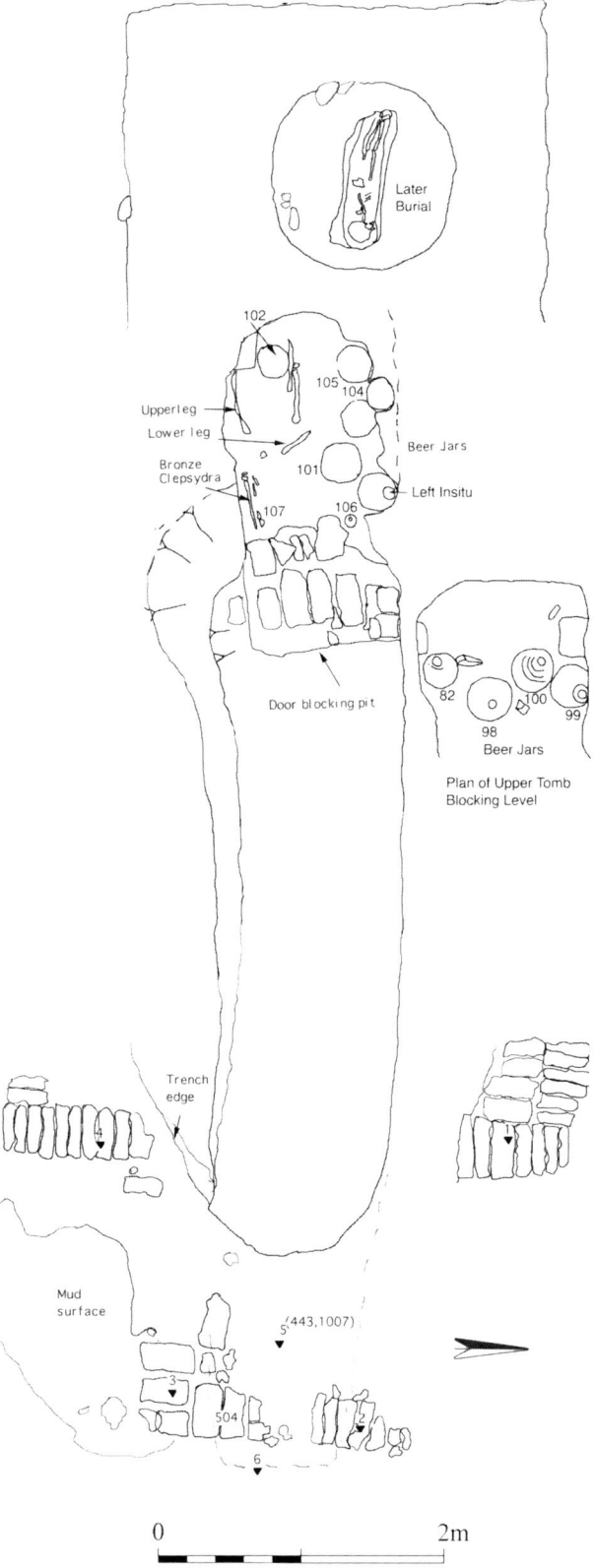

**Figure 1.9.** Site 159.2, T11, lower excavated levels (scale 1:50).

was also recorded by Dunham at the pyramids at Meroe in Beg.N.11, which dates from Generation 36, around the 2nd century BC (Dunham 1957, 72-73, fig. 44).

The only parallels available for this structure are those recorded at Karanog, where the superstructures were partly stone, and one at the site of Kedurma near the

Third Cataract (Edwards 1995, 40-47, fig. 5). In both these cases however, the superstructure was square and the chapel only a small eastern extension, of less than half the width of the square. The uniqueness of these chapels made it essential that further study was carried out, particularly due to their potentially early date.

Cut into the side of the brick superstructure was a Christian grave. The body was aligned slightly north of east-west, perhaps due to the difficulty of digging through the brick work, with the head to the east.

*Summary of Objects Registered 159.2/T11*
(See Figures 2.43-2.45, Plates 4.1 and 4.2)

159.2/T11/99C Vessel 1 - Black decorated with mark
159.2/T11/100C Vessel 2 - Black decorated with mark
159.2/T11/98C Vessel 3 - Red, plain
159.2/T11/82C Vessel 4 - Painted slipped, broken (re-numbered <1106>)
159.2/T11/101C Vessel 5 - Painted slipped with mark
159.2/T11/102C Vessel 6 - Red with inscription and mark
159.2/T11/103C Vessel 7 - Small red, burnished
159.2/T11/104C Vessel 8 - Small black decorated
159.2/T11/105C Vessel 9 - Black decorated (broken)
159.2/T11/106C Vessel 10 - *Lekythos*
159.2/T11/107C Vessel 11 - Sherds

**Tumulus 13**
This was a very small mound of ferricrete sandstone that covered a female child buried east-west, with head to the east. Unusually, it included jewellery and textiles, but is probably Christian. A number of other small burials around it suggests that this may have been the children's section of the cemetery.

*Objects Registered*
(See Plate 8.27)

159.2/T13.1/71S Earring, copper alloy
159.2/T13.1/77S Textile
159.2/T13.1/70S Various beads for necklace

## Conclusions

The work of the expedition covered all the sites noted as being directly affected by the road line after the Surface Survey in 1993. Twenty-six tumuli in six cemeteries were excavated, and six structures in three settlement sites. A total of 30 individuals was recovered and nine of the burials contained grave goods. The expedition was able to record all the material recovered from these sites in a suitable manner, but the tombs at Gabati raised serious questions concerning conservation. The tombs, of which only five out of a potential 50 with intact grave goods were opened, yielded an enormous quantity of very fragile textiles, leather, wood, ivory, copper-alloy and other metalwork. This prevented any serious attempt that season to excavate completely the site. It seemed clear after our return to Khartoum that it would not be possible to re-route the road due to the additional costs. The road construction was due to start in three to six months after the test excavations, and a suspension of the work around Gabati was likely until the following spring. Consequently, it was recommended that an excavation be carried out to complete the study of the site at Gabati as soon as possible, which resulted in the subsequent Rescue Season from 1994-1995 (Edwards 1998).

## The Publication

It had originally been intended to publish all material from Gabati in one volume. This was not practical to achieve and so the nine burials from the 1994 season are published here as part of the overall work of that season.

The importance of the excavations was partly to verify our dating of cemeteries based on the form of each cemetery and the surface finds which we had attempted in *Road Archaeology* Volume I, but also to safeguard the archaeological heritage inevitably threatened by the road building process. These aims have proved successful. The considerable improvement in identifying the burials from the Post-Meroitic and Early Christian periods in this area has proved invaluable for dating the cemeteries close by. The improvement in knowledge about the pottery and small finds available to non-royal burials away from the major centres in the Meroitic and Post-Meroitic periods has shown to be useful for estimating the contacts of these remoter populations with Meroe, 30km to the south.

A summary of the bone material was prepared by Francis Thornton and Joyce Filer. This has now been superceded by Rebecca Whiting's detailed analysis published here. Specialist studies on the leather and basketry have been provided by Barbara Wills and Willemina Wendrich. The leather analysis indicated that what had previously been identified as shoes were actually a form of loin cloth, which perhaps recalls those depicted as worn by sailors in ancient Egyptian wall reliefs. Wendrich's study documented cultural influences on the basket types and the kinds of grasses used. Laurence Smith, who also worked on the pottery and small finds from the Gabati excavation season, has demonstrated cultural influences from as far afield as Southern Egypt, and others associated with the Red Sea trade. His study presents a summary of the grouping of vessels and objects in each grave, and a brief consideration of their significance.

The illustrations' section attempts to represent key examples of the different excavations, and some of the more interesting finds. The site descriptions are in a standard format, and provide access to all the known scientific data on each site. Drawings show the key sections and contexts of the excavations and spot heights. These drawings have for the most part been included in the text.

The volume is a team effort as were the excavations, and the compilation of the text was worked upon by all team members.

## Acknowledgements

This SARS project was made possible by the assistance of the Sudan National Corporation for Antiquities and Museums, in particular the late chairman, Prof. Ali Hakim. The project was funded by the Haycock Fund of the British Institute in Eastern Africa and SARS. Help was given by the late Dr John Alexander and Dr Patricia Spencer in reading this text and Vivian Davies in support of the project. Assistance was also provided by the late Dr Friedrich Hinkel who provided the Sudan Archaeological Survey numbers and references for the project.

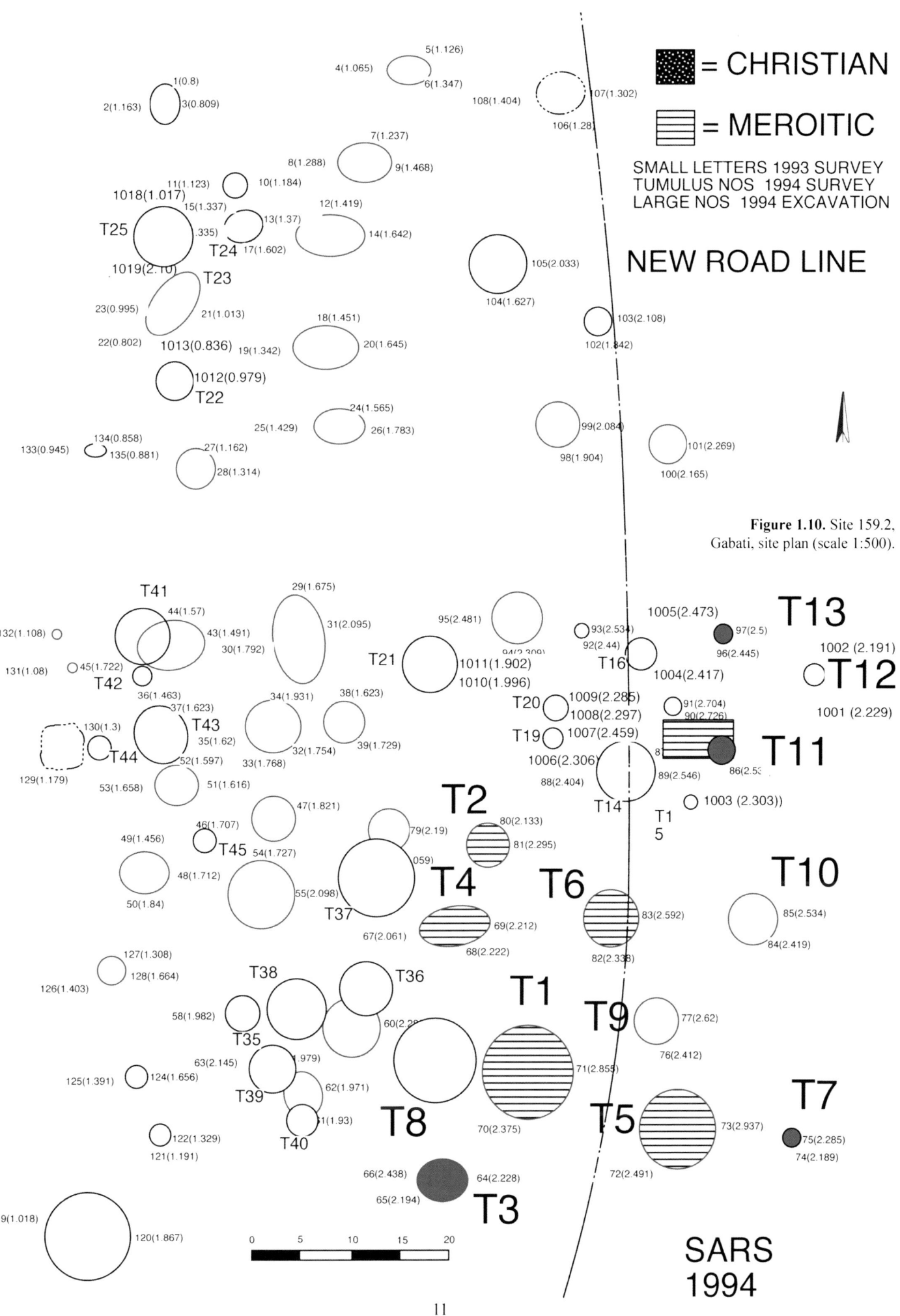

**Figure 1.10.** Site 159.2, Gabati, site plan (scale 1:500).

# 2. Site Descriptions

*Fathi Abdel Hamid Salih Khidr, Salima Ikram, John MacGinnis, Joyce Filer and Francis Thornton*

**SARS Survey Site Ref.**   101.4/T1
**Site UTM Grid Point**   N1874710 E578642
**Sudan Survey Number**   NE-36-O/3-E-19

### Description of site
A mound measuring about 7m north-south, 8.8m east-west and 300mm in height, covered with medium sized black stones and blown sand (Figures 2.1, 2.2, Plate 2.1). The tumulus is made up of black stones of different sizes mixed with grey soil. It was divided into four quadrants and excavated down to the original ground surface. A trench was made along the north-western quadrant section below the original surface revealing two stages of development; a Christian tumulus over an earlier Post-Meroitic monument. In the centre of the main tumulus T1.2 was found an east-west aligned, Christian skeleton head to east, in a pit; at its feet was a small sub-adult burial T1.6, aligned east-west, semi-flexed, facing south. In the north-eastern quadrant a pit was found filled with loose gravel, pebbles and fine sand. In the south-eastern quadrant at the lower level were found a very badly preserved skull fragment T1.1 and mammal bones. Under the fill to the south was T1.5, a semi-flexed skeleton, aligned east-west, on its left side, the head to the east facing south. The bone was in a very bad condition; no grave goods were found. The two Post-Meroitic burials were found further south. T1.3 a semi-flexed burial in very bad condition was aligned west-east with head to the west facing south. The deceased was buried with grave goods, including an iron band on a mat (Figure 2.2, Plate 2.3). To the south was a further burial T1.4, similarily aligned and very eroded (Plate 2.4). The superstructure and the make-up of the tumulus is of Christian date. It seems likely that the post-Meroitic period mound had been destroyed by water erosion. This is suggested by the section which showed several grey layers, Units 61 and 65, probably silts carried by water percolating through the mound, which had been sealed by the digging of the later mound. The Christian burial was associated with a layer of yellow sand, Unit 66. This may have been deliberately dug from an area to one side of the tumulus to purify the burial site, a practice known from Pharaonic Egyptian temples.
**Suggested Period**: Post-Meroitic and Christian

### Small Finds and Samples Report
**Object no.** 101.4/T1/2S
**Unit** 61
**Description** 1 flake of chert with cortex at one edge.
**Identification** A continuously retouched flake of chert.
**Dimensions** L: 13.3mm, W: 12.3mm
(Information on indentifications of all lithics from Mr H. Kenny)

**Object no.** 101.4/T1/9S
**Unit** 64
**Description** 1 bead of faience, cream in colour, cylindrical in shape, but somewhat asymmetrical, with slightly bevelled surface at each end (Figure 3.4).

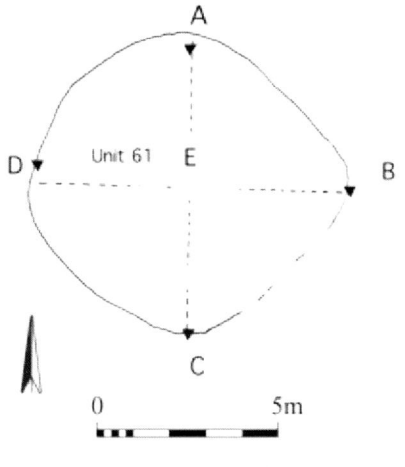

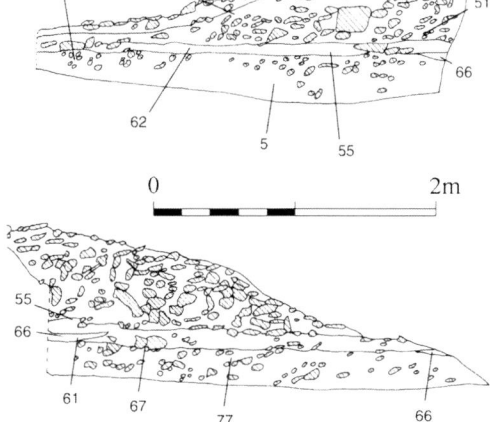

**Figure 2.1.** Site 101.4, Tumulus 1, plan (scale 1:200), sections AE and ED (scale 1:50).

**Dimensions** L: 3.2mm, D: 3.4mm

**Object no.** 101.4/T1/13S
**Unit** 64
**Description** 1 flake of chert with bulbar surface much eroded, with part of the bulb of percussion removed. Edges chipped.
**Identification** Wind-abraded denticulate
**Dimensions** L: 27mm, W: 27.2mm

**Object no.** 101.4/T1.3/47S
**Unit** 94
**Description** 9 sections and 1 fragment of a strip or band of iron (Figure 3.4).
**Dimensions** Sections generally *c*. 1.5-3mm thick
a) 43.8 x 12mm
b) 36.5 x 12.3mm *c*. 31 x 13mm
d) 39 x 12.8mm
e) 30 x 14mm
f) 7 x 15mm
g) 30.4 x 14.9mm

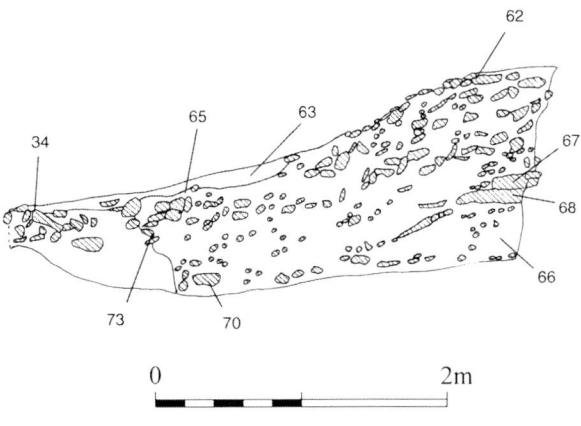
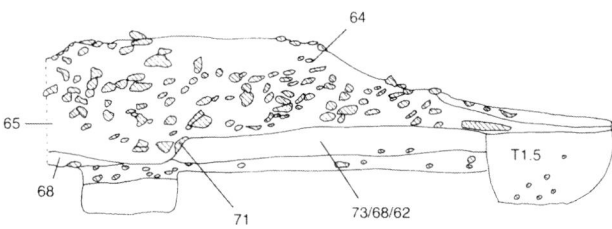

h) 19.3 x 10.3mm
i) 7.5 x 5mm

**Object no.** 101.4/T1.3/48S
**Unit** 94
**Description** 1 bead of faience(?), pale green in colour, of flat annular shape. A groove at the mid point of the hole, may indicate that the bead was made in two pieces (Figure 3.4).
**Dimensions** Th: 1mm, D: 4mm

**Object no.** 101.4/T1/Sample
**Unit** 94
**Description** Soil sample.

**Object no.** 101.4/T1.3/Sample
**Unit** 69
  **Description** Soil sample. Sample of deposit including remains of possible reed or other organic fibre. Possibly from matting. Associated with Skeleton 101.4/T1.3/41Or.

**Bioanthropological Report**
101.4 / T1.2 /38Or: A fully extended burial, aligned east-west, head to the east facing north. In good condition. Young adult, male (see pg. 124).
101.4 / T1.3 /41Or: A flexed burial, aligned west-east, head to the west facing south. Approximately 55% remains. Young adult, probable male (see pg. 124).
101.4 / T1.4 /42Or: A semi-flexed skeleton lying on its right side, head to the west facing south, in a poor condition. Young adult, probable male (see pg. 125).
101.4 / T1.5 /50Or: A possibly flexed infant burial with the head to the east, in a very poor condition (see pp. 125-6).
101.4 / T1.6 /51Or: A semi-flexed sub-adult skeleton (11-15 yrs), lying on its left side with the head to the east facing (south?). In a very poor condition (see pg. 126).

**SARS Survey Site Ref.** 101.4/T2
**Site UTM Grid Point = Unit 75 =** N1875182.40 E578850.68
**Ht** 368.97m
**Sudan Survey Number** NE-36-O/3-E-19

**Description of site**
A tumulus covered with black stones mixed with windblown sand (Figure 2.3). It measures 6.8m north-south and 7.8m east-west. Fragments of human bone and pot sherds were observed on the surface of the mound. The average size of the stones covering the mound is about 200mm. The make-up of the tumulus (Figure 2.4) is a layer of black stone flakes and small sized pebbles on red gravel and then grey and yellow sand. The superstructure of the tumulus was divided into four quadrants and these were removed in turn to reveal an oval-shaped pit at the original ground surface. The pit was filled with loose gravel and a whitish and yellow layer, Unit 75. The fill contains an extended skeleton aligned east-west, lying on its back, facing south (Plate 2.5). The present condition of the skeleton is good, the skull is fragmented. The skeleton is complete except for the small bones from the hands and feet. No finds were associated with it.
**Suggested Period**: Christian
**Burial Identity**: Adult? male.

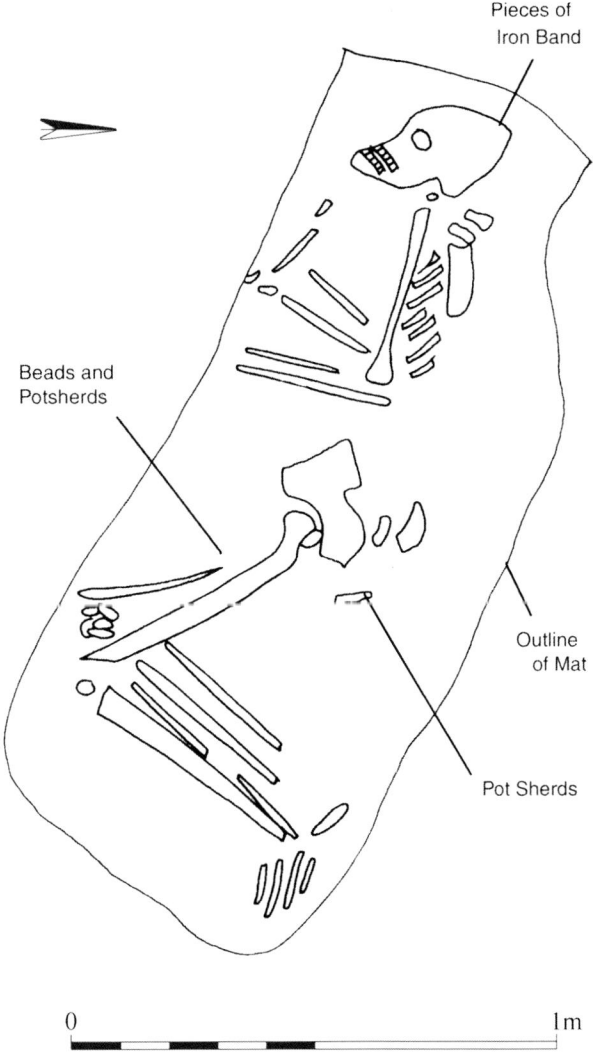

**Figure 2.2.** Site 101.4, Tumulus 1.3, sections EC and EB (scale 1:50), plan of grave (scale 1:15).

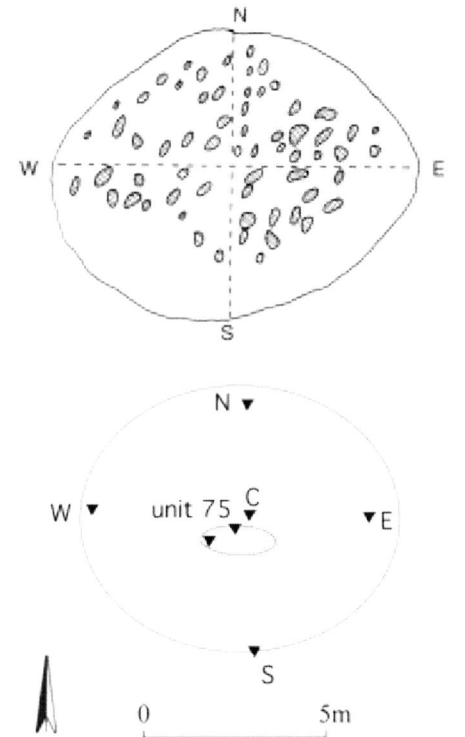

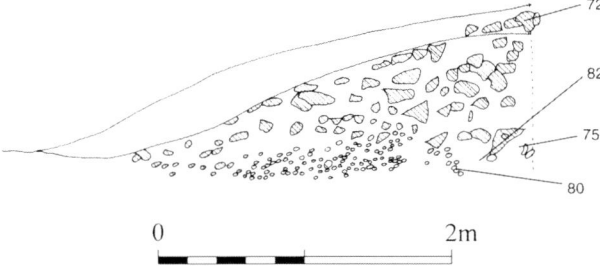

**Figure 2.3.** Site 101.4, Tumulus 2, plan before excavation, plan of burial cut (scale 1:200).

**Figure 2.4.** Site 101.4, Tumulus 2, north-south section, west face (scale 1:50).

**Small Finds and Samples Report**
**Object no.** 101.4/T2.1/Sample
**Unit** 79
**Description** Soil sample. Sample of deposit from near head of body in burial pit.

**Bioanthropological Report**
101.4/T2.1/39Or: Extended supine inhumation in shallow grave pit, lying on its back with the head facing south (Plate 2.5). The skeleton has a good level of preservation. Young adult, probably male (see pp. 126-7).

| | |
|---|---|
| **SARS Survey Site Ref.** | 101.4/T3 |
| **Site UTM Grid Point** | N 1875177.588 E 578852.186 |
| **Ht** | 368.172m |
| **Sudan Survey Number** | NE-36-O/3-E-19 |

**Description of site**
A tumulus covered with black stone blocks, measuring 350 x 200mm at the top centre, about 400mm in height, 4m north-south and 6m east-west. The tumulus make-up is a mix of stone blocks measuring 100mm to 150mm and loose gravel and soil (Figure 2.5). It was excavated down

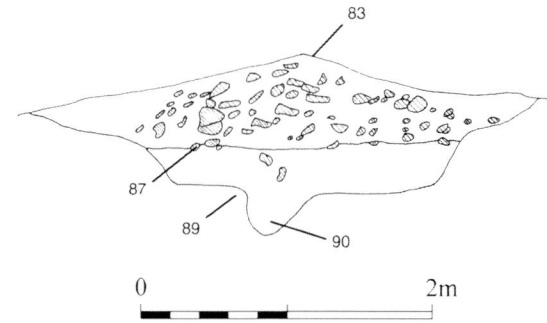

**Figure 2.5.** Site 101.4, Tumulus 3, section (scale 1:50).

to the ground surface where, just off centre in the western half of the tumulus, a grave was found, oval in shape and aligned east-west. The fill of the grave varies from hard to soft sand. It contained a fully extended inhumation, aligned east-west with the head to the east facing south (Plate 2.6). The skeleton is in good condition and almost complete.
**Suggested Period**: Christian
Burial Identity: Young adult, female.

**Small Finds and Samples Report**
**Object no.** 101.4/T3/31S
**Unit** 84
**Description:** Fragment of fine woollen cloth, plain pale brown tabby weave, warp and weft Z-spun, 16 threads per cm (Figure 3.4). (Information: S. Taylor.)
**Dimensions** L: *c*. 15mm, W: *c*. 5mm
**Object no.** 101.4/T3/34Or
**Unit** 85
**Description:** Seed. *Cassia* sp. (Identification: D. Fuller)

**Bioanthropological Report**
101.4/T3.1/40Or: A fully extended inhumation, aligned east-west with the head to the east facing south. Young adult, female (see pp. 127-8).

| | |
|---|---|
| **SARS Survey Site Ref.** | 102.1/T1 |
| **Site UTM Grid Point** | N1875421.524 E578939.65 |
| **Ht** | 365.88m |
| **Sudan Survey Number** | NE-36-O/3-E-19 |

**Description of site**
A tumulus about 6m in diameter, covered with black stone blocks. The tumulus make-up is a mix of stone blocks and loose soil. Down to the original surface on the western half of the tumulus just by the centre is an oval-shaped burial pit, aligned east-west and measuring 1.55m in length (Figure 2.6). The fill of the burial pit is a loose red gravel.

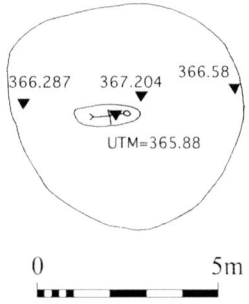

**Figure 2.6.** Site 102.1, Tumulus1, skeleton in relation to the tumulus (scale 1:200).

**Suggested Period**: Christian

**Small Finds Report**
No finds were associated.

**Bioanthropological Report**
102.1/T1.1/46Or: A fully extended adult inhumation (Plates 2.7 and 2.19), aligned east-west, lying on its back with the hands across the pelvis. The head is to the east facing north. Young adult, female.

**SARS Survey Site Ref.** 112.4 /T1
**Site UTM Grid Point** N1880280 E580445
**Sudan Survey Number** NE-36-K/22-U-5

**Description of site**
A shallow tumulus with a concentration of stone slabs in the centre and empty patches on the top, surrounded by a ring of stone slabs around the edges. The tumulus measured 3.25m north-south, 3.28m east-west and was 200mm in height (Figure 2.7). The structure of the tumulus was well preserved, comprising stones and earth, with larger stones (up to 500mm long) in the centre, decreasing to smaller ones round the edge (Plate 2.8). The outer two or three rings of stones were laid sloping inwards on top of each other. The

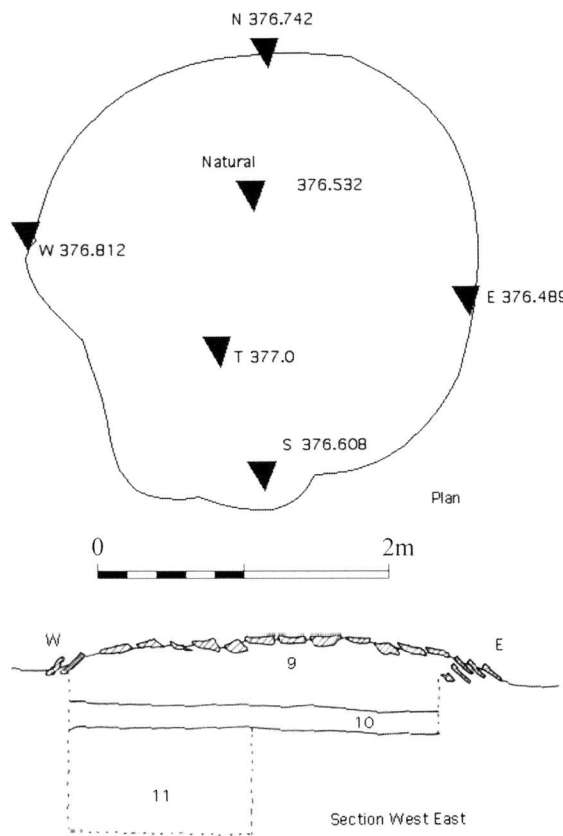

**Figure 2.7.** Site 112.4, Tumulus 1, top – east-west section, bottom – plan (scale 1:50).

removal of the upper layers of stones revealed a shallow stone-lined grave, slightly off-centre in the south-west quadrant. This measured approximately 700mm long x 300 mm wide x 200mm deep. It was cut into a layer of brown earth and stones that extended under the stone mound over the whole area of the tumulus. No skeletal material remained. It does not appear that this was due to robbing as the stone covering appeared to be intact, more probably the bone had been destroyed through natural factors.

**Contexts**
9 layer of black stones
10 layer of black stones mixed with brown earth
11 natural of white/brown stones and red sand

**Small Finds and Sample Report**
**Object no.** 112.4/T1.1/10Or
**Unit** Topsoil
**Description** Seed. *Balanites aegyptica*. (Identification: D. Q Fuller)

**SARS Survey Site Ref.** 112.4/T2
**Site UTM Grid Point** N1880280 E580445
**Sudan Survey Number** NE-36-K/22-U-5

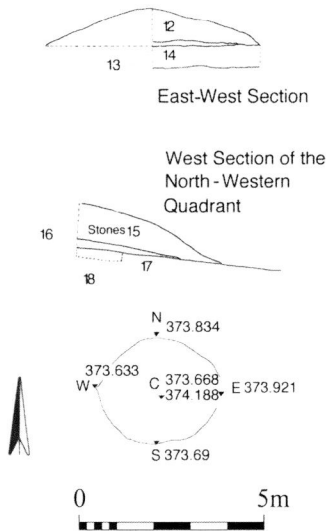

**Figure 2.8.** Site 112.4, Tumulus 2, top – east-west section, middle – west section of north-western quadrant, bottom – plan (scale 1:200).

**Description of site**
A shallow mound covered with black stone blocks and wind-blown sand (Figure 2.8). It measures 3.02m north-south and 3.52m east-west, with a height of 520mm. A certain amount of the uppermost stone covering was eroded but the structure of the tumulus was otherwise well preserved. The make-up of the tumulus consists of stone blocks, grey soil and hard red-coloured sand. In the south-east quadrant, approximately 400mm from the centre and 200mm below the surface of the tumulus, a small quantity of bone fragments was preserved (112.4/T2/3Or). These appeared to be in the stone structure rather than under it or in a cut grave, a circumstance possibly due to forces of erosion. A 300mm deep trench was dug in the very hard layer below the ground surface under the tumulus, but yielded no finds or other evidence.

**Small Finds Report**
112.4/T2.1/3Or context 13 bones
**Object no.** 112.4/T2.1/Sample
**Unit** 79
**Description** Sample of soil deposit from near head of body in burial pit.

**Contexts**
12 layer of black stones

13 layer of brown earth containing flat stones up to 200 x 400mm
14 natural of light brown stones and hard red sand
19 natural of gravel and very hard red sand

**Bioanthropological Report**
112.4/ T2.1/3Or: Bones placed 200mm below the mound surface in the south-eastern quadrant, just by the centre.

**SARS Survey Site Ref.**    112.4/T3
**Site UTM Grid Point**    N1880280 E580445
**Sudan Survey Number**    NE-36-K/22-U-5

**Description of site**

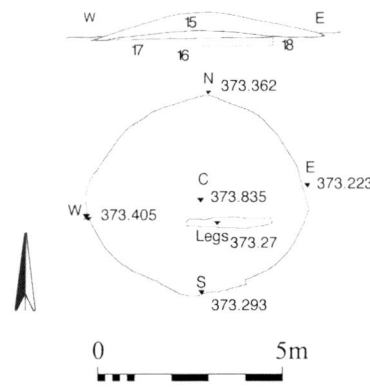

**Figure 2.9.** Site 112.4, Tumulus 3, top – east-west section, bottom – plan (scale 1:200).

A tumulus covered with small-sized stones, 5.27m north-south, and 4.01m east-west, with a height of 580mm (Figure 2.9). The tumulus was well preserved, being constructed of black stones. The removal, by hand, of the upper stone covering revealed a layer of brown earth containing a few small stones, in which a number of tiny light green and light blue faience beads were found (112.4/T3.1/2 S). The excavation of this layer revealed a grave cut into the natural red floor. This measured 700mm long x 250mm wide x 100mm deep and was aligned east-west, stretching out east from under the centre point of the mound. The grave contained only a very incomplete inhumation (112.4/T3.1/4Or), comprising fragmented pieces of an adult skeleton, insufficient to determine sex. We do not know for certain why, but the most likely cause is natural factors. Further faience beads were found in the grave fill and are included in the same small finds number as the ones above.

**Contexts**
15 layer of black stones; sealing 16
16 layer of brown earth containing small stones; sealing 17 and 18
17 hard red sand with white streaks; cuts 18, seals 16
18 grave, cut into 17 and sealed by 16, filled with dark brown earth mixed with sand

**Small Finds and Samples Report**
112.4/T3.1/2S context 16/18 light green and blue tiny faience beads (Figure 3.4).
112.4/T3.1/4Or context 18 bones
**Object no.** 112.4/T3.1/2S
**Unit** 016/018

**Description** 14 faience beads, of flat annular shape, but often somewhat irregular in shape and asymmetrical. 11 pale green, 2 pale greenish-blue, 1 turquoise.
**Dimensions: (mm)**

| Pale Green | | Greenish Blue | |
|---|---|---|---|
| length | diameter | length | diameter |
| 1.5 | 3.2 | 1.5 | 3.5 |
| 1.8 | 3.2 | 1.7 | 3.6 |
| 1.7 | 3.5 | | |
| 1.9 | 3.7 | Turquoise | |
| 1.2 | 3.7 | length | diameter |
| 1.3 | 3.0 | 1.5 | 3.5 |
| 1.8 | 3.8 | | |
| 2.4 | 3.2 | | |
| 1.5 | 3.2 | | |
| 1.7 | 2.9 | | |
| 1.2 | 4.1 | | |

**Bioanthropological Report**
112.4/T3.1/4Or: Fragments of an adult inhumation (Plate 2.9), not enough to determine the sex, in a very poor condition and fragile (see pp. 128-9).

**SARS Survey Site Ref.**    112.4/T4
**Site UTM Grid Point**    N1880280 E580445
**Sudan Survey Number**    NE-36-K/22-U-5

**Description of site**
A shallow mound situated on the slope of an outcrop, covered with black stones and wind-blown sand. It measures 5.40m north-south, 5.76m east-west and is 500mm high

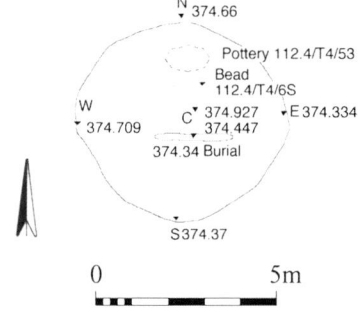

**Figure 2.10.** Site 112.4, Tumulus 4, plan (scale 1:200).

(Figure 2.10). Bone, potsherds and beads were found within the tumulus make-up at the north-eastern quadrant. The removal of the upper layer of black stones revealed a layer of grey sand and loose red-coloured gravel; a number of potsherds were found in this layer (112.4/T4.1/5C). Below this were the remains of a burial containing a very poorly preserved skeleton (parts of an arm, a leg and the skull, 112.4/T4.1/7Or) and a number of tiny blue faience beads (112.4/T4.1/6S). There was no cut, the burial lying on the cleared natural surface. The burial was evidently disturbed by either robbing or water action.

**Small Finds and Samples Report**
112.4 /T4.1/5C. Potsherds, 112.4 / T4.1/6 S. Beads, 112.4 /T4.1/ 7Or Skeletal material (see pg. 129).

**Object no.** 112.4/T4/6S
**Unit** 016
**Description** 59 faience beads, of flat annular to squat

cylindrical shape, but often somewhat irregular in shape and asymmetrical. Includes example of two beads fused together, one partially overlapping the other; 39 pale greenish turquoise; six brownish grey, four white, 10 bluish turquoise (Figure 3.4).

**Dimensions:** (mm)

| Greenish Turquoise | | Greenish Turquoise | |
|---|---|---|---|
| length | diameter | length | diameter |
| 2.3 | 3.2 | 1.8 | 3.7 |
| 1.8 | 3.2 | 1.9 | 3.8 |
| 1.9 | 3.3 | 2.1 | 3.0 |
| 1.4 | 3.0 | 2.2 | 3.0 |
| 1.8 | 3.2 | 1.9 | 2.7 |
| 2.0 | 3.2 | | |
| 1.9 | 2.9 | Brownish-grey | |
| 2.1 | 3.1 | length | diameter |
| 1.7 | 3.0 | 1.7 | 3.0 |
| 1.6 | 3.2 | 1.6 | 3.1 |
| 1.6 | 3.7 | 1.6 | 3.1 |
| 1.7 | 3.2 | 1.6 | 3.2 |
| 2.5 | 3.4 | 1.2 | 3.2 |
| 1.8 | 2.9 | 1.4 | 3.2 |
| 1.8 | 3.0 | | |
| 2.4 | 3.5 | White | |
| 2.0 | 3.2 | length | diameter |
| 2.8 | 2.9 | 1.5 | 2.8 |
| 3.0 | 3.2 | 2.5 | 3.2 |
| 2.6 | 3.0 | 2.1 | 3.3 |
| 2.0 | 2.9 | 2.0 | 3.4 |
| 2.4 | 2.9 | | |
| 1.8 | 2.8 | Bluish Turquoise | |
| 1.9 | 3.8 | length | diameter |
| 2.1 | 3.0 | 2.2 | 2.8 |
| 2.0 | 3.2 | 2.3 | 3.0 |
| 2.4 | 3.9 | 1.8 | 3.0 |
| 1.1 | 3.7 | 1.8 | 3.3 |
| 1.7 | 3.0 | 1.7 | 3.2 |
| 1.4 | 3.0 | 1.2 | 3.4 |
| 2.5 | 2.9 | 2.8 | 3.1 |
| 2.3 | 2.9 | 2.7 | 3.0 |
| 2.0 | 3.4 | 2.8 | 3.4 |
| 1.8 | 3.8 | 2.8 | 3.0 |

**Object no.** 112.4/T4.1/7Or
**Unit** 21

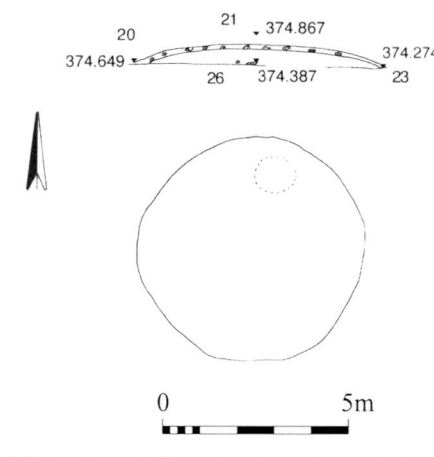

Figure 2.11. Site 112.4, Tumulus 5, section and plan (scale 1:200).

**Description** 5 ostrich (?) eggshell fragments (associated with bones 112.4/T4.1/7Or).

**Bioanthropological Report**
112.4/T4.1/7Or: Fragments of an adult, not enough to determine the sex; in very poor condition.

**Contexts**
20 layer of black stones, sealing 21
21 layer of brown earth containing flat stones up to 200 x 400mm, sealed by 20, sealing 23, 26
23 natural of hard red sand containing small stones, cut by 26, sealed by 21
26 grave (cut and fill), cut into 23, sealed by 21

Tumulus 112.4/T5 proved to be a pile of stones resembling a tumulus, but with a solid outcrop core (Figure 2.11).

**SARS Survey Site Ref.** 112.4/T6
**Site UTM Grid Point** N1880280 E580445
**Sudan Survey Number** NE-36-K/22-U-5

**Description of site**
A tumulus covered with black stones, 4.45m north-south and 4.4m east-west, with a height of 730mm (Figure 2.12). The tumulus is made from a mix of large and small stones and soil. Some fragments of pottery were discovered on the removal of the upper stone covering (112.4/T6/11C). Below this was a layer of grey dirt containing small stones, in which seed remains were also found (112.4/T6/13Or).

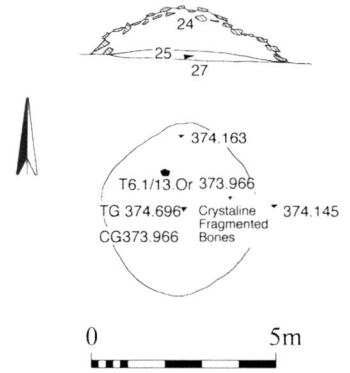

Figure 2.12. Site 112.4, Tumulus 6, top – section. bottom – plan (scale 1:200).

At the bottom of this layer, just above the natural, were the remains of a burial containing an extremely fragmentary skeleton (a few fragments only, very mineralised, 112.4/T6/12Or). There was no cut, the burial lying on the cleared natural surface. The burial was evidently disturbed, most likely by water action.

### Contexts
24 layer of black stones, sealing 25
25 layer of brown earth containing stones, sealed by 24 sealing 27
27 natural layer of hard red sand containing small stones sealed by 25

### Small Finds Report
112.4/T6/11 C context 24 pottery fragments
112.4/T6/12a Or context 25 bone fragments
112.4/T6/13 Or context 25 seeds (*Trignonella foenum graecum*) (Identification: D. Q Fuller)

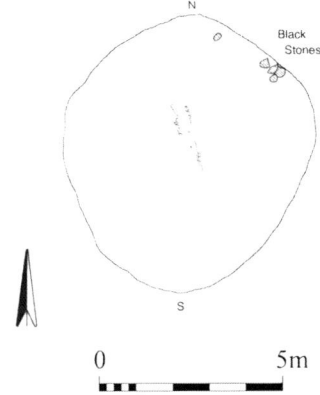

**Figure 2.13.** Site 112.3, X1, plan (scale 1:200).

**Object no.** 112.4/T6.1/12bS
**Unit** 25
**Description** Sample of concretion, rectangular in shape.
**Dimensions** (max.) L: 31.6mm, W: 15.1mm, Th: 18mm (Associated with bones 112.4/T6.1/12aS).

**Object no.** 112.4/T6.1/13 Or
**Unit** 24/25
**Description** Small yellow seeds.
**Dimensions** c. 2 x 2.25mm

### Bioanthropological Report
112.4/T6.1/12Or: A few fragments of extremely mineralised bone.

| | |
|---|---|
| **SARS Survey Site Ref.** | BM 112.3/X1 |
| **Site UTM Grid Point** | N1879695 E580155 |
| **Sudan Survey Number** | NE-36-K/22-U-3 |

### Description of site
Area strewn with potsherds (Figure 2.13). It yielded the following finds:

112.3/X1/8C Potsherds
112.3/X1/9Or Bones
112.3/X1/10S Broken stone quern

### Small Finds Report
**Object no.** 112.3/X1/10S

**Unit** Surface
**Description** Quern-stone. 1 segment of quern-stone (lower grinder) of sandstone. Grindstone Type 2. (Figure 3.2).
**Dimensions**: sector of circumference: 23mm, L: 113mm, Th (at base): c. 16mm

| | |
|---|---|
| **SARS Survey Site Ref.** | 112.5/S1 |
| **Site UTM Grid Point** | N1879780 E580645 |
| **Sudan Survey Number** | NE-36-K/22-U-5 |

### Description of site
A mound 9m in diameter and about 500-600mm high, situated on a flat plain surrounded by rock outcrops. The mound is covered with small-sized gravel, with about nine black stones on the top towards the northern half. After cleaning the gravel from the surface of the mound a thin muddy-coloured layer was revealed. This in turn covered a compact sand layer. The mound was divided into two sections along an east-west line. A fireplace comprising an ash-pit surrounded by three stones was revealed in the south-west part of the mound. On cleaning the compact sand layer about 100mm below the surface of the mound a hard muddy surface was traced. This only revealed pits in several different locations containing irregularly-shaped lumps of mud. A soft sandy layer occurred 200mm below, with similar mud-rubble filled pits containing irregularly-shaped lumps appearing in the north-east area.

| | |
|---|---|
| **SARS Survey Site Ref.** | BM 118/FS2.1 |
| **Site UTM Grid Point** | N1882714 E580228 |
| **Sudan Survey Number** | NE-36-K/22-U-17 |

### Description of site
Three round structures (112/FS2.1/S1, S2, S3), and nearby also rock carvings (Plates 2.10-2.13) including two Arabic womens' names, Zainab Fadl Almula and Naima Mohammed, and one unoccupied tumulus (118/FS2.1/T1). The structures (Figure 2.14) consisted of groups of stones around a roughly circular perimeter, with a well-preserved

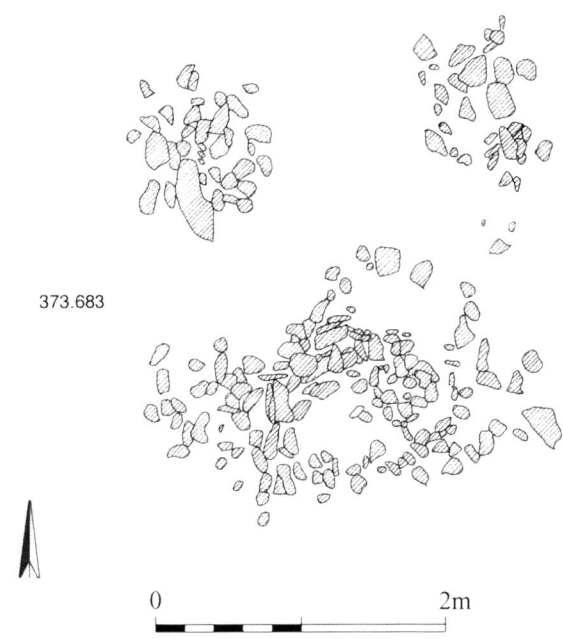

**Figure 2.14.** Site 118.FS.2.1 S3, plan of stone structure in House 3 (scale 1:50).

hearth in the middle with post-holes for a pot stand. The structures had compact floors (i.e. the desert floor cleared of large stones). Some Christian Group 1, and Groups 3 and 13 pottery was found.

**Small Finds Report: Site No. 118.F2**
**Object no.** 118F2/Site1/S1/14S
**Unit** 28
**Description** Bead of mid-green colour, anciently broken in half, of faience or glass paste. Sub-spherical in shape, with slight pitting on the exterior, together with an uneven, slightly 'ribbed' appearance to the surface. In the section, there is a hole tapering from *c.* 3.2mm at one end to 2.3mm at the other. The interior of the hole is black in colour (Figure 3.4).
**Dimensions** L: 8mm, D: 7.8mm

**Object no.** 118FS2/1/S3/Sample
**Unit** Structure 3
**Description** $C^{14}$ sample. Sample of burnt organic material from hearth.

**SARS Survey Sites Ref.** BM 153.4/T1 (Plate 2.14) and BM 153.8 – reports destroyed.

**SARS Survey Site Ref.** BM 154.5/T1
**Site UTM Grid Point** N1899449 E581786
**Sudan Survey Number** NE-36-K/22-F-27

**Description of site**
A small tumulus constructed of black blocks of stones in the middle of a rocky outcrop (Figure 2.15). Under the tumulus is the grave pit, an elongated oval-shaped shallow pit 1.9m long containing a prone extended burial aligned

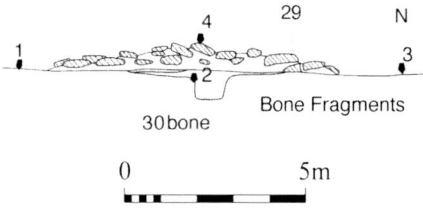

**Figure 2.15.** Site 154.5, Tumulus 1, section (scale 1:200).

east-west, head to the west, facing downwards (Plate 2.15). The partial remains of a burial, 154.5/T1.2, were also found (Plate 2.16).
**Suggested Period:** Christian
**Burial Identity:** Adult, male

**Small Finds Report**
154.5/T1/16C Pottery fragments
154.5/T1.1/28Or Shell

**Object no.** 154.5/T1.1/29S
**Unit** 30
**Description** Bead of ostrich eggshell, white in colour, generally flat annular, but somewhat asymmetrical, square with rounded corners.
**Dimensions** L: *c.* 1mm, D: 4.5mm

**Bioanthropological Report**
154.5/T1.1/3Or: A prone extended burial aligned east-west with the head to the west facing downwards. A shallow grave with animal damage to the uppermost part of the skeleton. The position of the body may indicate that the deceased was executed.
154.5/T1.2: Burial aligned east-west lying on its back facing north with head to the west. Partial skeleton only.

**SARS Survey Site Ref.** BM 155.4/S1
**Site UTM Grid Point** N1899906 E582014
**Sudan Survey Number** NE-36-K/22-F-33

**Description of site**
This is a large regular structure at the mouth of the valley, north east of BM 155. It is roughly rectangular (actually pentagonal, with a kink at the entrance in the south-east side), a little over 15m long and 12m wide (Figure 1.3). The walls are around 1.2m wide on average, with well built inner and outer facings of large stones (up to 600mm long x 400mm wide) neatly laid with a core of small stones, chippings and dirt. It is preserved up to six courses, about 700mm high (Figure 2.16a, c). The entrance is on the eastern side, with the walls turning in a short distance on either side of it. A certain amount of pottery and bone was found in the rubble from these walls, both inside and out.

The whole building had a laid floor 50-100mm thick consisting of gravel and stone chips set in mud with a few inclusions of ceramic and carbon fragments. Five storage pits were dug through this floor, Pit 1 (1.2m wide) in the eastern corner, Pit 2 (600mm wide) and Pit 3 (1m wide) just inside the north-eastern wall, and Pit 4 (1.4m wide) and Pit 5 (1.3m wide) in the southern corner. Pits 1, 4 and 5

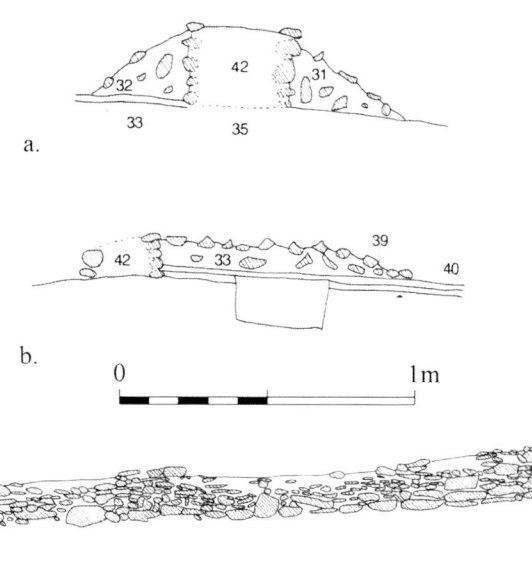

**Figure 2.16.** Site 155.4/S1, a – section through the west wall (scale 1:50), b – section through Room 1 (scale 1:50), c – western elevation, external wall face, House 155.4 (scale 1:200).

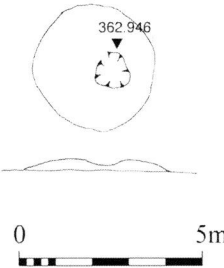

**Figure 2.17.** Site 155.4/S1/FP1, plan of fire pit south of House 155.4 (scale 1:200).

are discussed further below. Pits 2 and 3 both yielded bone and carbon remains, some pottery also coming from Pit 3.

Nineteen post-holes were uncovered (PH 6-24): three in the eastern part of the building (PH 6-8), one in the middle (PH 9), and five (PH 10-14) inside the main south-west wall. Of these, PH 13 and 14 were in the area that later became Room 2: whether or not they predate the construction of Room 2 it was not possible to tell. PH 11 was 500mm wide and may have been a pit rather than a post-hole.

One fire installation was excavated on the floor inside the structure (F2) but the main installation must have been the one outside the south-west wall (F1). This was well laid out, consisting of a low platform, 100-150mm high, of red sand raised above the desert floor (Figure 2.17). The platform was circular, 3.2m in diameter, and the sand had been baked hard by firing. A depression in the platform, off centre to the south appears to be the work of animals. A large number of substantial stones up to 300mm long was scattered on top of the platform, but with no obvious surviving pattern. The excavation of the platform yielded remains of charcoal, bone, camel droppings and pottery.

Possibly dating to the original construction of the building, there were two narrow internal walls (one coming perpendicularly out of the main south-west wall, one a continuation of the entrance wall: Walls 53 and 54). In this phase (Plates 2.20 and 2.21) the room (Room 2) was rectangular, measuring roughly 2 x 3m. These walls are of modest but not poor construction, 500 to 600mm wide, made with stones 100 to 400mm long. In some places they appear solid, in other places they are of casemate construction with a fill of pebbles and mud. These walls are built directly on the main laid floor of the building. There are two pits in this corner dug through the original floor (Pits 4 and 5). There is no direct stratigraphic connection between these pits and Walls 53 and 54: the pits could have been dug before, during or after the construction of Walls 53 and 54. But given how neatly Pits 4 and 5 fit into this earlier phase of Room 2, it is probable that pits and room all date to the same time, and probably very shortly after the construction of the main building. Pit 4 and Pit 5 each yielded a quantity of remains of pottery, bones and carbon.

In a phase subsequent to the original construction of the building, Room 2 was changed when a roughly circular wall was built in the southern corner, approximately 3.8m across (Wall 41, Plate 2.22). Bone, pottery and carbon remains were all found in the rubble, but once again it was not possible to determine whether this was rubble from the collapse of the main wall or the collapse of Wall 41.

At some stage after the initial construction of the building but prior to the construction of the late phase Room 1, a pit was dug just inside where the entrance to this room would subsequently be (Pit 1). The fill contained a number of potsherds.

On the north side the late phase extension turns back in from the entrance way and runs to the north east, thus delineating a room (Room 1) along the eastern side, approximately 2.5m wide, entered at its northern corner. This wall sits on the layer of sand covering the original floor of the whole building, indicating that it was built after the initial abandonment of the building but presumably before the collapse of the main walls. It is preserved only one to two courses high and was built of stones 200 to 300mm long roughly laid on top of each other (Figure 2.16b). Some potsherds and some carbon remains were found in the rubble overlying Room 1 but it was not clear whether they came from the collapse of the later wall or the original main walls. About a third of the way along the main outer wall of the room (from the eastern corner) were the remains of a drain passing through the wall at ground level. Almost certainly this drain was part of the original construction.

In summary, the main walls and the laid floor of the building were constructed at the same time. Later, but probably very shortly afterwards, Room 2 was laid out and housed two storage pits. A number of other storage pits and post-holes were also installed around the building during its main phase. There was one small fire installation inside the building and a much larger one outside. Subsequent to the initial abandonment of the building, but before its walls had collapsed, Room 1 was built along the eastern side.

**Contexts**
31 rubble outside main walls
32 rubble inside main walls
33 rubble inside Room 1
34 laid floor of Room 1
35 red natural sand under floor of Room 1
36 Pit 1, cut through floor 34
37 sand fill of Room 2
38 fill of Pit 1
39 later wall in NE corner of Room 1
40 sand above original floor
41 circular wall of Room 2 (later phase)
42 main walls of the building
43 Pit 2
44 fill of Pit 2
45 Pit 3
46 fill of Pit 3
47 rubble in Room 2
48 floor of courtyard
49 Pit 4
50 fill of Pit 4
51 Pit 5
52 fill of Pit 5
53 North-west internal wall of Room 2 (earlier phase)
54 North-east internal wall of Room 2 (earlier phase)
286 Fire installation 1

**Suggested Period** Christian?, Islamic?

**Small Finds Report**
155.4/S1/1C context 32 pottery in the inside rubble
155.4/S1/2C context 31 pottery in the outside rubble

155.4/S1/4Or context 31 bones in the outside rubble
155.4/S1/5Or context 32 bones in the inside rubble
155.4/S1/6C context 38 pottery
155.4/S1/7C context 33 pottery from fill overlying floor
155.4/S1/8C context 34 pottery from floor
155.4/S1/9Or context 34 C14 sample from floor
155.4/S1/10Or context 33 C14 sample from fill overlying floor
155.4/S1/11Or context 44 bone from Pit 2
155.4/S1/12Or context 44 C14 sample from Pit 2
155.4/S1/13Or context 46 bones from Pit 2
155.4/S1/14Or context 46 C14 from Pit 2
155.4/S1/15C context 46 pottery from Pit 2
155.4/S1/17Or context 47 bones in fill from Room 2
155.4/S1/18C context 48 pottery from courtyard floor
155.4/S1/19C context 47 pottery in fill from Room 2
155.4/S1/20Or context 47 C14 in fill from Room 2
155.4/S1/21Or context 50 C14 from Pit 4
155.4/S1/22Or context 50 shell/bones from Pit 4
155.4/S1/23C context 50 pottery from Pit 4
155.4/S1/24Or context 52 C14 from Pit 5
155.4/S1/25C context 52 pottery from Pit 5
155.4/S1/26Or context 52 shell(?) from Pit 5
155.4/S1/27Or context 52 bones from Pit 5
155.4/F1/84Or context 286 charcoal
155.4/F1/85Or context 286 bones, camel droppings
155.4/F1/86C context 286 pottery

**SITE NO. 155.4**
**Object no.** 155.4/S1/9Or
**Unit 34 Description** $^{14}$C sample of burnt vegetal material.

**Object no.** 155.4/S1/10Or
**Unit 33 Description** $^{14}$C sample of burnt vegetal material.

**Object no.** 155.4/S1/12Or
**Unit 44 Description** $^{14}$C sample of burnt vegetal material.

**Object no.** 155.4/S1/14Or
**Unit 46 Description** $^{14}$C sample of carbonised vegetal material.

**Object no.** 155.4/S1/20Or
**Unit 47 Description** $^{14}$C sample of carbonised vegetal material.

**Object no.** 155.4/S1/21Or
**Unit 50 Description** $^{14}$C sample of carbonised vegetal material.

**Object no.** 155.4/S1/24Or
**Unit 52 Description** $^{14}$C sample of carbonised vegetal material.

**Object no.** 155.4/S1/26Or
**Unit 52 Description** 1 snake egg.

**Object no.** 155.4/S1/84Or
**Unit 286 Description** $^{14}$C sample of charcoal.

**Object no.** 155.4/S1/85Or
**Unit 286 Description** Organic remains. 4 fragments of bone together with associated fragments: 3 camel (?) droppings.

**Object no.** 155.4/S1/Sample (1)
**Description** $^{14}$C sample of carbonised vegetal material.

**Object no.** 155.4/S1/Sample (2)
**Description** $^{14}$C sample of carbonised vegetal material.

**SARS Survey Site Ref.** 170.1/T1
**Site UTM Grid Point** N1906162.887 E582918.795
**Sudan Survey Number** NE-36-K/16-U-6

**Description of site**:
A low mound of black coloured stone blocks covered by wind-blown sand (Figures 2.18 and 2.19 top, bottom). Its diameter measures 5.5m north-south. The make-up of the mound is a mix of stones and soil. At ground level beneath the northern half of the mound is a shallow pit containing the burial of an infant, crouched, with the head to the north, on its right side, facing west. The fill of the burial pit is a loose soft soil mixed with sand.

**Bioanthropological Report**
170.1/T1.1/2Or: A crouched burial, with the head to the north, on its right side facing west (Plate 2.41). The left leg and pelvis are missing, possibly due to animal disturbance. Teeth recovered in area of the feet. Individual in early childhood (see pg. 129).

**SARS Survey Site Ref.** 170.1/T2

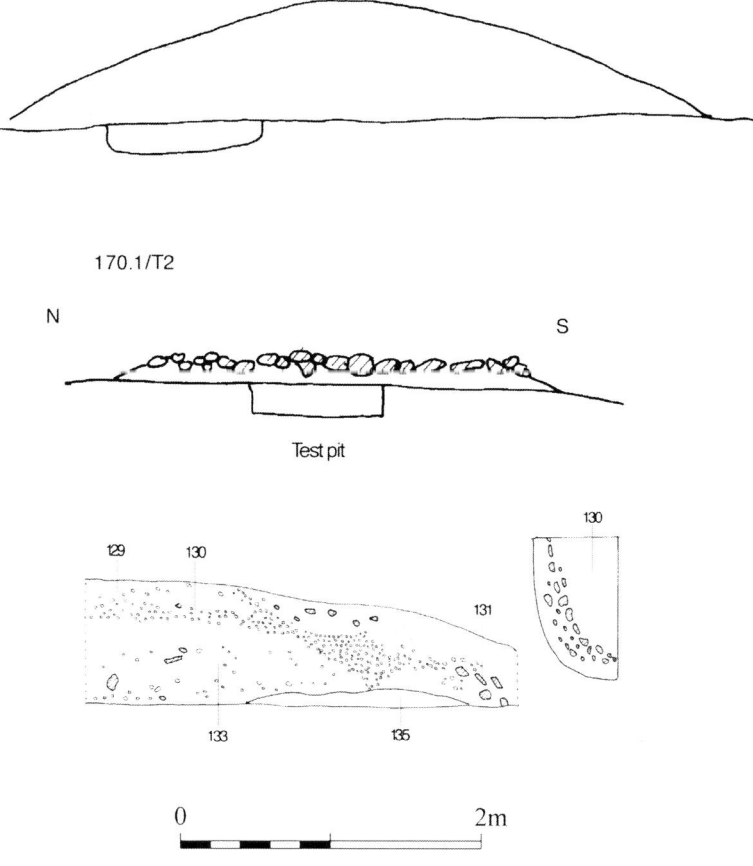

Figure 2.19. Site 170.1, Tumulus 1, top – section, middle – 170.1/T2, section, bottom – detail of T1 (scale 1:50).

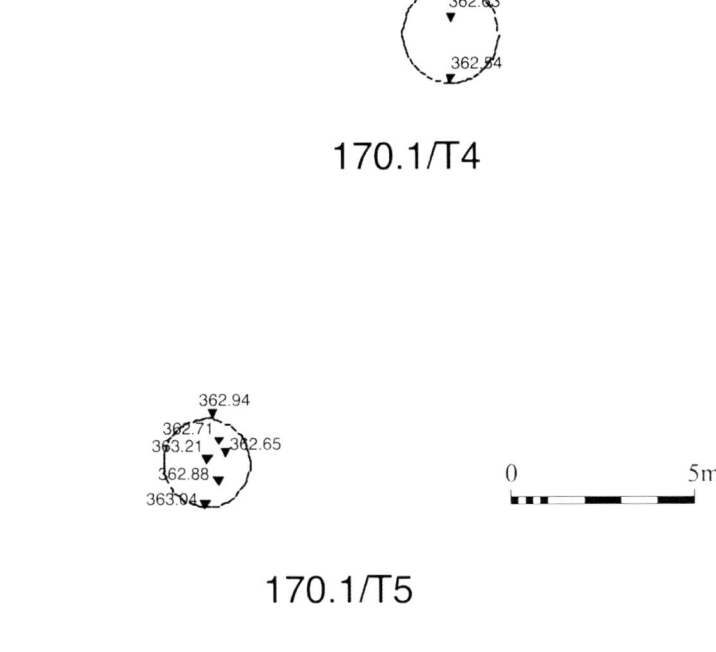

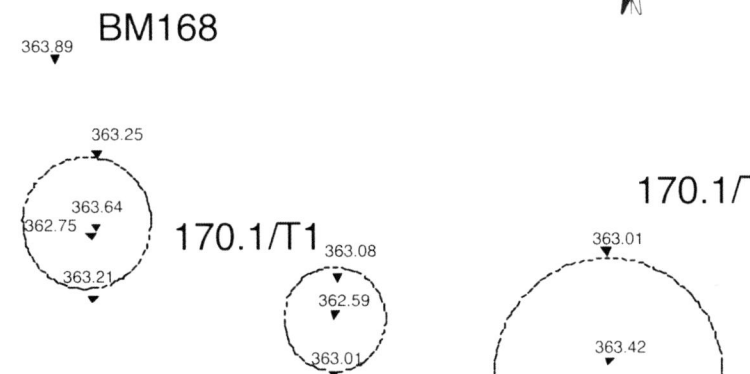

Figure 2.18. Site 170.1, general plan (scale 1:200).

**Site UTM Grid Point** N1906162.887 E582918.795
**Sudan Survey Number** NE-36-K/16-U-6

**Description of site**:
A small tumulus situated on the top of an outcrop made of black stone blocks covered by wind-blown sand (Figures 2.18 and 2.19 middle). Down to the original surface a semi-oval shallow grave pit was filled with a loose soil covering the burial. Its north-south diameter is 3.5m.

**Small Finds and Samples Report**
There were no finds associated with the burial.

**Bioanthropological Report**
170.1/T2.1/3Or: A crouched infant burial (0-1 yrs), head to the east, lying on its right hand side, facing north, relatively complete, skull very fragmentary (Plate 2.40) (see pp. 129-30).

**SARS Survey Site Ref.** 170.1/T3
**Site UTM Grid Point**
N1906162.887 E582918.795
**Sudan Survey Number**
NE-36-K/16-U-6

**Description of site**
A small mound of black stone blocks covered by wind blown sand, slightly raised.

**Small Finds and Samples Report**
No finds associated with the grave.

**Bioanthropological Report**
Tightly contracted (crouched) sub-adult burial of unknown age, lying on its left side with the head to the east, facing south. Forearms appear between the legs. Deciduous molars noted. Sex indeterminate. Clearly defined feet. Preservation condition is very soft and fragile.

**SARS Survey Site Ref.** 170.1/T4
**Site UTM Grid Point**
N1906162.887 E582918.795
**Sudan Survey Number**
NE-36-K/16-U-6

**Description of site**
A slightly raised tumulus of black stone blocks and wind-blown sand. Its diameter from north to south is 3.75m.

**Small Finds Report**
No finds associated with the grave.

**Bioanthropological Report**
170.1/T4.1/7Or: Tightly contracted burial, lying on its right side with a displaced skull to the east, facing west. Incomplete skeleton with disturbance, possible animal damage. Dental caries noted on mandible molars. Preservation is very poor. Puberty, 11-15 yrs (see pg. 130).

**SARS Survey Site Ref.** 170.1/T5
**Site UTM Grid Point** N1906162.887 E582918.795
**Sudan Survey Number** NE-36-K/16-U-6

**Description of site**:
A slightly raised small tumulus of black stone blocks and wind-blown sand. Its north to south diameter is 3.45m.

**Small Finds Report**
No finds in the burial.

**Bioanthropological Report**
170.1/T5.1/8Or: A crouched burial (Plate 2.17) lying on its left hand side with the body axis east to west, head to

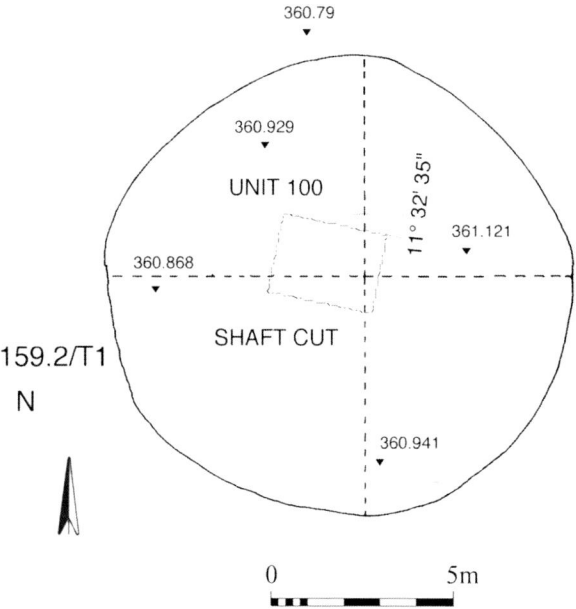

**Figure 2.20.** Site 159.2, Tumulus 1, plan (scale 1:200).

the east, facing south west. The skeleton, which is soft and fragile, is fragmented, incomplete and disturbed. Adult of unknown age and undeterminate sex (see pp. 130-1).

**SARS Survey Site Ref.** 166.2
**Site UTM Grid Point** N1905340.958 E 582794.709
**Sudan Survey Number** NE-36-K/22-A-16.

**Small Finds and Samples Report**
**Object no.** Village below site 166.2
**Unit** surface
**Description** 3 lithic artefacts (associated with sample of potsherds)
  1. Blade, triangular cross-section, with a thin cortex layer on the shortest side. One end is broken with a step fracture. The edge is chipped lightly and has small pressure flake scars (Figure 3.1, 9).
**Identification** Secondary blade with possible use retouch. Made on chert river cobble.
  2. Flake with thin cortex layer over much of the ventral surface. The dorsal surface is irregular, with part of the bulb of percussion absent. The piece is lightly chipped around its three edges (Figure 3.1, 10).
**Identification** Primary flake, with probable post-depositional damage on edges. Chert.
  3. Small fragment of flake, cortex on the ventral surface. Two edges have step fractures, whilst the third edge has light intermittent chipping.
**Identification** Broken primary flake, with haematite staining from long period on ground surface. Made on chert river cobble.
**Dimensions: 1.** L: 48.5mm, W: 26mm. **2.** L: 41.5mm, W: 30mm. **3.** L: 15.9mm, W: 15.8mm

**SARS Survey Site Ref.** 159.2/T1
**Site UTM Grid Point** N1902074 E582382
**Sudan Survey Number** NE-36-K/22-A-5

**Description of site**
This post-Meroitic tumulus was situated in the southern part of the cemetery (see Plates 2.18 and 2.25; for plan of Gabati area see Figure 2.47). It measured 8.8m east-west, 9m north-south and it was 800mm in height (Figure 2.22). The tumulus consisted of different sizes of sandstones, 100-300mm or smaller, with coarse gravel as a top layer. The actual mound sloped down gently from the centre to the circumference and contained a mix of loose brown gravel, silt and a layer of fine sand. The fill of the tumulus (Figures 2.23 and 2.24) contained red-burnished potsherds and snail shells, found 2-3m north of the east-west axis. At a depth of 700mm below the ground surface an earlier burial consisting of human bones associated with a half mud brick was found in a very bad condition. The tumulus fill also contained an ash pit and charcoal. The tumulus make-up covered a trapezium-shaped shaft situated in the centre of the tumulus, cut from the old ground surface. It was filled with a mixture of loose brown soil, white chalk and pink soil. At the bottom were uncovered the slabs blocking the burial chamber, which were laid in four layers occupying the entire space of the shaft bottom (Figure 2.23). The shaft was aligned 11.5° south of east-west descending from east to west towards the entrance of the single burial chamber. The entrance to the chamber was marked by two 'shoulders' cut into the western baulk (Figure 2.24). It measured 2.1m north-south at its western side and 2.7m east-west at the centre, with a depth of 1.5m at its western end. An oval opening gave access to a hollowed vaulted chamber, oval in plan, measuring 1.9m north-south and 1.3m east-west (Plate 2.25). It contained the skeleton of an adult male, lying in a flexed position, aligned north-south on his right side with the head to the south, facing east, the arms in a flexed position and the hands near the face. A collapsed or dismantled wooden bed was found *in situ* with four bed legs in the four corners of the chamber and the poles of the frame on each side (Plate 2.38). Pieces of ropes and leather webbing were found attached to the bed frame. A large pottery bowl was found at the southern edge of the chamber and another smaller bowl was found at the back of the head, 200mm east of the western edge of the chamber. Fragments of arrowheads were found in the area of the hands along with a stitched leather piece. Pieces of termite-eaten cloth and a leather object were found near the feet.
**Suggested Period**: Post-Meroitic.

Drawing Nos Figure 2.20, 2.22
Spot heights local datum BM 159, H 358.658m
BM Map Point: 17° 12'09.6928 N 33° 46'25.6458 E

**Small Finds Report**
Description of finds and samples
For the illustrations of the main objects, see Figure 8.1, Plates 8.1-8.4.
Handmade bowl (41C): Complete apart from a few chips. Fabric G12. (For fabric descriptions see the pottery report by Rose and Smith (1998)). Broad open black burnished bowl. It has incised decoration on the interior in the form of a 'W'-shaped design. H: 103mm, rim D: 340mm, max. D: 342mm.

Handmade bowl (42C): Complete. Fabric may be similar to G12. In form, it has a concavo-convex profile. The exterior surface exhibits a light burnish, on the lower part. Decoration consists of incised undulating loop motifs within a roughly rectangular panel on each side, with traces of white infill. H: 95mm, rim D: 107mm, max. D: 122mm.

(1) Textile: One band of textile from beside the head. Colour: reddish-brown. L: *c.* 50mm, W: 15mm.

(2) Samples of rope: Three samples of rope, from bed-frame. 2a: 28 pieces of rope. L: range *c.* 18-74.5mm, D: 2.5-4 mm. 2b: *c.* 40 pieces of rope from bed-frame. L: range *c.* L:65 - *c.* 55mm, D: 2.5-4.7mm. Includes two examples of knotted ends. 2c: *c.* 90 pieces of rope. Max. L: 50mm, D: 3.2-7.5mm. Includes one fragment of a knot.

(3) Sample of wood: One sample of wood, preserved in *c.* 20 pieces much attacked by termites. Pieces range in size: 102.6 x 16.2mm, to *c.* 15 x 1mm.

(4) Samples from leather pouch (?) : Fragments of leather from beside the ankles and lower legs. Sheet leather. Includes fragments with pairs of holes 1.8mm in diameter near edge, fragments curled and folded over and fragments of double twisted strips. Size: *c.* 100 x 55mm.

(5) Samples of textile: Three samples of textile fragments from headgear on body. Colour: reddish-brown. Very fine weave. Max. dimensions 44 x 32mm.

(6) Wooden bed frame: Part of wooden bed frame, at present curved into an arc-shape. Both ends are cut to form tenons to fit into the mortice sockets in the legs. Original diameter not preserved. L: *c.* 760mm, W: 45mm, T: 30mm.

(7) Wooden bed frame: Five sections.
1. strongly curved. Present L: *c.* 515mm, W: 40mm, T: 30mm. Has each end cut to form tenons.
2. curved. Present L: *c.* 565mm, W: 50mm, T: 40mm.
3. slightly curved. Present L: *c.* 395mm, W: *c.* 50mm, T: 30mm. Has one end cut to form a tenon.
4. slightly curved. Present L: *c.* 500mm, W: 54mm, T: 33mm. Has one end cut to form a tenon.
5. fragmentary section, L: *c.* 275mm, W: 55mm, T: 30mm.

(8) Wooden bed leg: Lower section, originally with a double bulbous section above foot. Remaining portion L: *c.* 310mm, W: *c.* 75mm, T: 80mm.

(9) Wooden bed leg: Three sections: comprising:
1. Section with rectangular mortice socket.
2. Section L: *c.* 180mm, cross-section *c.* 80mm square, with a socket *c.* 40 x 35mm.
3. Foot of bed leg: 2 bulbous sections and splayed-out foot. At the base, the cross-section is oval with max. D: of *c.* 95mm.

(10) Wooden bed leg: Three sections:
1. Upper part of bed leg. L: *c.* 190mm, cross-section *c.* 80 x 90mm. with two mortices.
2. Bulbous section of leg, cross-section *c.* 70mm square
3. Second bulbous section and splayed foot – L: *c.* 130mm, cross-section *c.* 100 x 85mm at foot.

(11) Sample of textile: Sample from strip of textile, striped horizontally in alternate blue and buff (originally white?) stripes. Evidence for eight stripes. Blue stripe W: *c.* 8.5mm, buff stripe W: *c.* 8.4mm. Max. dimensions *c.* 16.8 x 13.4mm.

(12) Sample of leather: Leather, with slightly roughened, 'braided' appearance of surface on one side. Rest featureless apart from irregular undulations. One piece with layer folded back on itself and then over again. Size: *c.* 100 x 50 x 1mm.

(13) Samples of textile and leather: Two samples of leather strips and textile fragments. Textile: 1: Colour: blue and white stripes each *c.* 8mm wide. Max. dimensions 26.5 x 18mm. 2: Colour: light brown. Max. dimensions 22 x 17mm. Leather strips, *c.* 24 x 4 x 1.5mm.

(14) Samples from leather garment:
a. Sample of leather 18.5 x 41 x 9mm.
b. Approximately 20-30 fragments of leather, including three pieces curled or folded over. Max. dimensions *c.* 45 x 20mm.
c. One sheet of leather with remains of thread through holes on one side, 65 x 50 x *c.* 9mm, one sheet of leather 22 x 14 x 9 mm, one piece folded over on itself, 28 x 18mm.

(15) Sample of leather: Two samples of leather, from either side of the bottom of a bed frame and over that frame.
a: section of leather sheet curved to fit over wooden frame. Appears to be one layer with approximately triangular shaped layer on top, with three rows of stitching.
b. curved section of a double layer of leather, comprising inner layer and an outer layer, shorter and with a rounded end. Has two lines of stitching *c.* 12.3mm from each edge and one in the centre. Size: 15a. 94.5 x 42.5mm. 15b. 68.7 x 63mm.

(16) Sample of leather: Section of double layer of leather, one end broken, other end with original edge, rounded at corners. Has row of stitching around one long side and the rounded end, *c.* 3mm from edge. Size: 58 x 45mm.

(17) Sample of textile: Colour: light brown. Sample of coarsely-woven cloth. Section folded over on itself and twisted. D: *c.* 47mm.

(18) Iron arrow head: approximately leaf-shaped with long tang. Tip and possible barb missing. L: 59.4mm, W: 21mm., T: 7.8mm.

(19) Samples from leather sheath: One end of leather sheath (with stitching). Fragments of leather: one fragment of leather comprising two sheets still attached in central area, and two fragments with possible stitch holes. Size: *c.* 28 x 21 x 0.9mm.

(20) Sample of rope: approximately 90 fragments of rope from the lower end of a bed. Three thicknesses: a. D: *c.* 6.2mm b. D: *c.* 5mm, c. D: *c.* 3.5mm.

(21) Samples of plaited leather: a. 150 fragments of leather strips up to *c.* 30 x 4 x 1mm in size. b. Five samples of leather strips including examples looped around plaited or twisted strips at right angles. c. Three samples of rope: max. dimensions L: 50mm, D: 5.5mm.

(22) Sample of leather: a. Two layers of leather sheet fragments. Majority slightly undulating, including four fragments with edges curled in, one fragment with edges folded over. b. Four fragments with textile adhering, one section of thong with two knots around it comprising a

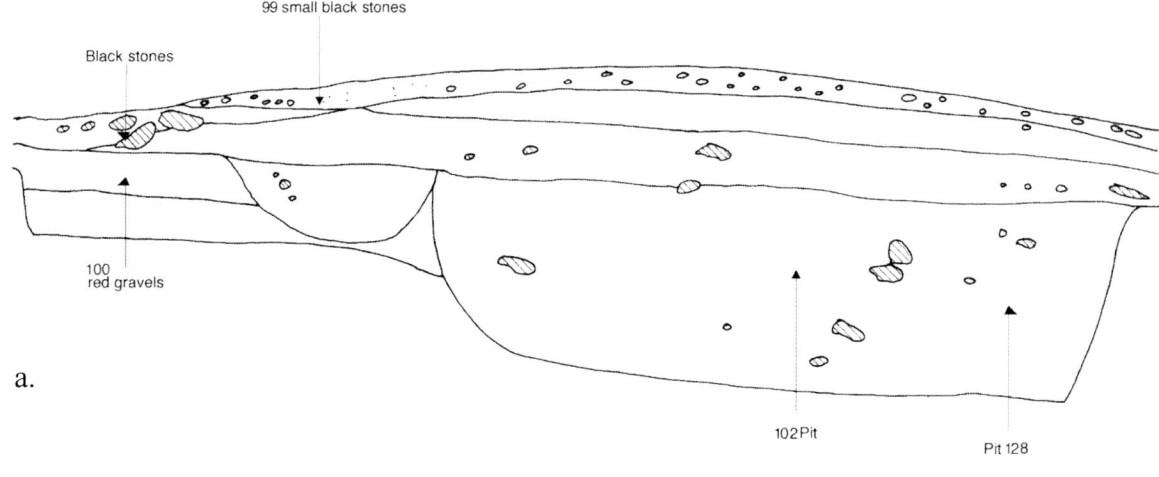

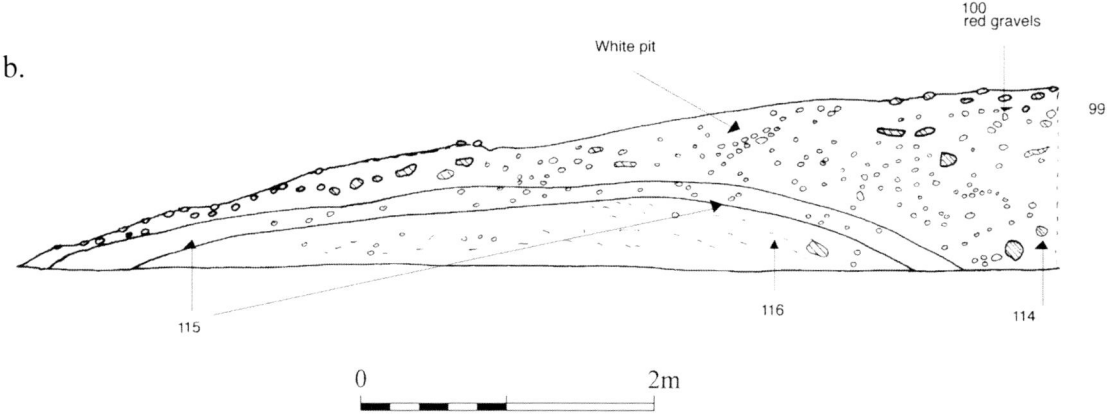

**Figure 2.22.** Site 159.2, Tumulus 1, a – east-west section, b – north-south section (scale 1:50).

looped strip, passed round a thong and the ends pulled back through loop. Size: *c.* 80 x 110mm (two layers).

(23) Leather thongs or strips: approximately 100 pieces, *c.* 7-34mm long x 4-5mm x *c.* 1mm. One fragment of a knot. One fragment of rope, L: 26mm, D: 5.5mm.

Iron spear or end of knife-blade (24): Corroded. One end of iron blade of narrow spear or knife. L: 78mm, W: 15.5mm, T: 6.5mm.

Sample of leather (25): *c.* 14 fragments of thin leather. Two appear to be folded over onto themselves. One piece with evidence for three thin slits *c.* 4.4mm x *c.* 0.5mm. Rest featureless. Max. dimensions *c.* 32 x 35.5 x 1mm.

Sample of bed-stringing (26): Rope and leather a. Sample of plaited rope – max. L: *c.* 50mm, general D: *c.* 5mm. b. Sample of leather with four doubled strips looped over plaited or twisted strip at right angles, having ends of rope passing round central plait, partly over, and partly in between the leather strips looped over it.

**SARS Survey Site Ref.**  159.2/T2
**Site UTM Grid Point**  N1902074 E582382
**Sudan Survey Number**  NE-36-K/22-A-5

**Description of site**:
This tumulus, a shallow mound about 7m in diameter and 300mm high, was situated in the southern half of the cemetery. The tumulus was constructed from a mix of loose gravel and soil, white and pink in colour. At the original ground surface the grave shaft was uncovered beneath the north-western quadrant of the tumulus, aligned east-west, trapezoidal in shape, measuring 2.5m long and 1.9-1.45m wide (decreasing from west to east), descending from west to east towards the entrance of the burial chamber (Figure 2.25). The cut was filled with a loose white and pink coloured gravel. Half of the shaft floor was sealed with stone slabs on the tomb entrance side. The entrance itself was blocked by stone slabs laid in layers and cemented in with mortar (Figure 2.26). A vaulted chamber contained the skeleton of an adult female, with a baby, lying flexed on its left side, head to the north, facing east with the hands near the face. Most of the body was covered by a shroud of coarsely woven textile, also some rust-coloured fabric and a mat or a thick cloth with pink threads (possibly oval in shape). It was also likely that the deceased was laid on a striped fabric. The skeleton was laid on a wooden bed woven with leather stitches, two bed legs were found on opposite sides of the tomb, north and south. At the southern part of the tomb 100mm from the wall was a rust-coloured bowl. There was also a scatter of small blue beads, shells, small ivory beads and three large beads near the ankles along with carnelian ones around the feet.

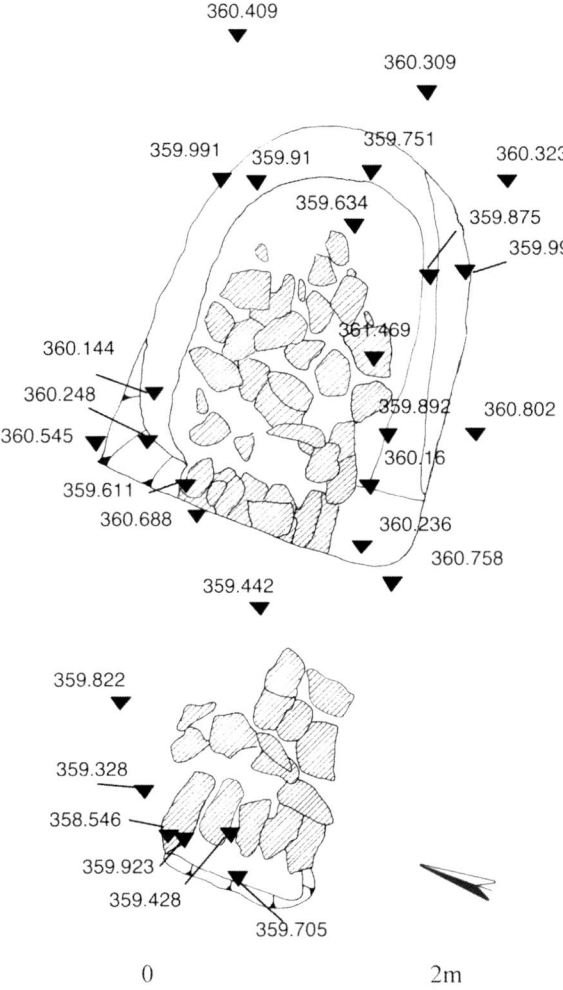

**Figure 2.23.** Site 159.2, Tumulus 1, tomb shaft, top – upper blocking stones, bottom – lower blocking stones (scale 1:50).

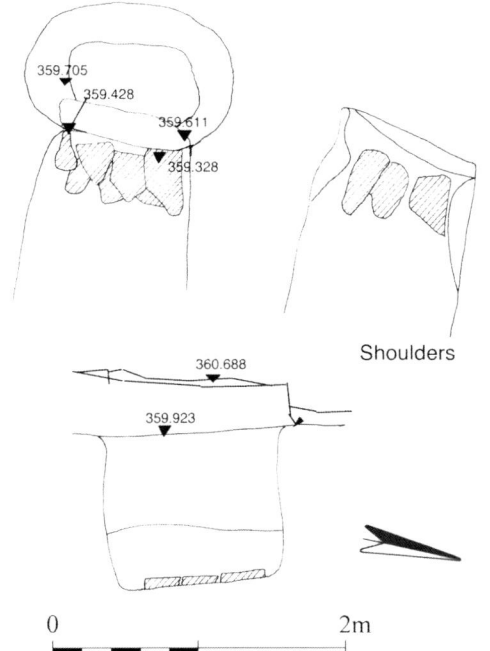

**Figure 2.24.** Site 159.2, Tumulus 1, top left – shaft cut and burial chamber, top right – lower two levels of blocking stones, bottom – section across the entrance shaft showing tomb entrance and lowest blocking stones (scale 1:50).

Part of the bed frame was pushed behind the deceased and laid out along the long axis of the tomb with a roll of plaited leather near it. More beads were found around the neck area.

**Small Finds Report**
Description of finds and samples
For the illustrations of the main objects see Figure 8.2, Plates 8.5-8.7.

(31C) Wheel-made oil-bottle: Complete. Fabric probably G18. The exterior is red slipped with decoration comprising shallow to moderate ribbing. H: 22mm, rim D: 76mm, max. D: 114mm.

(32C) Handmade bowl: Complete. Fabric probably G12. It has a strongly inturned upper body, with black-burnished exterior. Decoration comprises two vertical cordons with oblique incised lines and two sets of incised cross-like motifs. H: 102mm, rim D: 110mm, max. D: 171mm.

(1) Sample of textile: Colour: greyish-brown. Wrapped around foot. Dimensions not determinable.

(2) Sample of textile and leather: Textile. Colour: grey-brown. Dimensions not determinable. Leather: small fragments including a knot, *c.* 33 x 18mm.

(2a) Seven beads.
  spherical, opaque white quartz.
  spherical, semi-translucent quartz.
  cylindrical, semi-translucent glass.
  short cylindrical, semi-translucent glass.
  spherical, red-brown carnelian.
  barrel-shaped, purplish-brown with broad white central stripe, faience or glass .
  short cylindrical, dark blue with broad white central stripe, faience or glass.

(3) Sample of textile and leather: Sample of textile and leather fragments. Textile: Colour: grey-brown. Size *c.* 34 x 50mm.

(4) Samples of textile and tanned skin: Textile sample with leather sample.

(5) Biological sample, skin: See report on human remains (Judd 2012, 86).

(6) Sample of tanned skin/leather: Size: 55 x 28.5mm.

(7) Biological sample, baby bones: See report on human remains (Judd 2012, 86-87).

(8) Sample of textile and leather: Textile from around tibia and fibula of body, including small pieces of leather strips from bed-stringing. Colour: brown. Size: 120 x 70mm.

(9, 10) Samples of textile and leather: (9): *c.* 20 fragments of leather with rough surface and slightly corrugated. Max. L: *c.* 49mm. (10): Four fragments of textile. Colour: grey-brown and buff-brown. Size: *c.* 20 x 10mm. Five examples of looped and knotted leather strips, comprising two examples of leather strips or thongs of round cross-section, one with three strands, 3mm in diameter, twisted and looped into knot at one end and three twisted as above, looped around a second three-strand rope at right angles, together with *c.* 10-20 small fragments of leather.

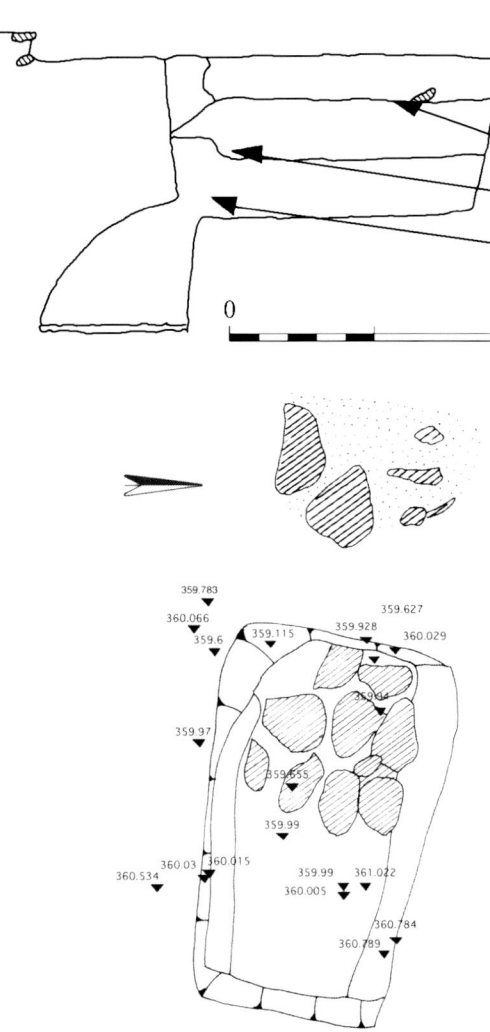

Figure 2.25. 159.2/T2, east-west section (scale 1:50).

Figure 2.26. Site 159.2, Tumulus 2, top – plan of blocking stones, upper level. bottom – plan of blocking stones, lower level (scale 1:50).

(11, 12) Biological samples, skin: See report on human remains (Judd 2012, 86).

(13) Sample of textile: One with colour: mid greyish-brown. Size; *c*. 150 x 70mm. One with colours dark grey and orange-brown. Size; *c*. 30 x 20mm.

(14) Sample of textile: Examples of textile threads. Colour: brown and pink. Size: *c*. 36 x 23mm.

(15) Biological sample, hair, scalp: See report on human remains (Judd 2012, 86).

(16) Sample of textile: Textile from in front of chin of body. Colour: dark brown. Size of sample: *c*. 120 x 65mm.

(17) Biological sample, skin: See report on human remains (Judd 2012, 86).

(18) Biological sample, hair, scalp: See report on human remains (Judd 2012, 86).

(19) Sample of textile: Textile from the bottom edge of bed. Includes borders of the cloth, together with threads from the centre. Colour: brown. Size of sample: *c*. 45 x 28mm.

(20) Sample of textile and leather fragments: 1: *c*. 25 fragments of leather. Max. dimensions 50 x 20mm. 2: Three to four small fragments of textile. Colour: grey-brown to buff-brown, *c*. 10mm square.

(21) Sample of textile and leather: 1: Sample of textile. Colour: light reddish-brown stripe >7mm wide, mid-blue stripe >10mm wide. Size: 23 x 35mm. 2: Small fragments of textile. Colour: greyish-brown. 3: Small fragments of leather. Size: *c*. 10mm square.

(22) Sample of textile and leather: 1: *c*. 10 fragments of leather, including one curved fragment. Max. dimensions *c*. 35 x 19 x *c*. 0.8mm. 2: Sample of textile with thin fragments of leather adhering in some areas to one side. Colour: buff-brown. Size: *c*. 80mm square.

(23) Sample of textile and leather: 1: *c*. 30 fragments leather. Size: *c*. 35 x 13 x 0.5mm. 2: fragment of leather curled over at edges with textile adhering to one end. Size: 42 x 46 x 1.5mm.

(24) Sample of textile: Colour: reddish-brown. Size: *c*. 45 x 40mm.

(25) Samples of textile: 1: one fragment of coarse weave. Colour: grey. 2: four fragments of fine weave. Colour: reddish-brown. Size: 25 x 17.3mm. 3: four fragments. Colour: buff with grey-blue stripe. Size *c*. 15 x 10mm.

(26a) Shell: Cowrie shell with back removed. L: 30mm, W: 20mm, Max. T: 8mm

(26b) Beads: 42 beads.
two cylindrical, greenish-blue faience.
two cylindrical, pale blue faience.
38 disk, white shell.

(27) Beads: 18 beads.
11 subspherical and ovoid, dark brown, made from husks of seeds.
seven disk, white bone.

Biological sample, bones of an infant (28): report on human remains in preparation.

(29) Sample of textile: Colour: striped mid grey, grey-blue, pale brown. Original dimensions indeterminate.

(30) Wooden bed leg: Form only preserved. Lower part of bed leg showing two indentations forming double bulbous profile. Remaining L: *c*. 160mm, cross-section *c*. 70mm square.

(31-32) Sample of bed pole and bed-stringing: Samples of

sections from pole from bed, preserved in form only. Five samples of plaited strips of leather from bed-stringing. Cross-plaited over/under/over, with portions of knotted ends. Strips Max. W: *c.* 5.5mm, Max. T: *c.* 2.2mm.

(33) Beads: 32 beads.
   two cylindrical mid amber-coloured, carnelian.
   two cylindrical banded buff and olive, agate.
   two subspherical opaque white, quartz.
   nine subspherical dark amber-coloured, carnelian.
   one subspherical opaque white, quartz.
   one cylindrical greenish glass.
   one assymmetrical cylindrical dark grey, 'stone'.
   one spherical amber-coloured, carnelian.
   two short cylindrical dark grey, 'stone'.
   four subspherical semi-translucent white, quartz.
   one subspherical grey, 'stone'.
   one barrel-shaped amber-coloured, carnelian.
   one tapering barrel-shaped white, quartz.
   one short annular white, bone.
   one asymmetric barrel-shaped translucent white, quartz.
   one cylindrical white, quartz.
   one biconical with square central cross-section, amber-coloured, carnelian.

(19C) Compartmented dish: Approximately two-thirds complete, recovered from fill of tumulus (Figure 8.2). Fabric G5. A dish of approximately oval shape, having two compartments. There is incised decoration on the outer rim and the central dividing wall comprising short, oblique lines and 'X'-like motifs. Original L: *c.* 190mm, W: 106mm, H: *c.* 63mm.

Spot heights local datum BM 159
H 358.658 m
BM Map Point: 17° 12'09.6928 N 33° 46'25.6458 E
Date Recorded 14.5.93+
**Suggested Period:** Post-Meroitic

**SARS Survey Site Ref.**    159.2/T3
**Site UTM Grid Point**    N1902074 E582382
**Sudan Survey Number**    NE-36-K/22-A-5

**Description of site**
This tumulus of ferrocrete sandstone flakes covered with wind-blown sand measures 2.4m east-west and 1.74m from north to south (Figure 2.27). The tumulus was divided into four quadrants which were excavated sequentially to reveal a layer of yellow-red gravel under the top layer of ferrocrete sandstone rocks, coarse pottery was found inserted in the fill at this early stage. The mound covered an extended north-south aligned burial pit filled with red silt and gravel and topped with a layer of different sized stones of which some were flat. The fill of the burial pit contained the skeleton of a supine extended adult male aligned north-south.

Spot heights local datum BM 159
Ht 358.658 m
BM Map Point: 17×12'09.6928N 33×46'25.6458E
Date Recorded 14.5.93+
**Suggested Period:** Christian
**Burial Identity:** Male

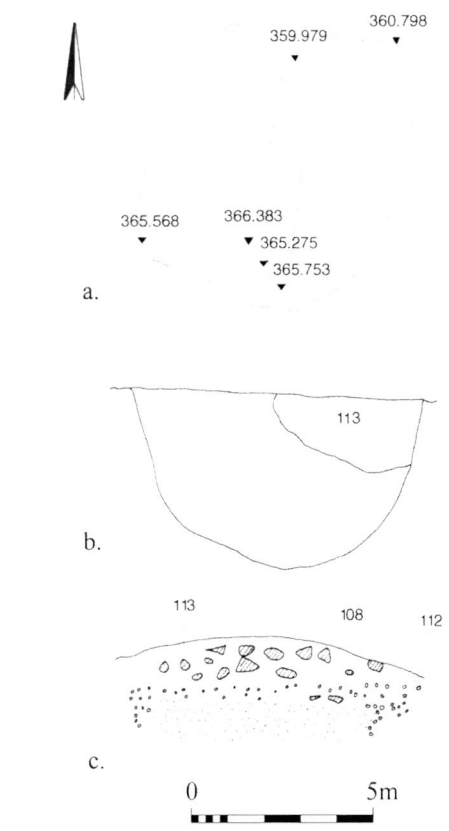

**Figure 2.27.** Site 159.2, Tumulus 3, a – plan, b – detail plan of excavation, c – section (scale 1:200).

**SARS Survey Site Ref.**    159.2/T4
**Site UTM Grid Point**    N1902074 E582382
**Sudan Survey Number**    NE-36-K/22-A-5

**Description of site**
This tumulus, a low mound 300mm high and 6m in diameter, was situated in the southern part of the cemetery. The mound was covered with small ferrocrete sandstone flakes, gravel and white chalk gravel. The mound was characterised by a ring of stones around its circumference (Figure 2.28). The make-up of the tumulus was compact grey soil mixed with gravel. Under the tumulus right in the centre was uncovered a sandy shaft measuring 800mm in diameter going down to the level of the burial chamber blocking. It appeared later that this shaft was associated with an earlier burial, possibly Meroitic, of an infant which was visible in the eastern part of the ceiling of the later post-Meroitic burial chamber. This precarious location prevented a proper excavation, but it was recorded from below. The tumulus was associated with a single shaft found 400mm below the ground surface. The shaft (Figure 2.28) was filled with loose pink-coloured gravel, and a bottom layer of flat stone slabs occupying a space of 1m from the entrance towards the centre (Plate 2.27). The shaft, trapezoid in shape, was aligned east-west sloping down gently towards the entrance of a burial chamber on its western part. The shaft sides measure 2.72m (north), 1.2m (west), 1.7m (south), 800mm (east) with a depth of between 1m (east) and 1.4m (west). A single step 500mm in height gave access down to a vaulted chamber through an oval entrance 400mm along its shortest dimension. A burial chamber, oval in plan, measuring 2.12 x 1.2m contained the

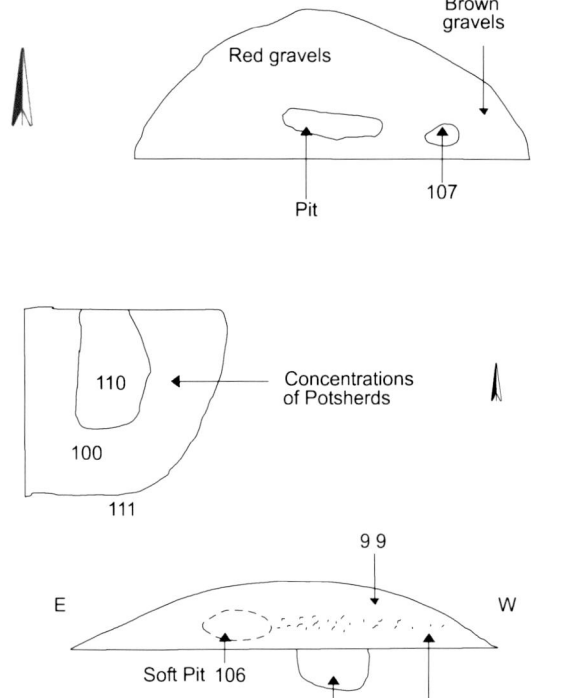

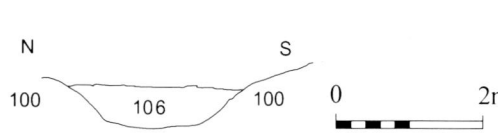

**Figure 2.21.** Site 159.2, Tumulus 3, top – plan of northern half of tumulus showing earlier pits (Meroitic burials?), upper middle – plan of south-eastern quadrant of tumulus showing earlier pits (Meroitic burials?), lower middle – east-west section, bottom – north-south section (scale 1:100).

skeleton of a sub-adult male(?) in a flexed position, lying on its right side head to the south facing east (Plates 2.24 and 2.26). The arms were flexed with the hands near the face (see Judd 2012, 87-88). Objects associated with the burial are two wooden bed legs, a jar and a gourd found at the southern part of the chamber between the skull and the wall. A metal ring and piece of textile were found near the face. Between the eastern wall of the chamber and the skeleton's back there were a wooden pole (part of the bed frame) and a single bead near the back of the skull (part of an earring?). A small basket and fragments of leather were found near the pelvis. A third wooden bed leg was found near the wooden pole. Small black beads were found around the flexed arms and large beads were found associated with an anklet lying by the ankles.

Spot heights local datum BM 159
H 358.658m
BM Map Point: 17×12'09.6928N 33×46'25.6458E
Date Recorded 14.5.93+
**Suggested Period:** Post-Meroitic
**Burial Identity:** Male(?)

**Small Finds Report**
Description of finds and samples
For the illustrations of the main objects, see Figure 8.3,

Plates 8.8-8.12.
(49C) Wheel-made oil-bottle: Intact apart from missing rim, Fabric G18. The exterior surface is red slipped and lightly burnished. Painted decoration comprises two sets of three red stripes on a broad cream band. H: indet., rim D: indet., max. D: 128mm.

(1a) Iron anklet: Somewhat corroded. Semi-circular iron element with approximately round cross-section. One end flattened and bent back to form a loop. D: 105mm. W: 7mm, T: 8mm.

(1b) Beads: 22 beads.
  one sub-spherical, brown and white banded agate.
  three ovoid, semi-translucent quartz.
  four sub-spherical, white quartz.
  one spherical, banded carnelian.
  two sub-spherical, banded carnelian.
  seven sub-spherical, carnelian, plain.
  three sub-spherical, semi-translucent quartz.
  one spherical, grey smoky quartz.

(2) Samples of leather: Max. dimensions reconstructed: c. 51.5 x 52.5mm.

(3) Beads: 32 beads (+ one fragment).
  26 dark brown-purple glass, including one oval, three square-sectioned
  22 annular, orange-brown glass.
  two annular, sub-spherical, orange-brown glass.
  one subspherical, orange carnelian (+ one fragment)
  one annular, green glass.
  one spherical, white quartz.
  one cylindrical, blue glass.

(4) Sample of textile: Nine fragments of textile. Colours: three dark brown, six mid-buff. Max. dimensions 15 x 10mm.

(5) Copper-alloy ring: Plain finger ring of thin metal with crescentic cross-section. D: 17mm.

Gourd (6): Section of gourd. Original D: c. 200mm.

(7) Wooden bed leg: Upper portion of square cross-section, lower part with double bulbous indentations and splayed foot. Remaining L: c. 360mm, cross-section c. 90mm square.

(8) Wooden bed legs: two bed legs, of same form as previous, upper part only coherent, remainder preserved in form only. Cross-section of upper part c. 80mm square. One has a mortice, c. 40 x 30mm. Remaining L: 190mm.

(9) Wooden bed leg: of same form as previous, the lower section only preserved, having double bulbous indentations and splayed foot. Remaining L: 240mm, cross-section c. 75mm square.

(10) Wooden bed frame: Section of bed-frame of approximately triangular cross-section, each side c. 35mm. Estimated original L: c. 480mm.

(11) Beads: 268 beads.
  one hexagonal bevelled from centre line to top and bottom, orange, carnelian(?)
  one annular, green glass.
  one annular, blue glass.
  one spherical, grey, stone(?)

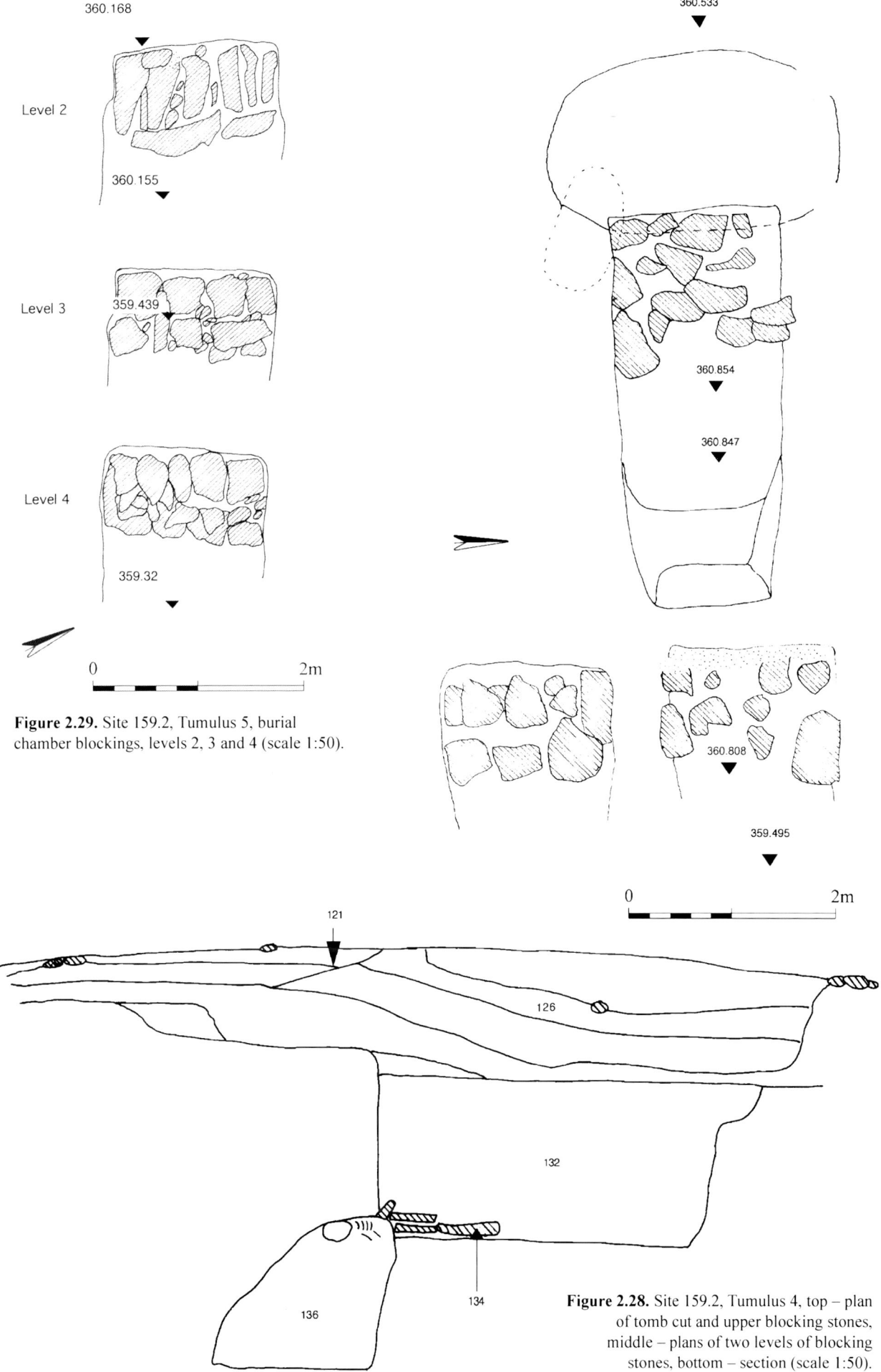

Figure 2.29. Site 159.2, Tumulus 5, burial chamber blockings, levels 2, 3 and 4 (scale 1:50).

Figure 2.28. Site 159.2, Tumulus 4, top – plan of tomb cut and upper blocking stones, middle – plans of two levels of blocking stones, bottom – section (scale 1:50).

12 annular dark brownish-purple glass.
one annular grey stone(?)
one barrel-shaped, dark brownish-purple glass(?)
c. 244 small annular bluish-green, glass, including two sets of two adhering to each other, six small annular greenish, faience.

(12) Copper-alloy spatula or *kohl* spoon: a straight handle of circular cross-section with, at one end, a shallow bowl, slightly splayed and flattened, the other end being rounded. L: 152mm, shaft D: 4mm, bowl W: 7.5mm.

(13) Sample of basketry: Parts of two small baskets. Made from strips or bundles of fibre, stitched together in a spiral pattern. Associated with a small sample of matting. D: c. 32mm.

(14) Beads: c. 20 small annular blue, glass or faience.

(15) Bead: Lozenge-shaped with flattened ends and approximately square cross-section, amber-coloured, carnelian.

(90S) Beads c. 170 beads.
two annular, dark purplish-brown glass.
one spherical, greyish-white, quartz.
one annular, dark bluish-green glass.
one annular, yellow glass.
one barrel-shaped, bluish-green glass.
two annular, orange 'clay'.
one annular, pale bluish-green glass.
one annular, greyish-brown, 'stone'.
c. 160 small annular, bluish-green glass.

(43S) Beads: (Unit 143)
two cylindrical, greenish-blue, faience.

**SARS Survey Site Ref.**     159.2/T5
**Site UTM Grid Point**     N1902074 E582382
**Sudan Survey Number**     NE-36-K/22-A-5

### Description of site
This tumulus, a low mound 10.12m in diameter, was situated in the south-eastern part of the cemetery and was marked on its surface by vehicle tracks. The top layer of the tumulus make-up was small sized ferrocrete sandstone flakes, gravel, large pebbles and wind-blown sand. A layer of soft fine yellow gravel, under which lay a layer of soft orange gravel, was also cleared. The bed layer of the make-up of the tumulus was a mix of soft fill and limestone pebbles. The tumulus covered a single shaft which appeared 250mm below the current ground surface (Figure 2.32). The shaft fill was a mix of hard orange gravel and different sized pieces of sandstone. The fill also contains the burial chamber blocking stones situated at the western end of the shaft. The shaft was trapezoid in shape, its long axis aligned east-west with a shoulder along its southern side measuring 540mm deep. The shaft measured 2.5-2.7m long, varying north to south, and 1.52-1.84m wide, varying east to west. The blocking stones (Figures 2.29 and 2.32, Plate 2.34) were sealed by red mud. A 600mm step led

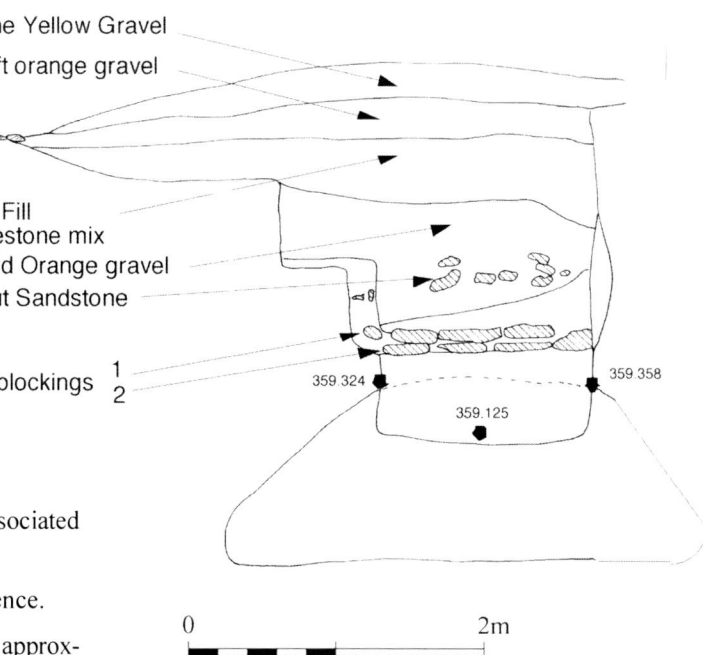

**Figure 2.30.** Site 159.2, Tumulus 5, section (scale 1:50).

down to an oval vaulted burial chamber measuring 2.8 x 1.8m in size (Figures 2.30-2.34, Plate 2.29). It contained the skeleton of an adult female, aligned east-west in a flexed position lying on its right side with the head to the south facing east (Plates 2.29 and 2.32) (Judd 2012, 88). The body was laid on a wooden bed; the legs of the latter were found at the four corners of the burial chamber, part of the bed frame (pole) was found behind the body at the western part of the chamber. Evidence of textile all over the body was clear, but had been attacked by termites. Textile was also covering pots and a grindstone. There was

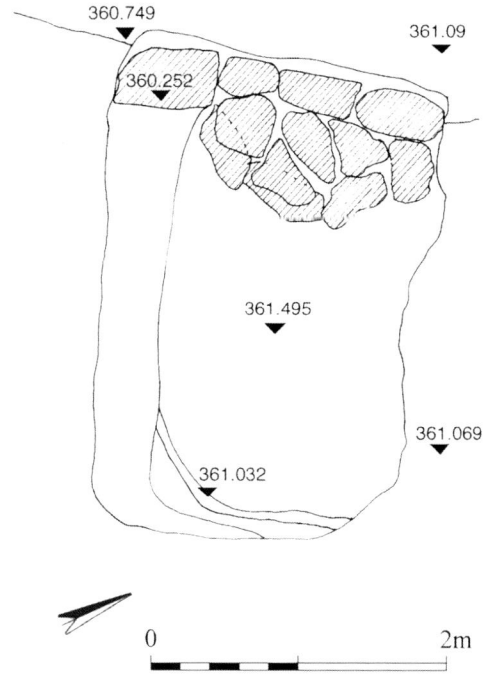

**Figure 2.31.** Site 159.2, Tumulus 5, plan (scale 1:50).

a concentration of leather around the pelvis and along the legs, and a blanket was also found under the head. A variety of beads was found around the wrists, near the chest, different colours around the shoulder, around the right ankle and later a huge amount came out of sieving. An ivory *kohl* pot provided with a lid (Plate 2.30) was found lying on its side 250mm, to the west of the pelvis. A pottery jar, pot-stand and bowl, and a mortar were found near the feet (Plate 2.33, cf. Plate 2.31). There was also a metal mirror near the face and a metal spatula in front of the abdomen. Small pieces of rope, wood fragments and leather thongs were found scattered in the chamber.

**Suggested Period:** Post-Meroitic.
**Burial Identity:** Female.

Spot heights local datum
BM 159 H 358.658m
BM Map Point: 17°12'09.6928 N 33° 46'25.6458 E
Date Recorded 14.5.93+

**Small Finds Report**
Description of finds and samples
For the illustrations of the main objects, see Figures 8.4 and 8.5, Plates 8.13-8.26.

(92C) Handmade pot-stand: Complete. Fabric probably G5. Bi-conical, with a dark purplish-red slip. Decoration, in white, comprises a series of wavy lines and heart-shaped motifs, together with an undulating horizontal line around the central stem, and an irregular white rim-band and an irregular circle on the base of the interior. H: 195mm, rim Diameters: 127mm and 147mm, max. D: 147mm.

(93C) Wheel-made spouted jar: Complete. Fabric G19. Red-slipped and burnished. Decoration includes motifs consisting of white rectangular patches having geometric designs executed in black lines. In the centre of each side of the vessel there is a vegetal motif. There is a deep groove below the rim and three, shallower, closely-spaced incised grooves around the body. H: 186mm, rim D: 97mm, max. D: 208mm.

(94C) Wheel-made 'table amphora': Intact, apart from one missing handle, apparently broken in antiquity. Fabric G18. The exterior is ribbed, with a red slip having a light burnish. H: 188mm, rim D: 57mm, max. D: 128mm.

(95C) Handmade bowl: Complete. Fabric probably G5 type. It has a strongly inturned upper body. The exterior is red-slipped and well-burnished, having a single horizontally pierced lug, approximately oval in shape. H: 77mm, rim D: 87mm, max. D: 135mm.

Handmade bowl (96C): Complete. Fabric G5. In form, an open bowl with sides sloping outwards, with a purplish red-brown burnished slip. Decoration comprises two sets of three white painted wavy lines. There are two holes through the body wall just below the rim. H: 53mm, rim D: 185mm, max. D: 187mm.

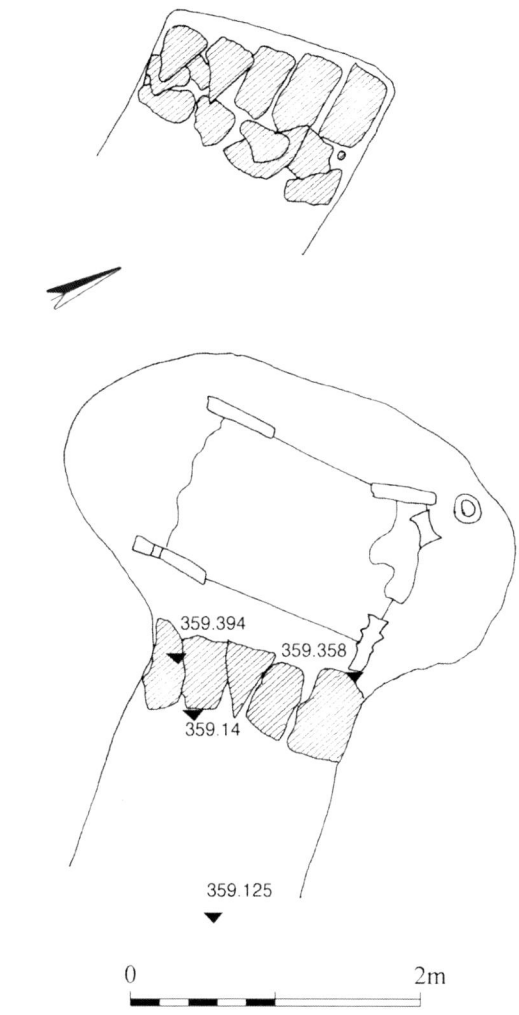

**Figure 2.32.** Site 159.2, Tumulus 5, top – plan of upper blocking stones, bottom – plan of lower blocking stones and burial chamber (scale 1:50).

(1) Beads: 36 beads:
  11 spherical, pale, mid- and dark amber-coloured carnelian.
  four spherical, mid amber-coloured and pale orange carnelian.
  one spherical, brown to dark brown carnelian.
  one short barrel, amber-coloured carnelian.

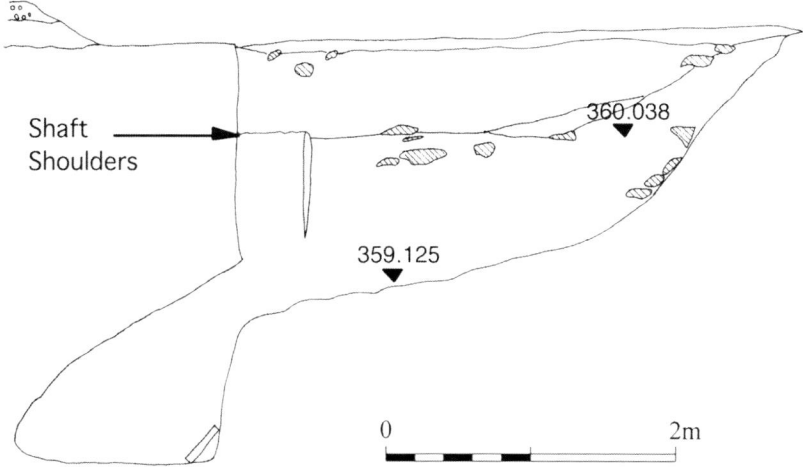

**Figure 2.33.** Site 159.2, Tumulus 5, section along east-west long axis (scale 1:50).

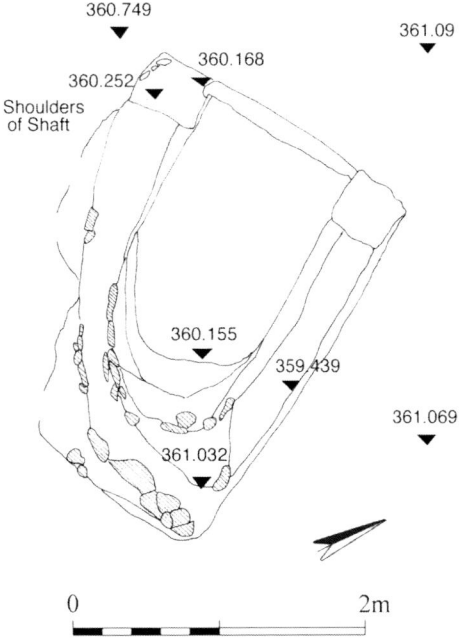

**Figure 2.34.** Site 159.2, Tumulus 5, plan (scale 1:50).

two short barrel, mid-dark brown with dark brown striae carnelian.
five spherical, opaque white quartz.
one spherical, semi-translucent white quartz.
two spherical, brown to light brown quartz.
three spherical, light and pale brown banded quartz.
two pear-shaped, semi-translucent white quartz.
one elipsoidal, light brown and white banded quartz.
one spherical, brown and white banded agate.
one long barrel, dark brown and white banded agate.
one annular, green and yellow glass. This has moderately convex sides and slightly concave top and bottom having a green background with irregularly spaced yellow roundels with seven or eight dark brown lines radiating from centre to circumference, and two irregular trapezoidal shapes outlined in dark brown on the top surface, together with one of similar shape and one of approximately triangular shape on the base.

(2) Beads: *c.* 688 beads.
one oval, blue-green glass.
33 cylindrical, white bone.
*c.* 636 annular, dark blue to blue-green glass.
one biconical, yellow-green glass.
one oval-section, white bone.
one cylindrical, orange 'clay'.
one annular, orange 'clay'.
one biconical, rose quartz.
two annular, light blue-green glass.
three annular, green glass.
one annular, yellow glass.
two cylindrical, yellow glass.
two cylindrical, mid-blue glass.
two annular, mid-blue glass.
one dark red, faience.
two cylindrical, white bone.
ten annular, blue-green glass.
one oval-sectioned, blue-green glass.

(3) Copper-alloy spatula and nail: Spatula comprising a narrow rod, slightly splayed and flattened at one end and rounded at the other. Bent around two right angles. Original L: *c.* 180mm. Shaft D: 4.7mm. Splayed part Max. W: 9mm T: in centre 1.4mm. One nail: L: 12.7mm, D: of head *c.* 7.7 cm, shaft top D: 2.6mm, shaft bottom D: 1.7mm.

(4) Sample of textile: Fragments of textile on right scapula and under chin of body. Colour: dark brown. Size of sample: 60 x 20mm.

(5) Ivory *kohl* pot: *Kohl* pot of turned ivory. Restored from four major pieces. Body: cylindrical base, with central depression and two concentric circular grooves in centre of underside, the base rises straight for *c.* 9.5mm and has a sloping shoulder with a step *c.* 1mm up to a groove below main part of body. The latter curves outwards and upwards from base to maximum diameter, the wall then tapering in slightly to 50.3mm diameter at the top. It has a central cavity 20.4mm in diameter at the opening and tapering at the bottom. Stopper: cylindrical base with shoulder sloping to flat top, with a deeply undercut flange above, leading to sub-spherical knob, having three sets of closely-spaced concentric grooves around it just below the top and at approximately one-third and two-thirds of its height. Body: H: 96mm, max. D: 65mm. Stopper: L: 87.8mm, max. D: 55.4mm.

(6) Copper-alloy mirror: Mirror, circular in form. D: *c.* 84mm, T: *c.* 1mm.

(7) Beads: 44 beads:
five short cylindrical, convex sides, dark brown glass.
29 ovoid, dark brown glass.
three subspherical, dark brown glass.
one short cylindrical, dark brown glass.
one cylindrical, light yellow-brown glass(?)
one lozenge-shaped, ovoid longitudinal cross-section, amber-coloured, carnelian.
one subspherical, light amber-coloured with dark amber-coloured stripe, carnelian.
one short cylindrical, striped brown/buff/brown, glass or faience.
one cylindrical, mid blue faience.
one annular, gold in glass.

(8) Sample of leather garment: Pieces of thin leather and some textile fragments. Leather generally flat, but includes some examples with turned-over edges. Max. dimensions *c.* 42.5 x 21 x 0.9mm.

(9) Beads: 36 beads. Forming short section of 'necklace' comprising three strings of dark blue-green annular glass or faience on either side of single white cylindrical bone bead.

(10) Beads: 286 beads.
three subspherical, light amber-coloured with irregular dark amber-coloured patches and striae, carnelian.
one sub-spherical, opaque white quartz.
one short cylindrical, dark red glass.
two short cylindrical, convex sides, dark grey, 'stone'.
one short cylindrical, with one end sloping, the other flat, dark grey, 'stone'.
one small cylindrical, dark red glass.

two annular, dark green glass.
one barrel-shaped, slightly asymmetric, dark green glass.
one short cylindrical, asymmetric, light yellowish brown glass.
two annular, light yellowish-brown glass.

(11) Samples from leather apron: Samples of thin leather, mainly flat but including a few examples of fragments folded over and with stitching. Max. L: 41mm, T: c. 0.9mm.

(12) Sample of textile: Colour: dark brown. Size: c. 20 x 15mm.

Sample of textile and thread (13): Colour: mid-brown. Textile fragments sizes: 7 x 9mm and 7 x 5mm.

(14-15) Wooden bed legs: Two bed legs with straight upper section passing to single bulbous section above broad splay foot. Remaining L: c. 410mm, cross-section, c. 120mm square at upper end, c. 140 x 130mm at base of foot.

(16-17) Wooden bed legs: Two bed legs. No. 17 having double bulbous section and moderately splayed foot. Remaining L: c. 480mm. Remains of joint or interlocking piece at upper end projecting c. 35mm. Cross-section approximately circular at upper end, D: c. 110mm, cross-section at foot: c. 90mm square.

(18) Sample of bed frame: Part of bed frame in poor state of preservation. Original L. indeterminate.

(19) Biological sample, hair: See report on human remains (Judd 2012, 88).

(20) Sample of leather: c. 30 fragments of leather, including two pieces with edges curled over, two pieces with stitched seam, one piece with stitch holes, one partly knotted piece. Remainder flat or undulating. Max. dimensions c. 54 x 29.5 x 0.9mm.

(21) Wooden comb: Teeth poorly preserved. Rectangular section of wood with thin teeth extending from both the long sides. Evidence for ten teeth with a maximum thickness of c. 2.5mm thick and a maximum length of c. 27mm. L: 84mm, W: 46mm, T: 6.6mm.

(22) Beads: 80 beads.
one cylindrical, white bone.
79 annular, dark green - black glass.

(22a) Bead: One white disk bead, bone.

(23) Sample of rope and leather: Rope, c. 80 fragments in total:
a. fragments of D: c. 4.4mm
b. fragments of D: c. 5mm. Leather
c. 100-200 small fragments, mostly flat or slightly undulating, some with evidence for being folded

over on themselves. Max. dimensions c. 24 x 8 x 0.7mm.

(54S) Stone slab: (Unit: 13). Slab of sandstone(?) from covering of burial pit, of approximately oval form. It has grooves on one face forming a cross-shape. L: 537mm: W: 300mm.

(97S) Bowl: Shallow bowl or mortar of stone (Greenstone?). It is approximately square in plan at the rim, with rounded, slightly projecting corners. It has rounded, convex sides and a base flattened on both the interior and exterior. L: 185mm: W: 187mm: H: 75mm.

**Test Excavations Begrawiya-Atbara 1994: 159.1, 159.2**
Other samples not from burial contexts

**Object no.** 159.2/T5/sample
**Description:** Sample of mortar (?) from sealing of Tumulus 5 burial chamber.

**Date:** Post-Meroitic

**SARS Survey Site Ref.** 159.2/T6
**Site UTM Grid Point** N1902074 E582382
**Sudan Survey Number** NE-36-K/22-A-5

**Description of site:**
This tumulus (Figure 2.35), a low mound that measured 400mm in height, 3m north-south and 5.6m east-west, was situated in the southern part of the cemetery. The tumulus was characterised by a ring of stones around the circumference. The top layer covering the tumulus make-up was a

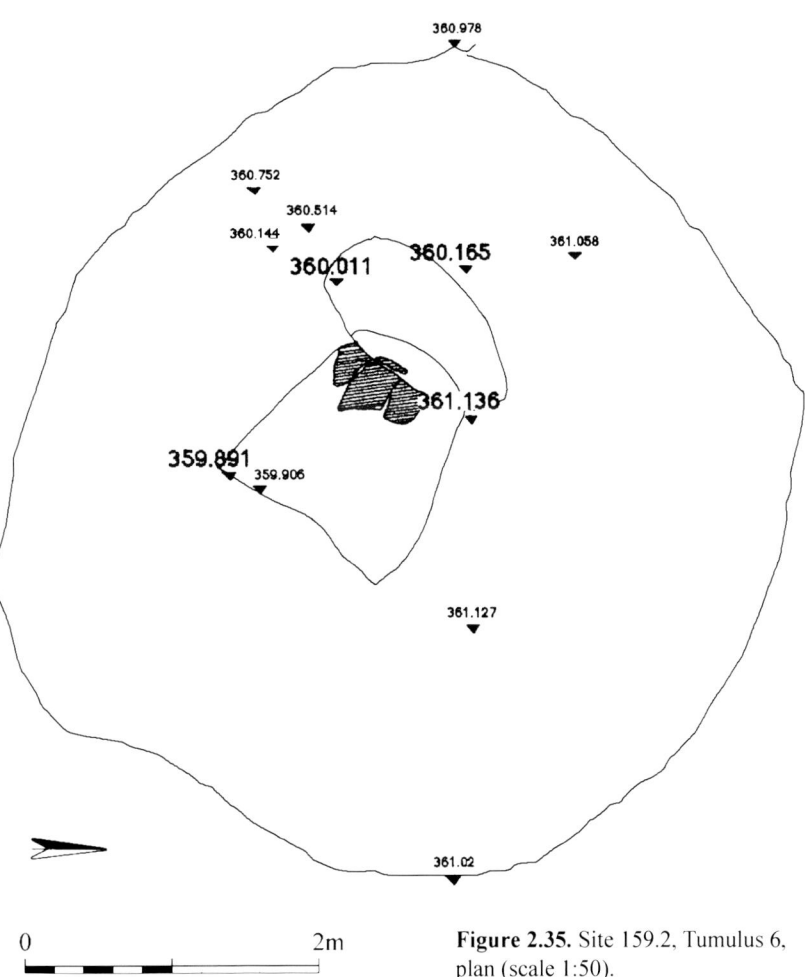

**Figure 2.35.** Site 159.2, Tumulus 6, plan (scale 1:50).

mix of small sized ferrocrete sandstone flakes, loose gravel and soil and fine wind-blown sand. The layer below the tumulus was a mix of coarse and fine gravel particularly in the centre of the shaft. A trapezoid-shaped shaft under the northern half of the tumulus was aligned east-west. The shaft was filled with loose gravel mixed with white and pink coloured soil, bits of wood and leather were found inserted in the fill 300mm below the surface, also fragmented wooden pieces were found between the blocking stones. The blocking stones of the burial chamber appeared at 600mm below the surface of the shaft fill forming a wall at the western side of the shaft, its thickness extending 140mm eastwards (Figure 2.36, Plate 2.35). Although the blocking stones were piled irregularly, the tomb was intact. The shaft measured at the bottom 1.96-2.26m long, varying south to north, and 1.7-2m wide, varying west to east. At the western side of the shaft lay a vaulted burial chamber with an oval-shaped bottom measuring 3.4 x 2.1m and with a height of 600mm at the entrance. It contained the skeleton of an adult male aligned north-south on its right side, in a flexed position with the arms at right angles to the body and the head to the south facing east (see Judd 2012, 88-89). The body was totally covered by layers of various qualities of textile, the top layer was a coarsely woven textile (shroud?). Leather was also found around the feet, apparently on a mat. Another thick folded blanket, stuffed with rubble, was used as a head rest. Apart from textiles and leather, no other grave goods were found associated with this tomb.

**Suggested Period:** Post-Meroitic
**Burial Identity:** Male

Sheet reference 159-161,93
Spot heights local datum BM 159
H 358.658m
BM Map Point: 17° 12'09.6928 N 33° 46'25.6458 E

**Small Finds Report**
Description of finds and samples
Samples of textile (1-20): Textile pieces from shroud, with some pieces from blanket interlayered. Colour: grey to buff-brown. Max. dimensions c. 260 x 130mm.

(21-22) Samples of textile: Textile pieces from blanket, with some pieces from shroud interlayered. Colour: mid grey-brown, orange-brown. Max. dimensions c. 260 x 130mm.

(23-24) Samples of textile: Textile pieces from shroud, with some pieces from blanket interlayered. Colour: grey to buff-brown. Max. dimensions c. 160 x 120mm.

(25-26) Samples of textile: Textile pieces from blanket, with some pieces from shroud interlayered. Colour: mid grey-brown, orange-brown. Max. dimensions 210 x 90mm.

(27-28) Biological samples, skin: See report on human remains (Judd 2012, 89).

(29) Sample of leather and textile:

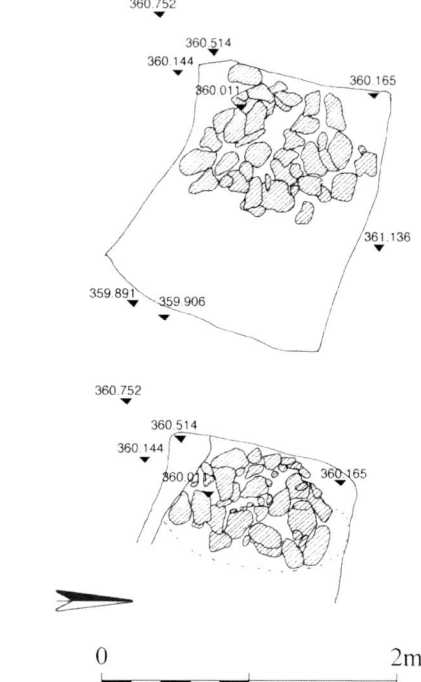

**Figure 2.36.** Site 159.2, Tumulus 6, top – plan of upper blocking stones, bottom – plan of middle blocking stones (scale 1:50).

Leather: from an apron. Includes fragments undulating and folded over on themselves and some fragments with stitch holes, c. 1mm diameter, in pairs c. 2.4mm apart, at intervals of c. 3mm, c. 2-3mm from edge. Max. dimensions of sample: c. 290 x 100 x 60mm. Leather average T: c. 0.8mm. Textile: four layers folded over. Size: 140 x 60mm.

(55S) Sample of worked wood: c. 68 pieces of worked wood comprising eight carved pieces and c. 60 smaller fragments, forming part of bier or bed frame. Includes three pieces with apparent square-cut corners, three pieces with parts of hollow socket, one piece carved; two bulbous sections separated by narrow section, and one piece with curved exterior profile. Remainder unidentifiable. Max. dimensions c. 160 x 55 x 30mm (see Figure 8.5).

(56S) Strips of leather: c. 12 strips of leather, five straight, remainder curled or knotted. Max. L: c. 120mm, max. W: 6mm.

**Suggested Period:** Post-Meroitic
**Burial Identity:** Male

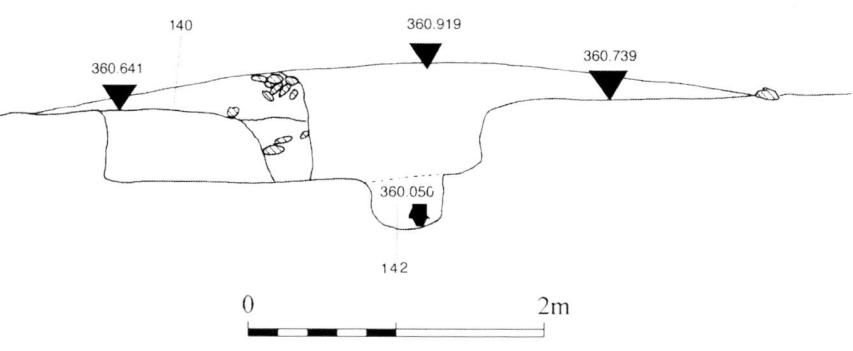

**Figure 2.37.** Site 159.2, Tumulus 7, north-south section (scale 1:50).

**SARS Survey Site Ref.**
159.2/T7
**Site UTM Grid Point**
N1902074 E582382
**Sudan Survey Number**
NE-36-K/22-A-5

**Description of site**
This tumulus (Figure 2.37) consisted of a mound covered with low ferrocrete sandstone flakes and was situated in the southern part of the cemetery. It measured 260mm in height at the centre and 4.2m in diameter north-south. The top layer of the tumulus make up consisted of different sizes of ferrocrete sandstone flakes: 100 x 100mm was the smallest at the circumference; bigger ones occurred towards the centre covered with wind-blown sand. The fill below the top cover was a brown gravel. This layer continued down to 400mm below the original surface. It covered a Christian burial pit, oval in shape and aligned east-west. The burial pit contained the skeleton of an adult male, fully extended in a supine position but slightly shifted to the right side, with the arms extended and the head to the east (Plate 2.28) (Judd 2012, 89). The body was on a layer of ferrocrete sandstone flakes but no grave goods were associated with it. Potsherds were found at the base of the tumulus make up: 159.2/T7/58C.
**Suggested Period**: Christian
**Burial Identity**: Male

Spot heights local datum BM 159
H 358.658m
BM Map Point: 17×12'09.6928N
33×46'25.6458E
Date Recorded 14.5.93+

**SARS Survey Site Ref.**
159.2/T11
**Site UTM Grid Point**  N1902074 E582382
**Sudan Survey Number**  NE-36-K/22-A-5

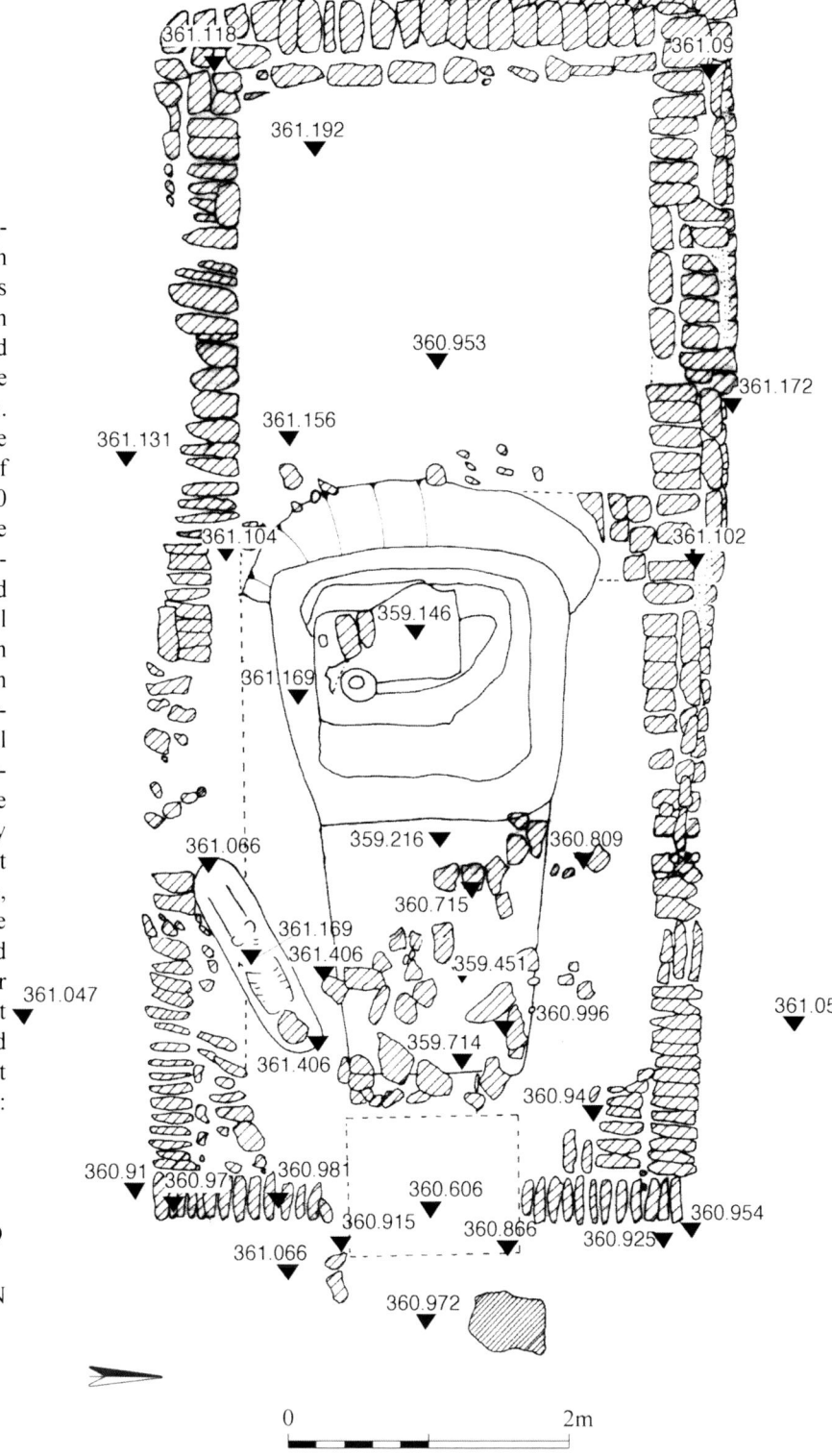

Figure 2.38. Site 159.2, T11, plan (scale 1:50).

**Description of site**
A small low ferrocrete sandstone flake-covered tumulus situated in the eastern part of the cemetery. On clearing off the top surface of the tumulus there was revealed an east-west aligned pit cut into the ground surface. The fill of the pit was loose pink coloured gravel and soil. It contained an extended east-west orientated skeleton, lying on its back with the head to the east facing north and the hands on the pelvis. The grave pit measured 2m east-west, and was 500mm by the head. No grave goods were associated with the burial. At the level of the surface from which the burial pit was cut there was uncovered, in the the tumulus area, a mud-brick structure, the north-eastern corner of a rectangular wall foundation (Figure 2.38).

The structure was divided into two rooms. The walls of the eastern room measured 5.1m-5.35m (south and north), by 3.9m (east). The eastern wall was pierced by a gap 1.2m wide in the middle. The northern part of the eastern wall measured 1.36m and the southern part 1.55m. The walls of the western room measured 3m (north), 3.9m (west) and 2.62m (south) and 3.38m (eastern dividing wall), the latter being 650mm thick.

The eastern room contained an east-west aligned shaft (Figures 2.41 and 2.42). The shaft fill was disturbed by two robber pits, one in the eastern part and the other at the western end. The western one went down to the entrance of the tomb. The back fill contained different sizes of stones, potsherds and mud bricks 370 x 170 x 100mm, which proved to be part of the tomb entrance blocking. The fill also contained loose pink coloured gravel. The shaft, 5m in length, descended westwards from the ground surface to a depth of 2m above the tomb entrance. Three globular jars and an incomplete broken one were found *in situ* in a row in front of the mud-brick blocking of the burial chamber (Plate 2.36). The burial chamber entrance was originally blocked by a mud-brick wall two courses thick, of which the upper part was removed by the robbers. The burial chamber was backfilled with white gravel and mud-brick rubble either following the robbery or due to the ceiling collapsing (perhaps also due to later robber activity). The burial chamber was cut into the white gravel of the western wall and under

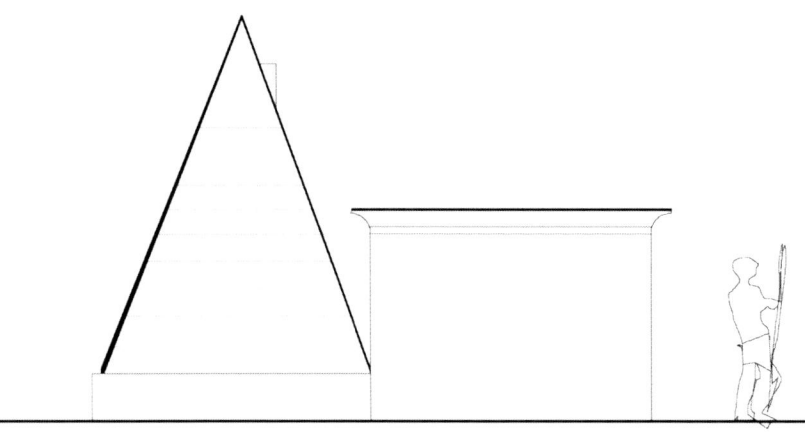

**Figure 2.39.** Site 159.2, T11, reconstruction of the pyramid and offering chapel.

the western room mud-brick foundations. The entrance to the burial chamber was of an irregular shape due to the robber pit having been cut through at one side. Two more inverted globular jars and large sherds were found buried in the soft fill near the entrance. It was clear that the tomb had been greatly disturbed in antiquity, and the chamber had been mostly rifled through. Due to a lack of time it was decided to back-fill the grave once the area uncovered had been recorded and the finds removed to storage.

The two rooms clearly represented the denuded foundations of a small mud-brick pyramid and associated eastern chapel (Plate 2.37). The threshold of the chapel was evident and an area of external mud surface recorded. From the finds and pottery it was clear the main burial dated from the 1st century BC, and the grave of the later *in-situ* skeleton must have been cut into the chapel after it had collapsed and been denuded to ground level. The small size of the later burial's mound suggested

this may have been constructed in the late Medieval period. The two robber pits must have been dug when the superstructure was still visible. A possible third pit was cut into the centre of the pyramid which may have contributed to its collapse. The robbers' ignorance of the design of the tomb suggests that at least one of the robbing attempts took place after this form of tomb ceased to be common place rather than during the Meroitic period. The fact that the width of the pyramid matched that of the eastern chapel suggests that the pyramid structure was elevated on a square base to at least the height of the chapel (Figures 2.39 - 2.40).

**Suggested Period:** Meroitic
**Burial Identity:** Meroitic

**Small Finds and Samples**

Description of finds and samples (Test Excavations only): The pottery recovered during the initial investigation of Tumulus 11 has been considered in detail typologically together with the rest of the ceramics from the Meroitic burials excavated in the Rescue Season (Rose 1998). Full listings of finds from this tumulus recovered from both seasons are given by Edwards (1998b, 13-15). Brief descriptions and illustrations of the complete and reconstructed vessels from this tumulus found in the Test Excavations are presented here for completeness, in order of their find number (see Plates 2.39, 4.1 and 4.2).

(98C) Wheel-made jar: Mostly complete, but restored from fragments. Surface over the base is eroded. Fabric G4? The exterior is red slipped and probably originally burnished. H: *c.* 460mm, rim D: 116mm, max. D: 320mm. (Figure 2.43, after Rose 1998. Base reconstructed)

(99C) Handmade jar: Complete but restored from fragments. Surface eroded. Fabric G5? The vessel is unslipped, the surfaces dark brown/black with some traces of burnishing. Decoration comprises a band of impressed motifs on the shoulder: a row of oblique oval impressions with two rows of overlapping comb-impressed lines below. Under this is a broad band composed of rows of oval impressions, *c.* 5 mm apart. H: 365mm, rim D: *c.* 100mm, max. D: *c.* 322mm. (Figure 2.43, corrected from Rose 1998.)

(100C) Handmade jar: Complete but restored from frag-

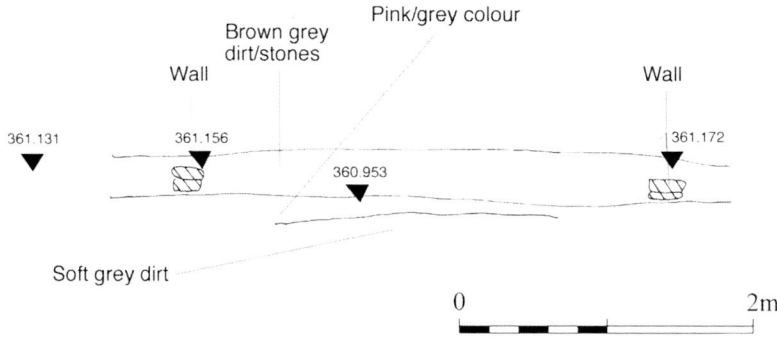

**Figure 2.40.** Site 159.2, T11, north-south section of pyramid structure (scale 1:50).

38

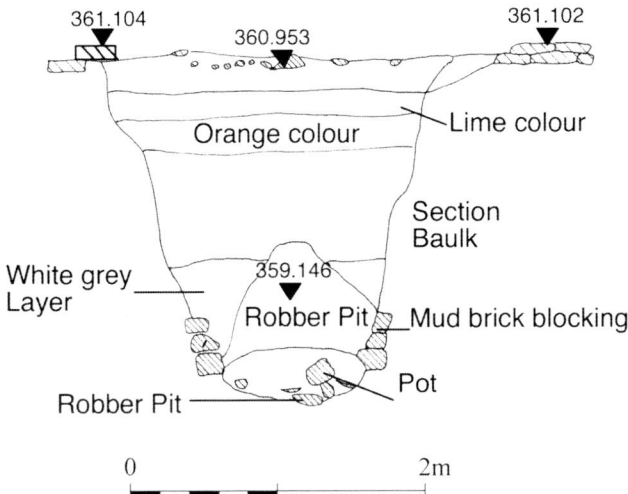

**Figure 2.41.** Site 159.2, T11, section of pit (scale 1:50).

ments. Surface eroded. Fabric G5? The vessel is unslipped, the surfaces dark brown/black with some traces of burnishing around the neck. Decoration below the rim consists of short oblique comb-impressed lines. Decoration on the shoulder comprises a row of oval impressions with two rows of overlapping comb-impressed curved lines below, under which is a wide band of oblique oval impressions, 2-5mm apart. There are three apparently deliberately-made holes in the body of the vessel. Three signs have been incised post-firing below the decoration on the shoulder. H: 375mm, rim D: 90mm, max. D: c. 310mm. (Figure 2.43 – after Rose 1998, and Figure 2.44).

(101C) Wheel-made jar: Complete apart from a short section of the rim missing. Fabric G1? Surface around the base eroded; decoration faint. The exterior surface has a dark yellow burnished slip. The neck and rim are red slipped and burnished. Painted decoration comprises a broad wavy band, divided internally by a second wavy band, all in red, on white, with stylised flower heads on short stalks in the troughs and peaks formed by the broad band. Two signs have been incised post-firing on the shoulder. H: 380mm, rim D: 102mm, max. D: 325mm. (Figure 2.43 – after Rose 1998, and Figure 2.44).

(102C) Wheel-made jar: Complete. Fabric G4? Exterior red slipped and burnished horizontally. An 'owner's mark' and a short Meroitic inscription have been incised post-firing around the shoulder. H: 443mm, rim D: 116mm, max. D: 330mm. (Figure 2.45 – after Rose 1998, pl. 46 and Figure 2.44).

(103C) Wheel-made jar: Complete. Fabric not visible, but G1, G2 or G4? Exterior red slipped and burnished horizontally. Three signs have been incised post-firing on the shoulder. H: 340mm, rim D: 88mm, max. D: 300mm. (Figure 2.45 – after Rose 1998, and Figure 2.44).

(104C) Handmade jar: Complete apart from some small chips on rim. Surface eroded. Fabric G5? Exterior dark grey and lightly burnished. Decoration, on the shoulder, is made by comb-impressions. This comprises a band of oblique impressions, double curved lines of rocker-stamp impressions with a broad band of cross-hatching below this, followed by a row of oval impressions. Under this are three rows of rather irregular horizontal comb-impressed lines. Pendant from the lowest of these are deep panels of vertical zigzags terminated by oval impressions with sets of shorter zigzags, and of vertical rows of paired oblique impressions in between these panels. Apparently deliberately-made holes are present in the shoulder and lower body. H: 354mm, rim D: 105mm, max. D: 280mm. (Figure 2.45 – after Rose 1998).

(105C) Handmade jar: Complete but restored from fragments. Fabric G5. Surfaces grey and lightly burnished. Decoration includes short oblique comb-impressed lines below the rim, and a band of comb-impressed rocker-stamped and impressed decoration on the shoulder. This comprises a row of oval impressions, with rows of horizontal zigzags underneath. Below are panels composed of two rows of vertical zigzags, terminated by a row of oval impressions. In between these panels are sets of shorter zigzags. H: c. 400mm, rim D: 102mm, max. D: 320mm. (Figure 2.45 – after Rose 1998).

(106C) Oil jar (*Lekythos*) : Complete, but surface eroded. Fabric indeterminate. There are traces of red slip around the neck and handle. It is asymmetrical about its vertical

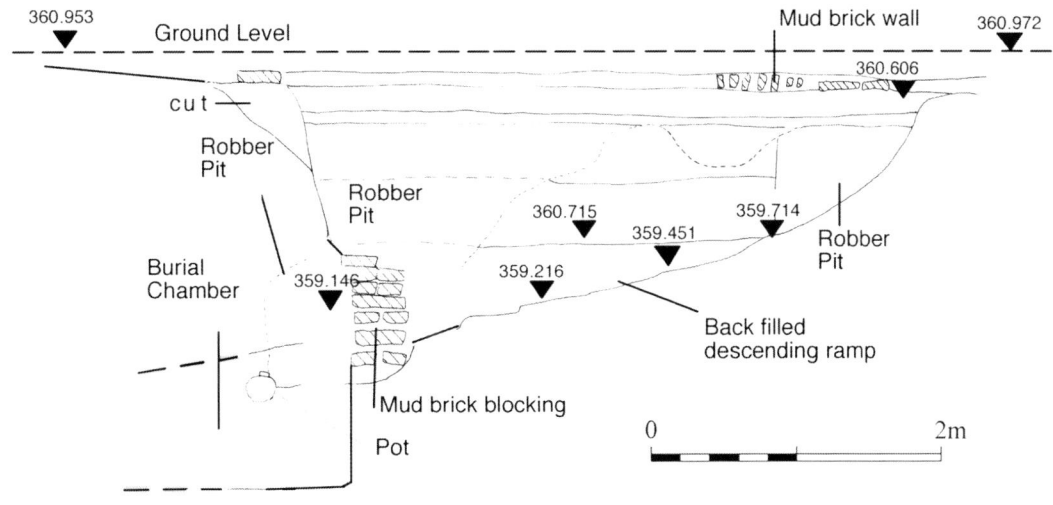

**Figure 2.42.** Site 159.2, T11. Section through the descendary and burial chamber (scale 1:50).

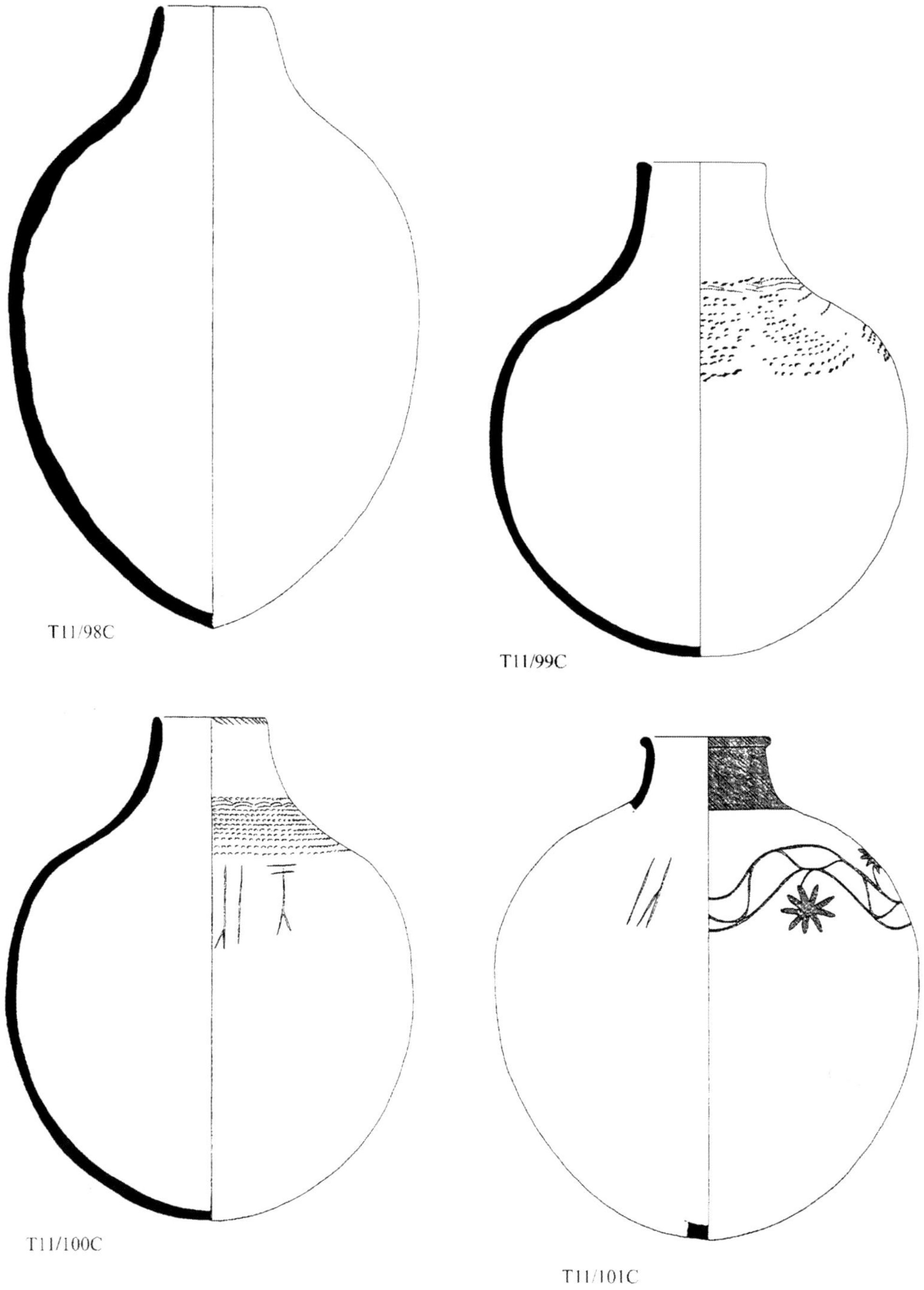

T11/98C

T11/99C

T11/100C

T11/101C

**Figure 2.43.** Site 159.2, T11. Complete and reconstructed jars recovered from test excavations of T11 (scale 1:4).

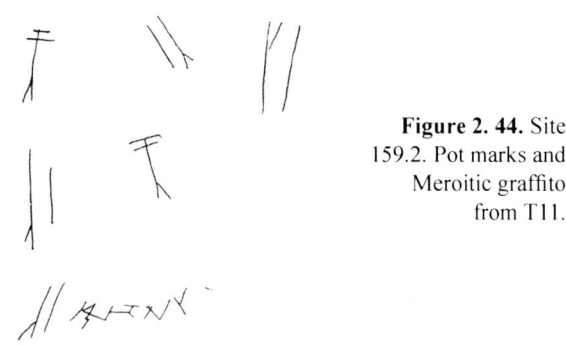

**Figure 2. 44.** Site 159.2. Pot marks and Meroitic graffito from T11.

axis. H: *c.* 100mm, D: *c.* 20mm, max. D: 81mm. (Figure 2.45 – after Rose 1998).

**Small Finds Report**
Other small finds not from burial contexts
**Object no.** 159.2/T11⁺/68S
**Unit:** 145
**Description:** Grinder or palette. Half of a flat disk of grey sandstone. It is smooth, but somewhat pitted on both faces. The edges appear to have been firstly chipped and then ground to the final shape.
**Dimensions:** D: 90mm, T: 15.7mm at the centre, tapering to *c.* 14mm at edge.
**SARS Survey Site Ref.** 159.2/T13
**Site UTM Grid Point** N1902074 E582382
**Sudan Survey Number** NE-36-K/22-A-5

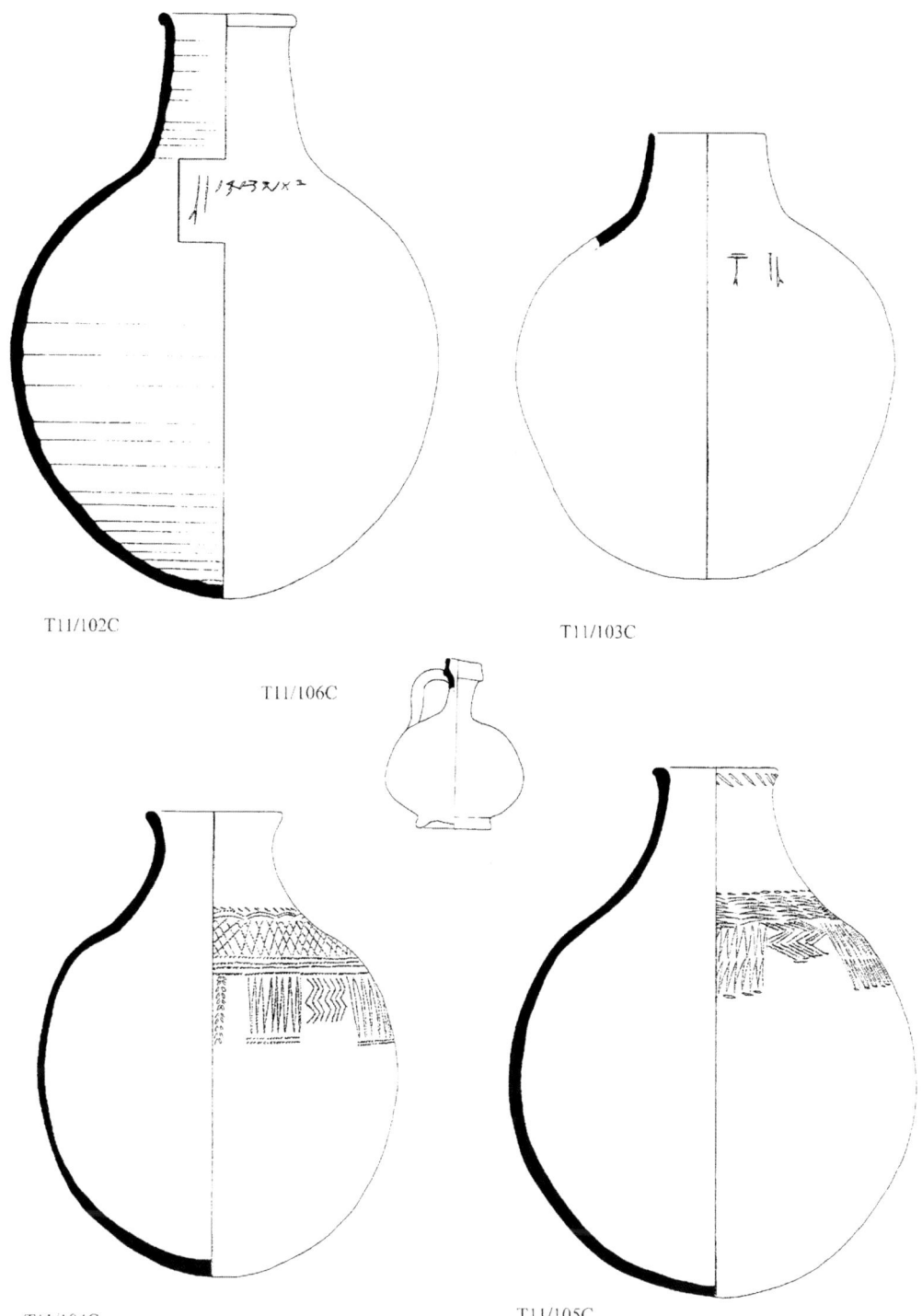

**Figure 2.45.** Site 159.2, T11. Complete and reconstructed jars recovered from test excavations of Tumulus 11 (scale 1:4).

## Description of site
This tumulus consisted of a low mound 3.6m in diameter covered in ferrocrete sandstone flakes. The make-up of the mound consisted of coarse brown gravel. It went down to the original surface of the ground, covering a pit filled with loose red-brownish gravel. The fill is contained within the burial pit of a Christian grave 500mm below the original surface of the ground (Figure 2.46). The burial pit was an irregular oval in shape, aligned east-west. It contained the skeleton – laying in the fill – of a child, lying fully extended, left arm at side, right arm across the body with the hand on the pelvis. The body lay with head to the east, facing south (Judd 2012, 92). No grave goods except a metal earring, fragments of textile and beads were found.

**Suggested Period:** Christian
**Burial Identity:** Christian

Spot heights local datum BM 159 H 358.658m
BM Map Point: 17° 12'09.6928 N, 33° 46'25.6458 E
Date Recorded 14.5.93+

## Small Finds Report
### Description of finds and samples
Sample of textile (13+/sample): Colour: greyish-brown. Size: *c.* 50 x 20mm.

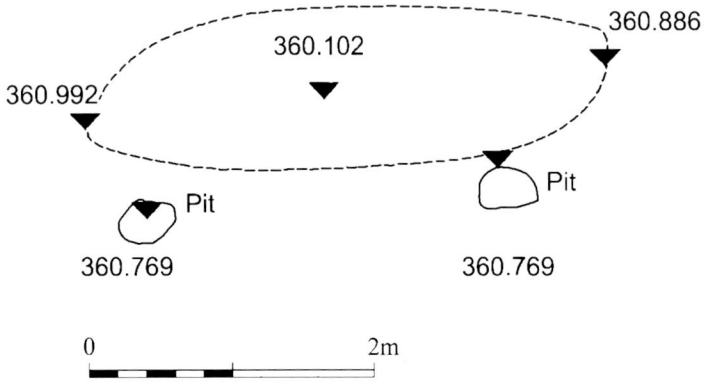

**Figure 2.46.** Site 159.2, Tumulus 13, plan of the burial pit (scale 1:50).

(67Or) Sample of textile: Colour: mid greyish-brown. Size: 20 x 8mm.

(70S) Beads: eight beads and four partial beads (Plate 8.27).
  five barrel-shaped, banded blue/white/blue, faience(?).
  one cylindrical, dark brown and white banded agate.
  one seven-sided, faceted on both halves and on centre line, amber-coloured carnelian.
  one barrel-shaped, yellowish-brown carnelian.
  four portions of biconical, striated amber-coloured carnelian.

(71S) Copper-alloy earring: One single loop earring of bronze or another copper alloy. Consists of a piece of metal with circular cross-section bent into a loop, with the diameter tapering at both ends. Max. external D: 2.28mm (see Figure 8.4).

(77S) Sample of textile: Colour: greyish-brown. *c.* 20 fragments of coarse weave textile from shroud. Max. dimensions *c.* 50 x 30mm.

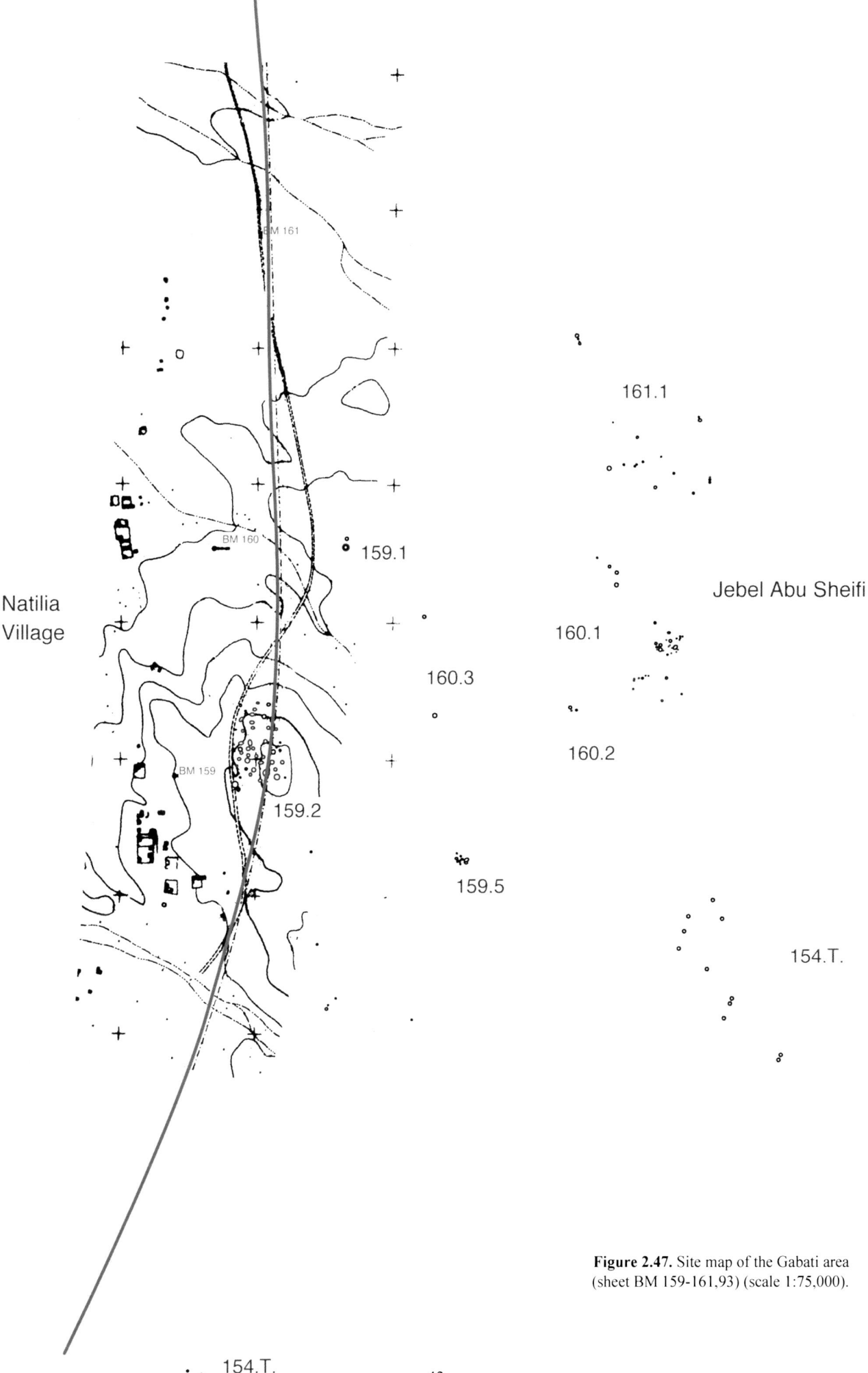

**Figure 2.47.** Site map of the Gabati area (sheet BM 159-161,93) (scale 1:75,000).

# 3. Small finds from the Surface Survey and from Test Excavations excluding Gabati Cemetery

*Laurence Smith*

The small finds from the Test Excavations at the site of Gabati at 159.2 were included in the main report on the site (Smith 1998a; 1998b). Material from the Surface Survey was collected from transects across each site identified, at approximately five metre intervals, with specific collections being taken where necessary, as in the case of the interiors of structures. This collection strategy was intended to provide an indication of the artefacts of all types and, potentially, all periods evident on the surfaces of sites. In general, it tended to reflect the frequency of material present, with that occurring in greatest abundance on the site surfaces being most strongly represented in the collections; this being pottery in most cases. Apart from the pottery, artefacts from the sites other than Gabati were relatively sparse in occurrence, represented a quite restricted range of material and can be encompassed by the following classes: stone artefacts, including chipped and ground tools, beads and a few metal items. Conditions for organic preservation were poor in contexts outside the Post-Meroitic burials at Gabati, so that the amount of organic material from the other sites was very small, and mostly in fragmentary condition. The present report presents a summary and brief discussion of the small finds and will concentrate on those from the Surface Survey and the Test Excavation sites, excluding Gabati. Reference will be made, where appropriate, to comparative material from the latter site.

## Chipped stone artefacts[1]

In the Test Excavations, very few examples of lithics were recovered, as might be expected given the nature of the sites and the late date of the main occupation at all of them. Consequently, the following summary and discussion will concentrate mainly on the lithics collected during the Surface Survey. All lithic pieces that appeared likely to be man-made were collected, but the artefactual nature of some pieces was not always certain, given that they had undergone a fair degree of abrasion and rolling. It must be noted that the strategy for the collection of the surface sample was not designed with the intention of providing a random sample specifically of lithic material, so that conclusions reached here can only be provisional.

The main characteristic of the overall sample is that *c.* 68% comprises flakes having few diagnostic characteristics. In the remainder, some can be recognised as primary or secondary flakes. There is one trimming flake, from 103.4, one ridge flake from 166.2, Site 3, and one likely primary flake from core preparation at 181.1. The majority of the flakes are chert, with a few examples probably being flint, fossilised wood, or milky and rose quartz. At three sites (138.2, 166.2, Site 3 and 222.1) some flakes were evidently heat-altered, but it is not clear whether this was accidental, or part of a deliberate treatment of the raw material.

Most of the raw material appears to have been from a pebble source, perhaps with a few exceptions including some larger flakes from 184.1. This indicates that the sites on which lithics were found are not sites where material was extracted from outcrops, at least on the basis of the samples taken.

The substantial use of a pebble source for raw material has been noted elsewhere in the Sudanese Nile Valley. It is characteristic, for example, of the 'Cataract Tradition' of Lower Nubia, distinguished by Shiner, in which the very great majority of tools are on Nile pebbles. In the Cataract Tradition sites the main rock types used are chert or jasper, agate, quartz and quartzite, all obtainable from the gravels bordering the river in that region. Other types used include petrified wood (Shiner 1968, 540). This latter raw material was also evident, together with the common materials: chert, quartz, agate and jasper, in the Shamarkian industry at Debeira West (Schild *et al.* 1968, 703, 710, 724). A similar range of raw materials is evident in the Neolithic at Geili, with a quartz pebble source being dominant, but with other types including basalt, agate, chert, fossil wood and flint being present (Caneva 1988b, 116) and at the Neolithic site of Kadero (Nowakowski 1984, 345, fig. 2). It seems, therefore, that in terms of the sources of raw materials present the lithics sampled in the Surface Survey and Test Excavations are compatible with those of other Nile Valley traditions, although the proportions of particular stone types appear to differ from some of the major sites within the same general region.

In general, the abraded nature of the lithics means that while there is quite extensive edge damage on the pieces, there is relatively little true retouch on the flake material. There is some possible intentional retouching on flakes from 101.9, 147.6, 168.1, and 211.2. One piece from 168.4 Site 1 exhibits possible '*sur enclume*' retouch and one flake from 113.2 has inverse retouch forming a steep backing to the piece.

The great majority of the flakes, together with such retouching as is present upon them, appear to be produced by percussion flaking. All have been made using the hard hammer technique, with little or no evidence for soft hammer being noted. There is very little evidence for pressure flaking, this only being noted on one flake, from a village below 166.2, which exhibits small pressure flake scars, although these may result from use.

In terms of the broader considerations of lithic technology, the sample does not present very conclusive evidence. There were flakes with prepared platforms recovered from sites 101.9, 184.1 and 222.1 (North side), the latter with some possible retouching, whilst one flake

---
[1] This section is based upon identifications kindly supplied by Mr H. Kenny and Dr T. Reynolds.

with a dihedral butt occurs in the samples from 101.9. These pieces could be produced using Levallois technology, as could flakes from 154.5, and 181.1. However, the frequency of pieces showing such characteristics is low, so that it is not possible to say that there is a full 'Levallois industry' present at these sites. This type of technology was not known previous to the Middle Palaeolithic but it continued to be used at much later periods, so that the presence of prepared platforms, at the frequency evident in the sample, cannot be used to date occupation at the sites unambiguously to the Middle Palaeolithic.

The number of flake tools in the sample is small. There is one truncation on a platform rejuvenation flake from 113.3 and a single endscraper from 165.1. Burins are represented by two '*Burins de Siret*', from 150.5 and 166.2 Site 3. Two possible notches were noted, one on a retouched secondary flake from 168.1, and the other on a '*machurée*' or 'bruised' flake from 110.4. The abraded nature of this latter piece means that the notch may have been produced by post-depositional crushing. Denticulates were present at 99.3 and 168.1, that from the former site being made on a truncated facetted piece, and that from the latter on a flake from a blade core. A third wind-abraded denticulate (find 101.4/T1/13S) was recovered from 101.4 in the Test Excavation season. Finally, one of the Levallois flakes from 181.1 was made into a truncated facetted piece. Examples of lithic artefacts are shown in Figure 3.1.

Considering the number of flakes, including the few identifiable tools on flakes, the evidence for blade technology is very limited. Apart from a broken bladelet core on a chert pebble from 101.9, virtually all the other evidence for blade and possible bladelet manufacture comes from site 166.2, Site 3, where a small number of pieces relating to blade and bladelet production occurred. These include a hinge-fractured flake from a bladelet core and a few tools, such as a borer made on a portion of an arched backed blade with abrupt retouch, and a proximal portion of a bladelet exhibiting retouch. This remains the strongest evidence for blade technology, and is the best potential evidence for bladelet production. However, it may be noted that it would still be possible to produce the short blades found at 166.2 on the types of cores available in earlier lithic technologies so that it cannot be certain, on the basis of the current sample, that a true blade industry is present.

It is evident that the main tool types present in the sample occur and re-occur in lithic industries dating to a wide range of chronological periods and extending over a large area geographically. Truncations, endscrapers, burins, notches and denticulates are known, for example, from sites near the Survey area, in the Mesolithic and Neolithic levels at Shaqadud, and at the Neolithic site of Kadero, whilst all except the burins are known from Geili, in the early Neolithic (Marks 1991, table 6-11; Nowakowski 1984, 345-346, fig. 5; Caneva 1988b, 121-133, fig. 21b). The tools from the 'Early Khartoum' Mesolithic site at Khartoum Hospital include endscrapers, borers, burins and backed blades (Arkell 1949, 42-48). Equally, the first three tool types occur in some of the Nubian Upper Stone Age and 'Epi-Palaeolithic' industries of Lower Nubia, such as the Gemaian, Qadan, and Shamarkian. Bladelets, including backed examples, are also present in industries such as the Shamarkian, and persist through at least into the Neolithic in Upper Nubia, as at Kadero (Shiner 1968, 544, 553, 563, 571, 573, 612, 617, 626; Schild *et al.* 1968, 697, 700-701; Nowakowski 1984, fig. 5).

Given this widespread distribution of the tool types the main method of comparison of lithic assemblages between sites, both chronologically and geographically, is through the relative proportions or percentages of the various types, as well as the percentages of the different raw materials and the presence or absence of technological traits (cf. Nowakowski 1984, 346-348; Caneva 1988b, 134-137). It must be noted that such comparisons are generally done of the basis of *excavated* assemblages, as is the case with those sites cited above. The low number of tools occurring in the sample solely from the *surface* collections from the individual sites and the long timespans during which these tool types were manufactured, means that their presence does not enable any attribution in terms of 'industry' to be made, nor does it allow any significant comparisons with lithic collections from other sites.

Cores form a relatively small component of the sample; only nine cores were collected, including the bladelet core mentioned above. The other eight include one possible Levallois core, a globular flake core used as a hammer stone from 168.1, and two 'change of orientation' flake cores, one from 165.1 and one on a chert pebble from 189.4. These cores can be characterised by a general lack of evidence for platform preparation, and appear to show a very basic lithic technology being employed. Only the 'change of orientation' cores could indicate a technology not found earlier than the Middle Palaeolithic, whilst the bladelet core indicates the presence of a technology that existed from the later Palaeolithic onwards.

Some of the few definite hammer stones recovered are described below under 'grinding stones', since typologically they can best be included there. They could have been used in the initial stages of forming artefacts such as stone spheres, of which one example was found, as well as in the chipped stone industries. One irregular piece from 165.1 was used for hammering, although it is not certain that this was necessarily in stone tool manufacture. Two pieces have evidence for utilisation, one being a pebble from 147.2 and the other a flake, from 154.3. The latter shows parallel striations near one edge, which resulted either from a poor polish (or perhaps the first stages of polishing) or from being used as a base on which a cutting action was carried out.

Taking the lithic samples from all the Begrawiya-Atbara sites as a whole, it is evident that there are only a few sites with evidence for specific lithic technology. The sites at 101.9, 165.1, 181.1, 184.1, 189.4 and 222.1 all have evidence which could indicate the presence of lithic technologies, including Levallois technology, in use from the Middle Palaeolithic onwards. Otherwise, apart from the fragmentary bladelet core from 101.9, only site 166.2 yielded substantial evidence for the blade and bladelet technology, which was practised from the Late Palaeo-

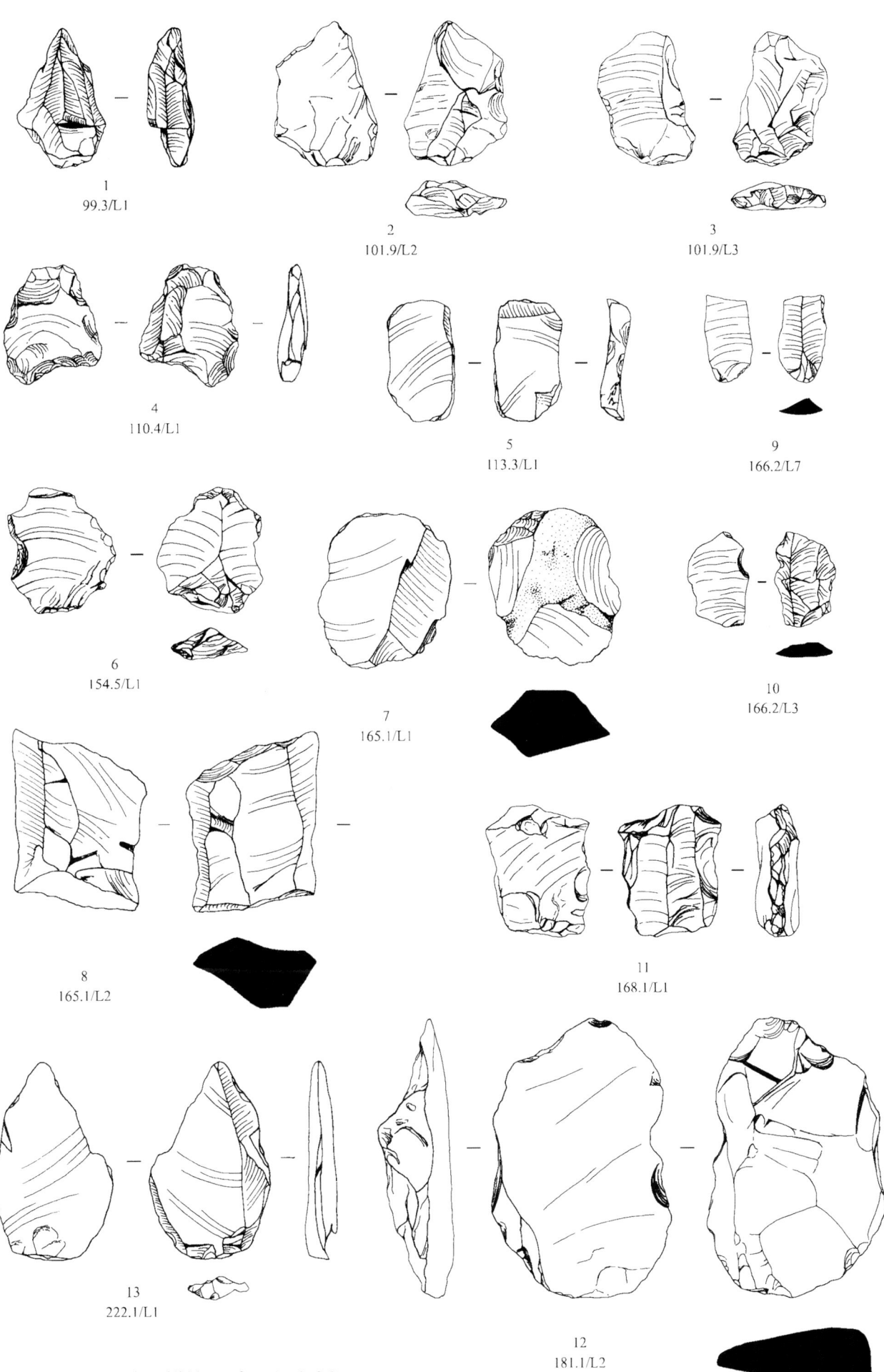

**Figure 3.1.** Examples of lithic artefacts (scale 2:3).

lithic onwards. It is of interest that no indubitably Lower Palaeolithic material, especially bifaces, was recovered, in contrast to a more recent road survey conducted across the Bayuda (Mallinson 1998). Here, in the course of a survey carried out in similar length of time, but over a slightly longer distance than the Begrawiya-Atbara Surface Survey, several classic Acheulean handaxes were found within the Wadi Muqaddam, albeit only at one site (Smith 1998d, 46, pl. 2). However, further sampling and excavation at the Begrawiya-Atbara sites may reveal evidence for Lower Palaeolithic occupation.

In conclusion, it must be noted that the lithic sample itself is mainly indicative of forms of lithic technology employed and the types of raw materials used. In the case of certain sites, it is likely that the chipped stone artefacts present are not inconsistent with the evidence from the other artefacts present and indicate the existence of a major occupation from a particular period or periods, as in the case of the Mesolithic at 222.1 or the Mesolithic to Neolithic at 165.1 for example. In some sites where the lithics were more apparent than featured sherds, this being particularly so at 166.2, the chipped stone material indicates the earliest probable period for which there is significant evidence for occupation. In most cases, since the lithic technologies evident from the sample, once established, remained in use for long periods and the proportion of identifiable tool types is small, the lithic sample on its own is not closely diagnostic chronologically, and cannot give clear indications of the cultural affinities of the sites.

## Grinding Stones

The complete and, more usually, fragmentary examples of grinding stones collected exhibited a moderate degree of variability. Following the basic method of classification of such material used by Arkell (1949, 56 ff.) it was possible to order the collection by dividing it into a relatively small number of morphological types. This classification is based on, firstly, a distinction between 'lower grinders' or 'quern stones' and 'upper grinders' or 'rubbers'. In many cases this morphological classification can probably be equated with an actual functional difference between the elements, particularly in the case of 'querns' with clearly well-hollowed surface(s) and larger 'upper grinders' with pronounced convex surface(s). However, in some of the smaller grinding stones with only slightly convex or concave faces, the functional distinction between quern and upper grinder cannot be assigned unambiguously from morphology. Therefore, the present classification must be taken as descriptive of the shapes of grinders present. Secondly, the classification involves the shape of the piece in plan, essentially whether it appears closer to a rectangle, oval or circular plan, as far as could be ascertained from their present state. Thirdly, the shape in transverse cross-section is considered, the main criterion for classification being the degree to which each face of the piece is concave, flat or convex.

On this basis, 20 Types were distinguished. These are illustrated in Figures 3.2-3.4.

1. Oval quern with both faces concave;

2. Oval (?) quern with upper surface concave and lower surface convex to flat;
3. Oval quern with upper surface flat and lower surface slightly convex;
4. Quern, as previous, but with lower surface flat;
5. Circular (?) quern with upper surface concave and lower surface convex;
6. Quern formed from small irregular stone block, of approximately oval plan, with deep depression worn into upper face (not illustrated);
7. Rectangular grinder with oval cross-section, slightly asymmetric;
8. Rectangular grinder, similar to previous, with one face flatter and a slightly asymmetric cross-section;
9. Oval grinder, with oval cross-section, slightly asymmetric;
10. Oval grinder with oval cross-section, one face convex, the other flat;
11. Oval (?) grinder with oval cross-section, one face strongly convex, the other flat;
12. Circular grinder with symmetrical oval cross-section, both faces being moderately convex. One specimen of this type, which may be oval or rectangular in plan, has a relatively sharp edge, but the most usual variety has a somewhat rounded edge, seen most clearly in cross-section;
13. Circular (?) grinder with sub oval cross-section, having somewhat squared-off corners.

In addition to those pieces reasonably securely identified as querns or grinders, seven further Types were distinguished:
14. Disk with circular depression symmetrically in the centre of each face;
15. Disk with circular depression in centre of each face, one deeper than the other;
16. Disk with circular pecked area in centre of one face;
17. Sub-spherical 'hammer stone' with irregular chipped surfaces;
18. 'Hammer stone', similar to previous, but closer to a cylindrical form, with one chipped end;
19. (Hemi) sphere;
20. Rectangular (?) block with irregular deep pecking on one surface.

It has not been possible to date closely the majority of the types of grindstone on the basis of characteristics of the artefacts themselves. Most of the grinding stones collected comprise forms that have tended to remain the same over long periods (cf. Caneva 1988b, 141). However, there are some types that can be more closely dated than others and these will be discussed first. Types 14 and 15, the disks with circular depressions in each face, both came from site 222.1. They are generally similar in form to artefacts of sandstone from the Khartoum Hospital site of the Khartoum Mesolithic, described as 'grinders with depressions in the face', exhibiting circular depressions in the illustrations. These grinders are considered by Arkell to be possibly palettes for ochreous paint. Such artefacts occurring at Khartoum Hospital are differentiated from 'unfinished stone rings', partly because examples occurred in which the depressions are not opposite each other, and partly because there are examples where the depression was made after the grinder broke (Arkell 1949, 62-63, pl. 33, 1-3).

It is not possible to distinguish between 'grinders' and 'unfinished stone rings' in the case of the specimens from 222.1. Types 14 and 15 are similar to examples of 'disk grinders' shown in Arkell's plates 33, 1 and 36, 1 and in plates 33, 2 and 36, respectively, since the depressions are placed centrally and opposite each other and seem to have been executed before breakage. In these respects they also appear to be comparable to examples of 'stone rings' in the early stages of manufacture shown by Arkell (1949, pl. 34, right hand end). No traces of ochre were noted on either Type 14 or 15. In cross-section, Type 14 is very similar to artefacts from Wadi el Kenger, although the degree of similarity in plan is uncertain, owing to the incompleteness of the Type Specimen. Type 15 is very similar in the portion extant to a second specimen from el Kenger in both plan and cross-section. Both the Kenger specimens are classed as 'unfinished stone rings' by the excavator, who considers this class of artefact to be known mostly from Mesolithic contexts (Caneva and Gautier 1994, 76, fig. 9). Stone disks with depressions on each face are quite widely distributed, being present outside the Nile Valley in Shaqadud Midden. However, it may be noted that here they are in levels assigned culturally to the Khartoum Mesolithic/Neolithic transition, as well as the full Mesolithic (Marks 1991, 112, fig. 6-8, tables 6-7, 6-10). A Mesolithic dating for Types 14 and 15 would fit best with the ceramic sample recovered from the Site 222.1, since all the featured sherds collected were classed as Group 8b or 8d, dated to the Mesolithic (Mallinson *et al.* 1996, 135; Smith 1996, 192-193).

Concerning the querns and other grinders, generally very comparable types can be found at sites of quite wide-ranging date. There may be some differences in the detail of form between examples from sites of a particular period and those from the Survey and Test Excavations, but it is not always certain how significant these differences are. Such variations cannot be assumed to indicate that the Survey and Test Excavation specimens are of a different period from the comparative examples.

Type 16, the disk with a circular 'pecked' area in the centre of one face (Figure 3.4), can be paralleled in the Mesolithic, being present at the Khartoum Hospital site (Arkell 1949, 61, pl. 32, 1). While disks with a pitted area on one face continue into the Khartoum Neolithic, as at Shaheinab (Arkell 1953, 47, fig. 11), the example from the Surface Survey was found at Site 222.1, in association with the few 'unfinished stone rings', together with the pottery of Groups 8b and 8d, so that a Mesolithic date would be most probable.

Several forms of grindstone have a concave upper surface with a flat or convex lower surface as in Type 2 shown in Figure 3.2. There are approximately flat grindstones from the Khartoum Hospital site, considered to be more characteristic of the Mesolithic at this site than the 'saddle quern' type. One example given by Arkell is a quadrilateral fragment with maximum thickness of 400mm, comparable to specimens of Types 3 and 4 (Figure 3.2), but the Khartoum Hospital example was subsequently used as an upper grindstone and as a grinder. Several of the other types distinguished by Arkell are not fully comparable to those from the Survey, the fragments being thinner, of 25mm maximum thickness. These could be from flat oval grinders that are thinner than Type 3 and 4 specimens. One example from Khartoum Hospital, with a rounded edge and flattish faces is of very similar thickness to the maximum thickness of Type 3, *c.* 32mm, although the Type 3 specimen lacks the heavy pitting evident on the former. One type present at the Khartoum Hospital site comprises a thicker form, of maximum thickness 47mm, with a more or less rounded edge, but with a deeper concave face than Types 3 and 4. The latter Type appears closest to one example of grindstone with an upper face only slightly concave, although the Khartoum Hospital specimen has greater maximum thickness (Arkell 1949, 70-71, pl. 42, 4, 5; pl. 43, 1-4).

If the identification of Types 3 and 4 as 'querns' is correct, it appears a comparable form is not present in the Neolithic of Geili where the usual 'querns', as reconstructed from fragments, have a flat smooth edge to the upper surface with a deep oval impression in the centre (Caneva 1988b, 141, figs 32, 6-9, 34, 1-3). Concerning parallels from later periods, grinders with two flat, or only slightly concave, faces occur at Jebel Moya. These are closest to the Survey Type 4, although the type as it occurs at Jebel Moya is more regular and thinner than the Survey specimens recovered. Many were from the Jebel Moya East Cemetery, in use during Gerharz's phases II and III, dated to between the 3$^{rd}$ and near the end of the 1$^{st}$ millennium BC (Addison 1949, 170-171, 178, pls LXIV, type VIII.A.b, LXXII, B; Gerharz 1994, figs 12 and 13, 329-331). Stone 'querns' or 'rubbing bases' having shapes in plan of which Types 3 and 4 could reasonably be fragments occur at Soba; these could be dished or with flat surfaces, depending on the amount of wear (Shinnie 1955, pl. XXII, top right, and bottom; Allason-Jones 1991, 149).

In the cases of Types 1, 2 and 5, querns with one or two strongly concave faces and a more or less distinct form of 'rim' to the piece (Figure 3.2), there are general parallels in the Mesolithic, in particular grindstones from the Khartoum Hospital site illustrated by Arkell (1949, pl. 44, 2, 5, 6) which appear similar to Types 2 and 5 especially. Other examples also with a 'rim' show evidence of subsequent use as an upper grindstone, which has affected the appearance of the 'rim' (Arkell 1949, 71-72 pl. 44, 1, 3, 7), so that these do not appear such close parallels for the Survey Type specimens. Type 5, in particular, could be a fragment of a grinder comparable to that from the Neolithic of Shaheinab, with one face concave and the other convex (Arkell 1953, 46, fig. 10). However, none of the three Types is exactly comparable to the cross-sections of querns illustrated from the Neolithic of Geili (Caneva 1988b, fig 32, 6-9), although Type 2 is closer to a quern from Kadada, described as having a well-defined round rim (Caneva 1988b, 143; Geus 1984, 60-61, no. 10).

An example with a rounded rim very similar to that on Type 2, but having more squared-off edges on the lower surface, was found in the Post-Meroitic burial of Tumulus

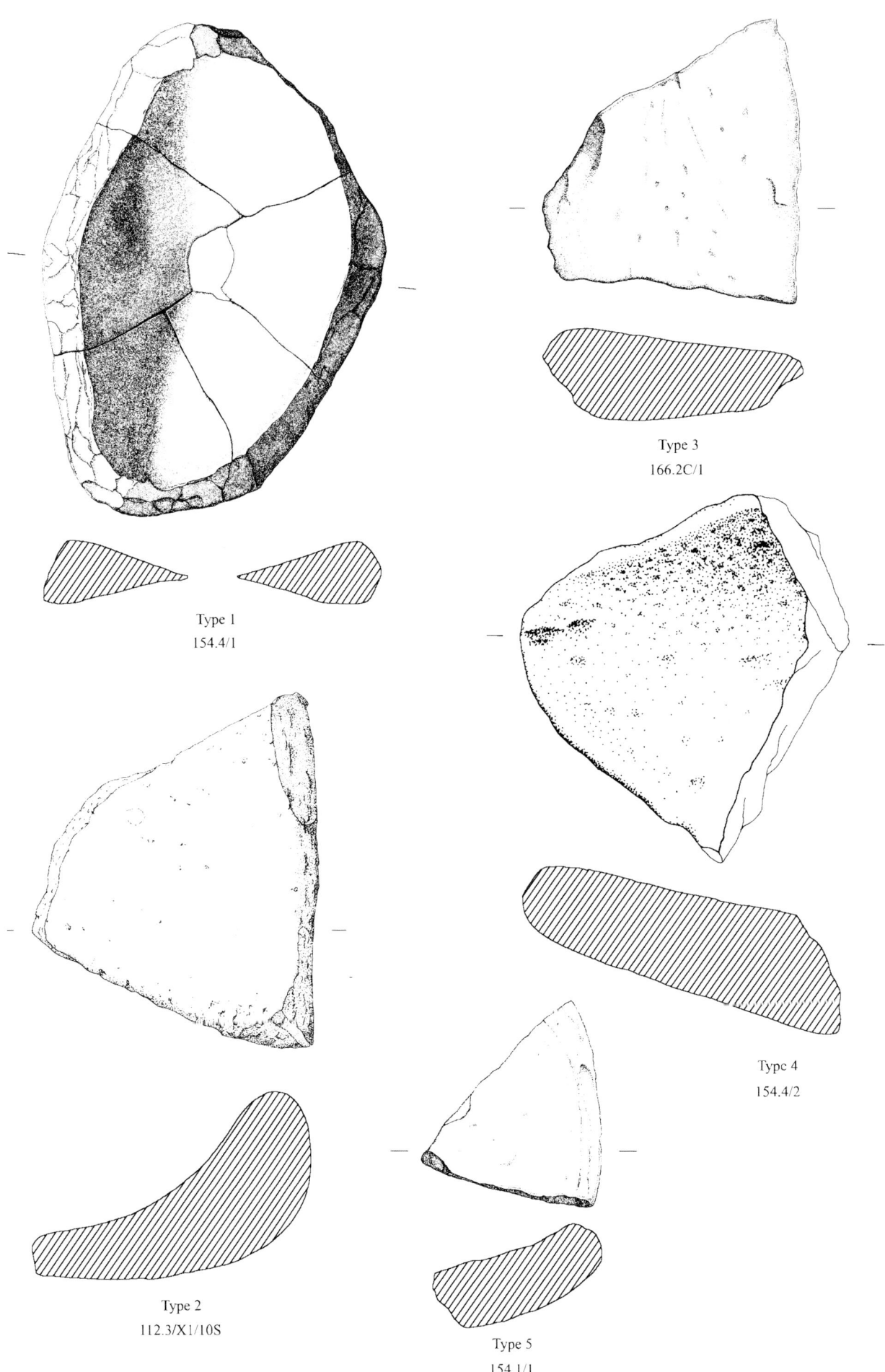

**Figure 3.2.** Grindstones, Types 1-5 (scales 1:2 Type 1, 1:4).

124 at Gabati (Edwards 1998c, 97; 1998d 128, fig. 5.13). Type 1 is reasonably well paralleled in shape by querns found at Soba, which are said to be identical to the 'saddle quern' termed *murhaka*, in use until recently in Sudan (Shinnie 1955, 60, pl. XXII, top). It is not clear from the illustration as to whether the Soba types were used from both faces, as was the Type 1 specimen from Site 154.4. This type appears comparable in general form and size to the description of querns in use during the late Medieval period at Qasr Ibrim given by Adams (1996, 103). A general later Christian date would be most appropriate for the Type 1 specimen, since virtually all the featured sherds collected from the site were assigned to the Late or Terminal Christian to Islamic periods (Mallinson *et al.* 1996, 80).

The various forms of grinder, rectangular to oval in plan and generally ovoid in transverse cross-section, comprising Types 7-14 (Figure 3.3), include shapes similar to those present at the Khartoum Hospital site, but which do not seem extensively paralleled there. For example, Type 7 is of similar subrectangular shape, relatively thin in cross-section, to a rubber with incipient keel illustrated by Arkell (1949, pl. 37, 2) although the Type is probably less strongly keeled and is in porphyritic stone rather than sandstone. Although most examples from the Survey and Test Excavations are fragmentary, in general they do not exhibit the elongated ovoid shape in plan seen in many Khartoum Hospital site rubbers (Arkell 1949, pl. 37). Some Types have a very similar transverse cross-section to Khartoum Hospital examples with an incipient keel, being slightly convex on one face and more strongly convex on the other, such as Type 11. However, none of the Survey Types has the strongly developed keel seen on several of the grinders from the Khartoum Hospital site (Arkell 1949, pl. 31, 4-8). Type 11 is closest to an example from the earlier Neolithic at Geili in cross-section, and appears likely to be the end of a piece of similar oval form (cf. Caneva 1988b, fig 33, 3). It can also be reasonably well paralleled at Jebel Moya in the oval shape in plan and in cross-section (Addison 1949, pl. LXIV, VIII A.a). No close similarities in form were noted from the, later, contexts at Soba or Qasr Ibrim, on the basis of examples illustrated.

Concerning the remaining Types, the pieces with traces of battering or chipping, considered to be hammer stones (Types 17 and 18, Figure 3.4) must of necessity have been present throughout the prehistoric and into the protohistoric periods, as long as stone tools were still in even occasional use. Type 19 is an almost perfect hemisphere, the flat face having an irregular surface indicating that that it is likely to be a broken spheroid. Such stone balls are known from wide-ranging periods. They do not appear to be the same as the spheroids known from the Mesolithic of the Khartoum Hospital site, which are less regular, and tend to have a flattened area and /or a keel, through being used as grinders (Arkell 1949, 51, pl. 25, 4-6). Spheroids occur in the Neolithic contexts at esh-Shaheinab (Arkell 1953, 52) and at Shaqadud, although apparently the spheroids at the latter site are rare, and approximate less to true spheres than does Type 19 (cf.

Marks 1991, 112, table 6-7). Such artefacts are present in much more recent periods, over 350 being recovered during the two campaigns at Soba (Allason-Jones 1991, 149; 1998, 78). They have been considered to be jar stoppers or, more usually, pounders or grinders. The latter is the function ascribed to examples from the Late Christian period, 14$^{th}$-15$^{th}$ century AD, at Qasr Ibrim where they have been found in association with the flat quern stones (Adams 1996, 103). A date in the Christian period for Type 19 would be consistent with the likely date of the area in which it was found (Mallinson *et al.* 1996, 58).

## Ground and Polished Stone

Only two stone items from the Surface Survey and Test Excavations were found that were not evidently querns or rubbers. One retained clear evidence of having been deliberately ground and polished, in contrast to those artefacts exhibiting a probably unintentional 'polish' due to use in a grinding action. The first is a fragment of stone from 155.3/2/T5, which has evidence for four faces quite smooth and polished. Two faces meet at a rounded corner, or ridge, whilst the other two form the short sides of the block, being at an angle. The rock type is most likely to be basalt.[2] Apart from the presence of the faces, the piece lacks any definite characteristics so that the type of artefact from which it comes is not readily identifiable. It may be a fragment from an architectural element, or statue base. It may be from an implement such as a strongly 'keeled' grinder, even though it appears to be rather shorter than might be expected.

The second artefact (211.2/1S) is a portion of a polished stone ring, made most probably of porphyritic basalt, or possibly gabbro (Figure 3.4). It has an interior diameter of about 58mm and is about 20mm wide. The cross-section is approximately triangular, with the interior face of the ring being the base and the outermost edge the apex. On the interior the surface is slightly bi-convex in profile, which would indicate that it was manufactured by drilling from each face before being polished.[3] The type of artefact from which the extant fragment comes cannot be definitely determined. In cross-section, it is similar to some of the smaller types of the discoid mace head known from the Neolithic. Nevertheless, judging by examples from Kadada illustrated by Geus (1984, 60, nos 2, 4, 6) even the smaller ones are wider than the Surface Survey specimen, ranging from about 35mm to about 43mm wide, and have relatively smaller central holes, about 17-32mm in diameter. It clearly differs in shape from other mace heads from early contexts, such as those from Shaheinab (Arkell 1953, pl. 23). The diameter of the central hole seems equally to militate against the ring being from a staff head. Another possibility could be something of the nature of a digging stick weight, but this does not appear to be consistent with the finely polished surface of the piece. From the foregoing, it would be most likely to be a portion of a bracelet, armlet, or anklet. The internal diameter would be most reasonable for that of a bracelet,

---

[2] Thanks are due for identification of stone types to Dr Judith Bunbury, Department of Earth Sciences, University of Cambridge.

[3] I am indebted to the late Dr J. Alexander for this point.

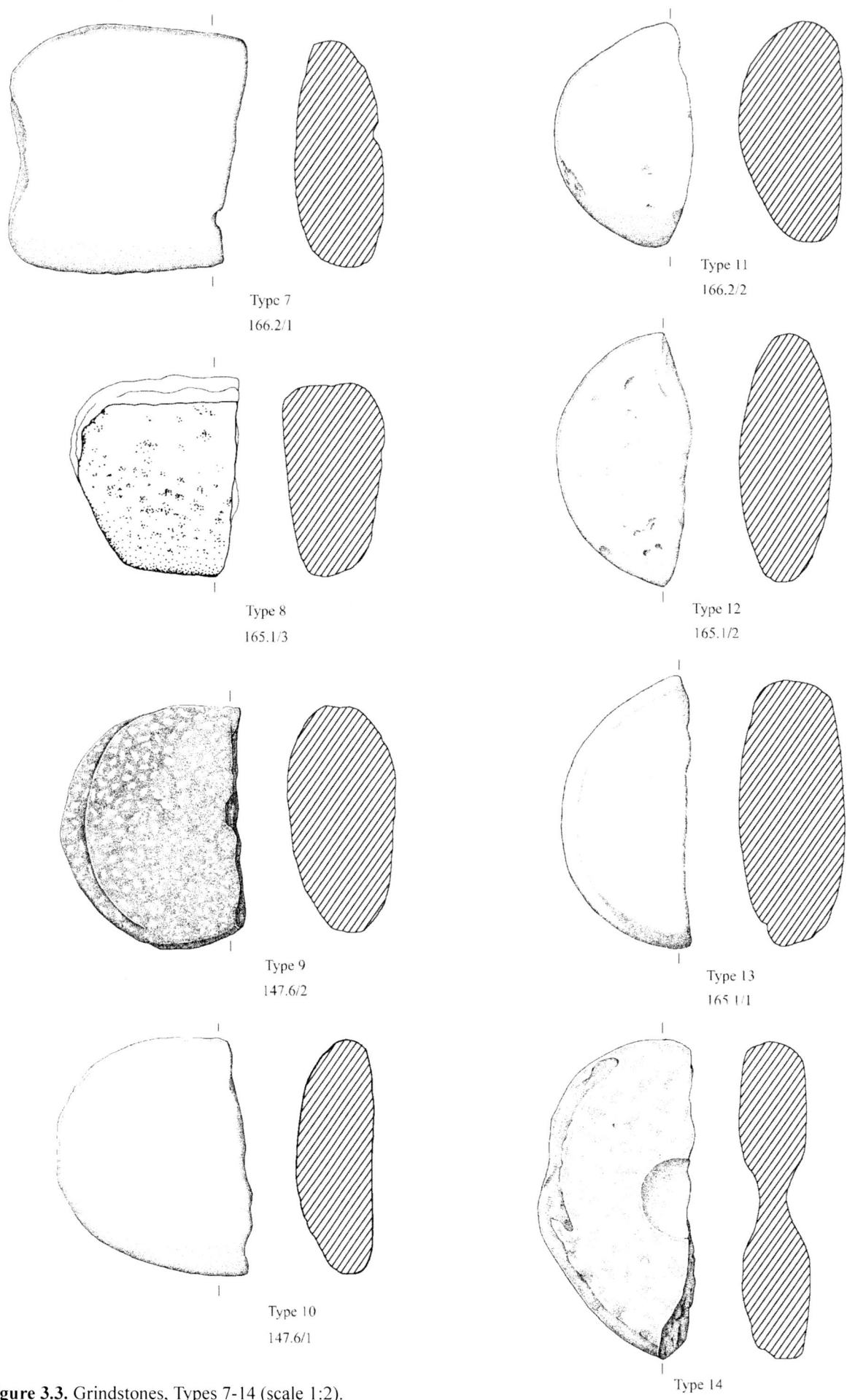

**Figure 3.3.** Grindstones, Types 7-14 (scale 1:2).

even though the width of the ring is a little greater than might be expected.

Taking the identification as a bracelet as the most probable, a date for the object on stylistic grounds may be suggested. The site on which the bracelet was found (211.2) had nothing but late to Terminal Christian or Islamic period pottery present amongst the featured sherds (Mallinson et al. 1996, 131). However, it cannot be assumed that the bracelet is of the same date as the ceramics. Bracelets with triangular cross-section are known from the contexts of the late Christian to Islamic date, but these often are in materials other than stone. For example, glass bracelets with such a cross-section are known from Quseir el-Qadim in the late Ayyubid to early Mamluk levels (Meyer 1992, 75, 178, pl. 20, 554-559). They also occur in the later levels of Qasr Ibrim, being abundant in those of the Islamic (Ottoman) period (Adams 1996, 183).

Bracelets or armlets in several varieties of stone are present at the latter site in the late Christian and Ottoman levels. Although in stone, the majority appears to be in types such as alabaster, schist and limestone, rather than in porphyrite or gabbro. The cross-sections tend to be D-shaped or wedge-shaped and not triangular. Hence, these comparisons do not confirm a specifically Late Christian/Islamic date for the piece.[4]

The closest parallel found for the bracelet comes from the site of Soba, where a fragment of a stone ring in diorite, classed as an 'armlet', was found (catalogue number 277). This piece is slightly greater than the Surface Survey example in internal diameter, at 65mm, and is somewhat thicker but the width is very similar, being 22mm. In cross-section it presents a triangular section with rounded corners, with the base forming the internal face of the ring; the latter is biconvex. In these respects it is close to the Survey specimen, differing mainly in being nearer to an equilateral triangle in shape. A second armlet, of chlorite schist or chlorite shale, was also found (catalogue number 279). This is of larger internal diameter (70mm), and has an irregular triangular cross-section, closer in shape to an equilateral triangle than the Survey bracelet but of similar thickness, at 14mm. (Allason Jones 1991, 147, fig. 71). The Soba armlet 277 may be presumed to be most likely from a generally 'Christian' context, but the dating of the piece itself is not very secure. It was recovered from the fill of Room m30 in Building D (the 'Mud Brick Building'), and may relate to Phase V of that structure, dated to c. AD 1200 at the earliest. However, it is not clear that the filling of this room was necessarily of the same date as similar fills in other rooms sealed by Phase V floors. (Welsby and Daniels 1991, 34, 115, 354). The context of Find 279 is the fill of another room in Building D, m26, which similarly may date to Phase V (Welsby and Daniels 1991, 353).

These parallels indicate that the bracelet could be assigned to a period before AD 1200, but do not provide any very close dating for the item. Comparisons were sought with other material to see whether there were reasonable parallels with more securely dated specimens. One of the types of stone ring (Type o.1) from Jebel Moya has a similar form of triangular cross-section, with somewhat 'squared-off' corners to the interior face. This example has width and thickness very close to the Survey specimen, viz 22mm and 14mm respectively, although it is smaller, being only about 52mm in interior diameter. Three further types of stone ring from Jebel Moya (Types l.1, n.1 and n.2), have generally similar triangular cross-sections to the Surface Survey specimen, although all have more rounded corners to the interior face, and Type l.1 is closer to an equilateral triangle. Type n.1 is closest in overall size, with an interior diameter of 58mm. (Addison 1949, pl. LXIV, Type VI). A bracelet in a light-coloured stone, or possibly ivory, was recovered from a Meroitic grave at Geili. It has an internal diameter of c. 51mm and is 17mm wide, with a biconvex profile to the interior face (Caneva 1988c, 200, 383, fig. 33, 1). Although the material of which this bracelet is made is not the same as that of the Surface Survey specimen, the form and size is quite similar.

The bracelet differs from the forms of bracelet, including examples in stone, from the Post-Meroitic ('X-Group') period, as illustrated from excavations at Ballana and Qustul (Emery and Kirwan 1938, 192, pl. 38, C; Farid 1963, 101, fig. 57; Williams 1991, 128-130, fig. 145 a, b). Few stone bracelets/armlets were recovered from Meroe West and South cemeteries; one example was from the South cemetery. This bracelet is from tomb S45 tentatively assigned to Generations 6-8, giving a date of c. 664-623 BC. It is of serpentine, with an approximately oval cross-section. Other specimens were of ivory (Dunham 1963, 47, 153, 408, fig. 226, C; 1955, 2; Welsby 1996, 207). Fragments of stone bracelets were found on the Meroe City site, in component 5 of the 79/80 area, and in the middle and top component of the '50-line' (Shinnie and Bradley 1980, 192). One final example was noted from the New Kingdom cemeteries of Lower Nubia, from unit 185/238A at Fadrus. This comprised a bracelet in stone, 'possibly granite', with a 'somewhat triangular' cross-section. It has an internal diameter of only 32mm (Troy 1991a, 135-136; 1991b, 266, pl. 28, 8).

From the foregoing, it appears that the closest parallels for the bracelet found in terms of material and form are the examples from Soba, Geili and Jebel Moya. Given that the latest dating for the latter site is now considered to be around 100 BC (Gerharz 1994, 331) thus falling within the earlier Meroitic period, whereas it is currently thought there is insufficient evidence to demonstrate a Meroitic period occupation at Soba (cf. Welsby 1998b, 114; 1998c, 272), it is difficult to make any definite attribution of date. However, a date in the Meroitic or Christian period appears most likely, and may relate to burials on the site 211.2 earlier than that of the featured pottery on the surface.

## Metal: band

Seven strips of metal (101.4/T1.3/47S) were excavated from burial 3 in Tumulus 1 at Site 101.4 during the Test

---

[4] Thanks are due to Dr P. J. Rose for allowing me to examine bracelets from Qasr Ibrim and to the Egypt Exploration Society for permission to cite this comparative material.

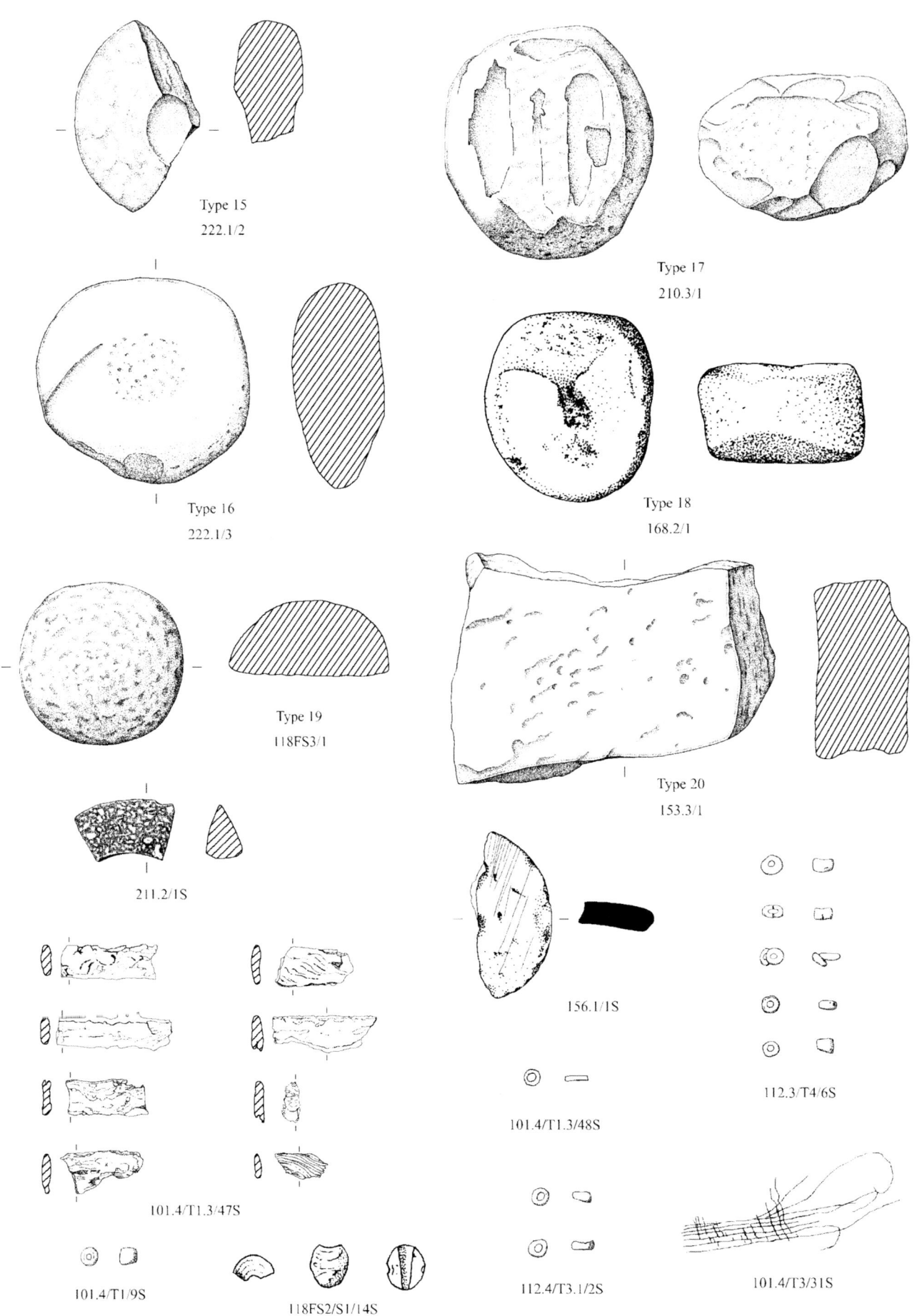

**Figure 3.4.** Grindstones, Types 15-20 and other small finds (scales 1:2, beads 1:1).

Excavations. The strips, together with a couple of small fragments, are largely replaced by corrosion, but from the colour, it is probable that they are of iron. The largest piece is 43.8 x 12mm, with the rest mostly between about 30-39mm long and 12-15mm wide. They are between 1.5 and 3mm thick (Figure 3.4). All have rather irregular edges, and it is not clear as to whether originally they made one continuous strip widening towards one end, or formed two or more elements. If the former were the case, the single strip would have been a minimum of about 237mm in length.

The type of object the strips came from is not immediately apparent. It is possible that it could have been a knife or dagger blade, but there is no evidence for a tang, and if the strips formed one piece the blade would seem to be very long in relation to its width. In general, the strips do not seem likely to be from a knife or dagger of the types from late Meroitic to Post-Meroitic contexts, such as those from Gamai or Qasr Ibrim Cemeteries 192 and 192A (Bates and Dunham 1927, pl. LXVII, 17-28; Mills 1982, 11, 37, pl. VIII, 2.25, pl. XL, 7.6). If a single piece, the strips could be of appropriate length to be a knife of the type illustrated from Abka, Site 416, dated to the Meroitic or Post-Meroitic ('X-Group') period (Säve-Söderbergh et al. 1981, 169, pl. 89,1) which has a blade 320mm long, but they still appear too narrow in proportion, since the blade of the Abka specimen is about 27mm in width. They are not clearly similar to examples of these implements from closer to the Survey area, such as those illustrated from the necropolis at Meroe (Garstang et al. 1911, 47, pl. LIV, 1, 6-9), or those from certain tombs in the West and South cemeteries. If more than one object is represented, the strips could be reasonably similar to three blades in a Post-Meroitic tomb G1, although they are still somewhat narrower than any of the G1 tomb blades, two of which still clearly retain a tang (Dunham 1963, 334, 336, Fig. 187, B). Three fragments (b, d and g) could have formed part of a blade, somewhat similar to the portion of a knife or spear blade from the Post-Meroitic burial T1 at Gabati (Smith 1998a, 113, fig. 5.4 no. <T1/24>), but the remaining strips do not appear to be compatible with being from a blade of this form.

If the strips cannot be unambiguously identified as blade(s) from a weapon or weapons, the other most likely possibility is that they were part of the fitments for a container, although none retains any clear evidence of any means of attachment. They could, for example, be from a 'strap' such as that illustrated from the 'Ballana Period' Cemetery 193 at Qasr Ibrim (Mills 1982, 47, 58, pl. LXI, 120.1) or they could be a more fragmented example of an object such as that from Soba consisting of a bar of rectangular section having two further bars attached. The width and thickness of the bars is reasonably close to those of the metal strips from the Test Excavations, at 12mm and 3mm or 4mm respectively. The Soba piece is likely to be more recognisable as a specific implement than these strips, being compared to the jewellers' balances from Ballana (Allason-Jones 1991, 135, fig. 66, no. 112; Emery and Kirwan 1938, pl. 105, A, C, D).

# Beads

### Metal

Two small metal rings, of bronze or other copper alloy were recovered from the surface near a tumulus during the Surface Survey. The identification of these rings is not certain, but it is likely that they are examples of metal beads. These are closest to type I.B.2.b (short cylinder) with VIb perforation in Beck's classification. Whilst the great majority of beads recovered archaeologically are in shell, stone, glass or faience, examples in metal are known. The most valuable would generally be in precious metals, such as silver or gold, but some occur in other metals.

### Glass/Glass paste

The first glass bead found (101.4/T1/9S) is cream in colour, is generally cylindrical in shape, with rounded edges and has slightly bevelled ends (Figure 3.4). In size, the bead is 3.2mm in length and 3.4mm in diameter. It can be classed according to the system of Beck (1928) as I.C.4.f.b, the 'standard cylinder with two convex ends'. The glass paste, or faience, bead recovered during the Test Excavations comprises a mid-green bead from 118FS2, Site 1 (Figure 3.4). This specimen (118FS2/S1/14S) was broken in half anciently, so that the perforation can be clearly seen, tapering from one end to the other. This bead can be classified as Beck Class I.C.1.a, with Type III perforation.

### Faience

Three finds of faience beads were made during the Test Excavations at the sites other than Gabati. One was from the burials in Site 101.4 Tumulus 1 (101.4/T1.3/48S), being a pale green disk bead of Beck Class I.A.1.b, termed 'barrel disk', 1mm long and 4mm in diameter (Figure 3.4). A further two sets of faience beads were found in Tumuli 3 and 4 at Site 112.4 (Figure 3.4). Although the number of beads from the former burial was considerably lower than from the latter, all the beads from both tumuli are evidently similar, and seem likely to be of the same manufacture, being rather irregular in form. The beads from 112.4 can, consequently, be summarised as in Table 3.1.

### Eggshell

Bead (154.5/T1.1/29S) is of eggshell, most probably of ostrich eggshell. It is badly eroded, so that its original form is uncertain. It is most probable that it is an example of a 'standard' ostrich eggshell disk bead, of Beck Class IA.1b or IA.2b.

### Comparanda

In general the beads described above are not very distinctive types. Metal beads survive from various periods. Examples of similar shape to the Survey specimens are known from A-Group graves further north, the Surface Survey beads being closest to type E1 in the classification of Nordström. Few examples in copper or copper alloy were recovered in the Scandinavian Joint Expedition

**TABLE 3.1.** BEADS FROM SITE 112.4.

| Find Number | Colour | Beck Class (Range) | Lenth x Diameter (range, in mm) | Number |
|---|---|---|---|---|
| 112.4/T3.1/2S | Pale Green | I.A.1.b; I.A.4.f.b | 1.2 x 3.7-1.3 x 4 | 2 |
| | Pale Green | I.B.1.b-I.B.2.b-I.B.4.f.b | 1.3 x 3-2.4 x 3.2 | 9 |
| | Greenish Blue | I.B.2.b; I.B.4.f.b | 1.7 x 3.6; 1.5 x 3.5 | 2 |
| | Turquoise | I.B.1.b | 1.5 x 3.5 | 1 |
| 112.4/T4/6S | Greenish Turquoise | I.B.1.b-I.B.2.b-I.B.4.f.b | 1.4 x 3-2.6 x 3 | 35 |
| | Greenish Turquoise | I.C.4.f.b | 2.8 x 2.9-3 x 3.2 | 2 |
| | Greenish Turquoise | I.A.1.b | 1.1 x 3.7 | 1 |
| | Greenish Turquoise | II.B.4.f.b | 1.4 x 3 | 1 |
| | Brownish Grey | I.B.1.b-I.B.2.b-I.B.4.f.b | 1.2 x 3.2-1.7 x 3 | 6 |
| | White | I.B.2.b-I.B.4.f.b | 1.5 x 2.8-2.5 x 3.2 | 4 |
| | Bluish Turquoise | I.B.1.b-I.B.2b-I.B.4.f.b | 1.2 x 3.4-2.8 x 3.4 | 8 |
| | | I.C.1.b | 2.8 x 3-2.8 x 3.1 | 2 |

(SJE) survey, but two occurred in A-Group graves at Ashkeit (Nordström 1972, 125, 175, 177, pl. 52). Similar ring-shaped metal beads occurred in C-Group cemeteries, forming types E1 and E5 in Säve-Söderbergh's scheme. Most of the specimens found in the SJE survey are of gold or silver, from C-Group, Kerma and Pangrave interments. Only two examples of type E1 were found, at sites 262 and 266 in the Abka district, one being of copper and the other probably of copper wire (Säve-Söderbergh 1989, 80, 235, 237, fig. 29). Such bead forms continued into the New Kingdom period being known, for instance, from a few sites investigated by the SJE. Types E1 and E2 appear closest to the Surface Survey specimens, although only one probable example in copper was noted (Troy 1991a, 83, fig. 18).

Simple ring-shaped beads in metal re-occurred during the 25$^{th}$ Dynasty to Meroitic periods. The Surface Survey specimens are closest in shape to Dunham's bead type IVi, described as a 'ring-bead'. Several specimens of this type were found in Meroe West and South cemeteries, for example, although ring beads noted in the publication are most often in gold, such as those from tomb W634 and W33. Nearly all surviving examples of bronze beads are described as 'ball-beads' or 'band-beads'. The latter appear to be more nearly equal in length and diameter than the Surface Survey beads (Dunham 1963, 43, 222, fig. R). Only one bead form in metal was noted from the earlier investigations at Ballana and Qustul, this being in silver, and nearer to a globular than a ring shape (Emery and Kirwan 1938, pl. 43, Type 17) During a more recent excavations at Qustul, a few examples in gilded or silvered glass and in lead, silver or pewter were found, although no types in copper or bronze were noted (Williams 1991, 130-150, figs 63, 65). It is apparent that the simple form of the beads precludes definite dating on stylistic grounds, but they would be consistent with a range of periods, to which the site from which they were recovered may be assigned on other grounds.

The cream glass bead 101.4/T1/9S is reasonably close in shape to some examples known from Post-Meroitic/'X-Group' contexts. It is closest to Type 28 from Emery and Kirwan's typology of beads from Ballana and Qustul. Those in this form recovered during the original excavations at the site are described as being in coral and faience, but a few examples of this type from the investigations of the Oriental Institute of Chicago are of quartz, presumably of a white colour (Emery and Kirwan 1938, pl. 43; Williams 1991, 130, fig. 62, n, p, table 22). No exactly similar bead was found in the excavations at Gabati, although one example classed as I.C.4.f.b, albeit with a more pronounced convex longitudinal profile, was found in the Post-Meroitic burial T5 during the Test Excavations. The Gabati bead is in glass, but of dark brown colour rather than white (Smith 1998b, 230, fig.10, no. 51). No entirely comparable beads are illustrated from Meroe town site. The closest in shape is no. 2235, of white glass with red decoration (Shinnie and Bradley 1980, figs 65-68), although this is slightly longer in relation to its diameter than 101.4/T1/9S. The overall form of the bead could come within the range of the Type Ib beads in the shape classification of those from the Meroe West Cemetery, but the Type as illustrated has a more regular convex longitudinal profile than 101.4/T1/9S. In this respect, the latter is closer to Dunham's Type IIf, although this is longer in proportion to the diameter (cf. Dunham 1963, fig. R). However, from the parallels found, a later Meroitic to Post-Meroitic date for the bead would be reasonable.

Concerning the small faience beads from Sites 101.4, 112.4 and 154.5 these are, again, of a general type that had a long history in Nubia. Faience beads of green to greenish-blue colour were present in Lower Nubia as early as the 'A-Group' period (Nordström 1972, 125). The form and size of the faience beads from the Test Excavations can be reasonably well paralleled in New Kingdom contexts in Lower Nubia. Beads of similar shape to 101.4/T1.3/48S in green and blue are known from New Kingdom sites in the Debeira area (Troy 1991a, 84, fig. 18, Type F19). The beads from 112.4 can be paralleled by Types F1 and F4, the latter described as 'somewhat irregular', which occur in blue, green and white colours from Debeira, including the cemetery at Fadrus, and at Ashkeit (Säve-Söderbergh and Troy 1991, 83-84, 253,

296-297, 309, fig. 18). Such types continued through the Meroitic period, in Upper Nubia, similar beads of blue faience being present, for example, in the burial of Beg. N29 dated to AD 150-167 (Dunham 1957, 7, 170, fig. 110). They are present in Post-Meroitic contexts in sites as widely separated as Qustul and Ballana, and Gabati, occurring in the colours blue, green and blue-green (Williams 1991, 130-136, 138-150, fig. 62, m; Smith 1998b, 230, fig. 9, no. 7; Edwards, 1998g, 225-227).

The general form of the beads, particularly those from 112.4, is similar to that of those termed 'drum beads' at Soba. Numbers of these, of comparable size, ranging in colour from green through turquoise to pale blue, were recovered in the 1989-1992 campaign, largely from Area MN3 and Mound Z. The former site is dated by ceramics to between the 6$^{th}$ and 9$^{th}$ centuries AD. The main occupation at the latter site is dated from about the 5$^{th}$ century AD through the Transitional to the Early Christian period, but material relating to later use is dated as late as the 9$^{th}$ to 12$^{th}$ centuries (Allason-Jones 1998, 69, fig. 28, nos 136, 137; Welsby 1998c, 272-273, 1998d, 21).

The range of periods during which bead types forming reasonably close parallels for the Test Excavation beads occur, means that the latter cannot be other than consistent with the dating for the sites ascertained through different classes of evidence. Bead 101.4/T1/9S could be contemporary with the main burial at 101.4 (Post-Meroitic) or, potentially, be residual or re-used from an earlier period, most probably the Meroitic. The site of 112.4, from where the largest number of the faience beads was recovered, had only a small number of diagnostic sherds. Most of these are from Groups 2d and 13, considered to date to the early Christian and Post-Meroitic to Christian Periods respectively. The most probable Groups to which the majority of the non-featured sherds could be assigned were 1a, 1h, 2b, 2d and 2f. The pottery thus indicates that a date in the late Christian Period is probable for the burials, a period within which the beads could reasonably have been in use, given the similar forms present at Soba.

Bead 118FS2/S1/14S is subspherical and rather irregular, coming closest to Beck Class I.C.1.a in overall form. Beads of this general shape in glass have sufficiently close parallels from several periods. There are two bead forms of similar shape from the Meroe West Cemetery, forming Types Im and In of Dunham (1963, fig. R). It is not certain from the illustration whether the longitudinal profile of the former Type meets the perforation, but it does appear to do so in the latter. Although resembling 118FS2/S1/14S in shape, both the Type illustrations of these Meroe beads are smaller in size. Other beads of spherical shape, including examples in glass, were noted as being numerous in the excavations of Meroe City site (Shinnie and Bradley 1980, 172, fig. 66). However, the form I.C.1a was quite rare amongst the sites excavated by the Scandinavian Joint Expedition, only a single example of similar size and colour to 118FS2/14S being noted in the corpus of beads from the Late Nubian cemeteries. This example was from Site 434 at Gamai, being associated with pottery of Ware Group AI and of Wares R25, R32, R35, W25, W26, U1 and H9 and in Adams' classification (Säve-Söderbergh *et al.* 1981, 40-41, 178, Table 12). These would give a date within the Meroitic or earlier Post-Meroitic ('X-Group') periods in the Lower Nubian chronology (cf. W. Adams 1986, 418-419, 436, 454, 468, 514-515, 526).

The form is present at Gamai in the tombs excavated by Bates and Dunham, although the Type as illustrated appears more truly spherical than the Test Excavation specimen (Bates and Dunham 1927, pl. LI, Form 1). Several subspherical forms were distinguished in the classification of the beads from Ballana and Qustul, including Types 3, 5 and 12, with Types 3 being closest to 118/FS2/S1/14S. The parallels are not exact, since the illustrated Types 5 and 12 from Ballana and Qustul are approximately half the size. Type 5 occurs in blue glass, otherwise specimens of this Type and of Types 3 and 12 are made from a various sorts of stone (Emery and Kirwan 1938, pl. 43). Several examples of form I.C.1.a in glass were present in a Post-Meroitic context in Tumulus 5 at Gabati, although these beads were dark brown in colour and slightly smaller than 118/FS2/S1/14S (Smith 1998b, table 7).

Beads of comparable shape to the Test Excavation specimen continued in use into the Christian period, since it is similar to those classed as 'globular' beads and to some 'drum' beads from Soba. Seven examples of such beads in green glass were found at the western end of Mound B and a further, unfinished, specimen was present in Area MN3, dated by ceramics to 6$^{th}$ to 9$^{th}$ centuries AD (Allason-Jones 1991, 145; 1998, 66, fig. 28, no. 117; Welsby 1998d, 21). On the basis of the sites examined for comparative material, it appears that beads similar to 118/FS2/S1/14S were present at least from the Meroitic to the Christian periods, but that the closest parallels for the combination of shape, material and colour occur in the Meroitic or Christian periods. This would be consistent with the dating of the pottery Groups from the site, these being Groups 3 and 13, considered to be most probably of the Post-Meroitic or Christian periods.

The ostrich eggshell bead comes from one of the longest lasting artefactual traditions in Nubia. Such beads occur on sites ranging in time from the Khartoum Mesolithic and the Neolithic, through the C-Group, Kerma, New Kingdom and the Meroitic periods and continue into the Post-Meroitic and Christian periods (Arkell 1949, pl. 9, 3-5; Caneva 1998c, 167, fig. 19, 6-8; Säve-Söderbergh 1989, 76-78; Troy 1991a, 81, fig. 18; Shinnie and Bradley 1980, 174; Williams 1991, 132-136, 138-150, fig. 64, l; Smith 1998b, 230, fig. 9, no. 17; Shinnie 1955, 53-54, fig. 28, 1). It can only be said that the presence of the bead in the burial at 154.5 identified as Christian on other grounds would be quite reasonable, given that these beads were clearly being manufactured during this period (cf. Allason-Jones 1991, 143).

## Ceramic (Re-used sherd)

The only ceramic artefact found not classifiable as a vessel sherd was a ceramic disk (156.1/1S), from amongst the 'nonfeatured' pottery from Site 156.1 (Figure 3.4). Just under half the disk is extant, being close to a semi-circle in shape, of maximum diameter 62mm. The concave-convex cross-section and the dark brown core of the ceramic showing at the edges, indicate that the disk was cut from a vessel body sherd, in a vegetable tempered fabric in the range exhibited by Fabrics 1a, 1b or 2a. The concave surface is quite rough, with vegetable temper voids evident, the convex face is smoothed and exhibits a number of parallel and sub-parallel striae, some extending most of the way across the surface. There is no evidence for a central hole in the surviving part of the disk, so it is not clear as to whether it could have been a spindle whorl similar to that from Post-Meroitic burial T27 at Gabati, also cut from a sherd but in this case one with an incised design (Edwards 1998c 76-79, 1998d 124, fig. 5.10).

It is apparent that such pottery disks were in use over a considerable timespan, since numerous examples were found during the excavations at Meroe City site and at Soba, including examples with no perforations, or with one, two, or three perforations. Those noted at Soba range from about 20mm to about 80mm in diameter. Despite the relatively large number both at Meroe and at Soba, it has not been possible to determine the uses to which such disks were put. Depending on their size and the presence and number of perforations, and of decoration, they could be gaming counters, buttons, weights, loom weights, pot lids or jar stoppers (cf. Shinnie and Bradley 1980, 189; Allason-Jones 1991, 149, 151, fig. 73, nos 366-381; 1998, 74, fig. 30, no. 197). It may be suggested that since the specimen from 156.1 is equivalent in size to the larger disks from Soba, it could have been a lid or stopper (cf. Allason-Jones 1998, 74), although this function would not explain the striae. However, one striation appears to be truncated at the present edge of the disk, and so may possibly relate to scraping of the surface carried out in the manufacture of the original vessel or to a previous period of use, perhaps as a base for cutting or scraping, before the sherd was finally cut down to make the disk.

## Textile: fragment

The generally poor preservation of organic material at the sites apart from Gabati resulted in only one example of textile (101.4/T3/31S) being recovered from the other sites investigated during the Test Excavations. A fragment, c. 15 x 5mm in size, of fine woollen cloth of plain tabby weave was found in Tumulus 3 at Site 101.4. It is pale brown in colour, this probably representing the natural colour of the material.[5] The piece is undecorated (Figure 3.4).

The small size of the fragment means that it is not possible to tell the type of garment or other object from which it comes. It may be noted that woollen shrouds were used in burials during the Meroitic period, but were not common (Caneva 1988c, 212-214). Within the region of Lower Nubia, woollen textiles are particularly characteristic of the Early Christian period at Qasr Ibrim and are known from Christian period burials in cemeteries such as those at Ashkeit, Debeira and Abka (N. Adams 1986, 24-26; Bergman 1975, 8, 77, 78). In Upper Nubia, also, shrouds appear to have been woven from wool during the Christian period. Shrouds, or items of clothing together with shrouds, made from vegetable fibre, probably flax, and from animal fibre, probably wool, were found in Christian period burials at Soba (cf. Vogelsang-Eastwood 1998, 178-179).

However, textiles of natural brown colour, undecorated, in tabby weave were present in a number of the Post-Meroitic graves at Gabati, as well as in a few of the Christian or 'medieval' period burials. These textiles indicate the range of items from which the fragment 101.4/T3/31S could have come. Most prevalent at Gabati are either shrouds or blankets, the latter often used as shrouds for wrapping the body. In one grave textile of similar type to the fragment from 101.4 has been identified as forming a tape or band of material (Taylor 1998, tables 8-9; Edwards 1998c, 80-94). On the basis of the Gabati textile finds, it appears that 101.4/T3/31S is most likely to be from a shroud, although the possibility of its having come from a smaller piece, such as a type of sash, cannot be excluded.

## Conclusions

In general, the small finds from the sites other than Gabati are restricted in number and type. The majority are not closely datable stylistically, with some representing traditions that lasted over several millennia in Nubia. Consequently, most do not allow a refining of the chronology of the sites from which they come over that derived from pottery dating. In some cases, such as the bracelet fragment from Site 211.2, the small finds can indicate that a site could have occupation or use extending over a longer period than that apparent from the pottery alone, which should be taken into account in future investigations.

Although the small finds are not of great significance in terms of dating, they do demonstrate the various levels of technology employed in the production of the different artefact classes, such as the prepared platform flakes and the bladelets amongst the lithic material, or the use of glass and faience amongst the beads. The final aspect on which the small finds can give direct information, related to the previous aspect, is that relating to the use of raw materials. This can be seen, for example, in the stone types present within the lithic collection and the indications that most probably pebbles were utilised as the main source of the raw material. It can be seen in the collections of artefacts from later periods in the use of man-made materials, especially the glass and faience amongst the beads, and in the small number of metal items. Further investigations at those sites examined in the Surface Survey and Test Excavations considered in the present report may provide a similar range of material to that found at the one site, Gabati, which will further extend the sample of the material culture of all periods from the Begrawiya-Atbara area.

---

[5] This information was kindly given by Ms S. Taylor.

# 4. Report on the pottery from the Test Excavations: Sites 101.4, 112.3, 112.4, 118FS2, 153.8, 155.4, 165BM, 166.2 and 170.1

*Laurence Smith*

The ceramics from the Test Excavation season at the site of Gabati (159.2) have been discussed together with the material from the rescue season undertaken at that site (Smith 1998c; Rose 1998). The present study will be concerned with the ceramics from the further nine sites from which pottery was recovered, investigated during the Test Excavations. Study of this material followed largely the same scheme as that developed for the Surface Survey carried out in 1993. The material was divided into featured sherds, or diagnostics, and non-featured, or undiagnostic sherds. The former comprised rims, bases, other identifiable vessel parts such as handles, and decorated body sherds, whilst the latter comprised all undecorated body sherds. In the case of the diagnostics, sherds were classed into a number of groups according to similarities and differences in fabric, surface treatment, style and motifs of decoration and form, where this could be determined. The greater part of the present report presents descriptions of each of the groups, concentrating on those into which the featured sherds have been placed. The comparative evidence leading to the dating of each group through the identification of links with other bodies of material is put forward. The dates currently assigned to each group are summarised in Table 4.1.

## Fabric

Fabric of the sherds was examined at 10x magnification in a freshly-broken cross section of the sherd. The general scheme of fabric description is based on that presented by Peacock (1977, 26-33). General features noted in the fabric on which the fabric classification was based were as follows: consistent fired colour, the degree of hardness, the absence, presence or abundance of vegetable temper (usually as indicated by voids within the fabric), the type and abundance of mineral inclusions. Further features of the inclusions themselves were also used, namely the degree to which grains of each mineral type were similar in size, or of varying sizes (sorting) the degree to which grains appeared angular or smooth (roundness) and the extent to which grains approached a spherical shape (sphericity). All of the comparisons of sorting, roundness and sphericity were carried out by means of standard charts (Bullock *et al.* 1985, 30-32), whilst estimates of the size of grains and their frequency within the cross-section have been assigned to descriptive terms corresponding to approximate sizes in millimetres for the former and percentages for the latter.

Identifications of mineral type have been limited by firstly, being made at maximum magnification of 10x, and, secondly, by being made on the appearance of grains in 'hand specimen', including such features as colour, degree of translucency and reflectance, angularity and habit. Consequently, it has been possible often only to identify the mineral types to rather broad classes of minerals, since the range of features available in other methods of examination, particularly thin-section analysis, were not observable in the solid specimen.

The main difference between the classification of the material in the study of the Surface Survey pottery and that from the Test Excavations is that in the former case the fabric description was subsumed under the description of the group. It was found that this was restrictive, since it meant that, in practice, the classification of the fabric could not be done independently of the other aspects of the pottery (such as surface treatment and decoration) which also determined the group to which individual sherds were assigned. Furthermore, it meant that a fabric description always had to be given for each group, whereas in some cases, fabrics of groups were closely similar, the groups being distinguished on surface treatment, decorative style and vessel form, rather than fabric. In view of this, it was decided to assign material to fabric types designated independently of the groups, in the study of the Test Excavation material. The fabric types were assigned in the first instance to groups already designated during the study of the Surface Survey ceramics. In general, each fabric distinguished did, in practice, correspond to a separate group. However, in some cases, fabrics belonged to two or more groups, which were separated on the basis of other criteria. The procedure adopted was that each new fabric distinguished formed the basis of a new group, but that if examples of sherds with the same fabric were found that differed significantly from the known ones in aspects such as surface treatment, decoration, etc., these were separated into further new groups.

The system of classification adopted for the fabrics involved distinguishing five main classes of fabric viz:

1. 'siltwares', with coarse porous texture, usually having high vegetable temper,
2. 'siltwares' with fine, compact texture, usually very low in vegetable temper, or lacking such temper,
3. fabrics considered likely to be from 'residual' clays rather than alluvial clays, on the basis of their lacking the range of inclusions present in the siltwares and often having a finer texture, with a low-iron matrix, thought to be probably kaolinitic clays,
4. fabrics considered likely to be from 'residual' clays but with a high iron matrix,
5. 'marl' wares.

Each of these was given a number from 1 to 5, preceded

**TABLE 4.1.** GROUPS PRESENT IN CERAMICS FROM TEST EXCAVATIONS AND THE PERIOD
OR PERIODS TO WHICH THEY ARE DATED.

| Group | Assigned Period | Group | Assigned Period |
|---|---|---|---|
| 1a | Late to Terminal Christian/Islamic | 5c | Egyptian Late Period |
| 1b | Late to Terminal Christian/Islamic | 6a | Late Neolithic ('III$^{rd}$ Millennium') |
| 1c * | Late to Terminal Christian/Islamic | 6b | Late Neolithic ('III$^{rd}$ Millennium') |
| 1d | Late to Terminal Christian/Islamic | 7a | Neolithic |
| 1e | Late to Terminal Christian/Islamic | 7c | Neolithic |
| 1f | Late to Terminal Christian/Islamic | 8b * | Mesolithic |
| 1g | Late to Terminal Christian/Islamic | 8c | Mesolithic |
| 1h | Late to Terminal Christian/Islamic | 8d | Mesolithic |
| 1I | Late to Terminal Christian/Islamic | 9 * | Meroitic (?), Christian (?) |
| 1j | Late to Terminal Christian/Islamic | 11 | Egyptian Late Period |
| 1k * | Late to Terminal Christian/Islamic | 13 | Post-Meroitic to Christian (?) |
| 1l * | Late to Terminal Christian/Islamic | 15a | Late Neolithic (?) |
| 1m * | Late to Terminal Christian/Islamic | 15b | Late Neolithic (?) |
| 1n * | Late to Terminal Christian/Islamic | 16a | Late to Terminal Christian/Islamic |
| 1o * | Late to Terminal Christian/Islamic | 16b | Late to Terminal Christian/Islamic |
| 1p * | Late to Terminal Christian/Islamic | 16c | Late to Terminal Christian/Islamic |
| 2a | Christian | 17a | Meroitic to Post-Meroitic |
| 2b | Christian | 17b | Meroitic to Post-Meroitic |
| 2d | Christian | 18a | Meroitic to Post-Meroitic |
| 2e | Christian | 18b | Meroitic to Post-Meroitic |
| 2f | Christian | 19 | Christian (?) |
| 2g | Christian | 20 | Late Neolithic (?) |
| 2h * | Christian | 21 | Napatan to Meroitic (?) |
| 3 | Post-Meroitic to Christian | 22a | Late Neolithic (?) |
| 4d * | Meroitic | 22b | Late Neolithic (?) |
| 4e * | Meroitic | 22c | Late Neolithic, Napatan (?) |
| 4g * | Meroitic | 22d | Late Neolithic, Napatan (?) |
| 4h | Meroitic | 23 | Christian (?) |

\* indicates Groups to which only non-featured sherds were assigned.

by 'F', to differentiate clearly the designation from those for form and decoration. The individual fabric types distinguished within the main classes of fabric were then assigned a number from 1 onwards, depending mainly on the order in which they were recognised. This resulted in a two part designation for each fabric type, for example F1:1, F5:2. Such a procedure was adopted so that the classification would group together fabrics appearing basically similar to each other, rather than using a system of numbering every fabric sequentially as it was recognised, in which fabrics of similar basic appearance could be widely separated in numbering solely because they were not recognised sequentially during the course of the study.

Descriptions of the fabrics present amongst the material from the Surface Survey and in the Test Excavations are set out below. In each case where featured sherds occurred in a particular fabric, the group or groups to which the fabric belongs is given. The fabrics distinguished on the basis of the Test Excavation material alone are indicated by the letters 'TE'. Definitions of the terms for grain size and abundance are shown in Table 4.2. Descriptions of those fabrics encountered only in the Surface Survey pottery are omitted here, being given in the first volume (Smith 1996).

## *F1:1 (Groups 1a, 2b, 1h)*

F1:1 is an earthenware, usually of black fired colour, often with narrow orange-brown outer zones. It is soft, with a laminated fracture and a porous texture. Frequent impressions from vegetable temper, frequent very fine mica and frequent fine quartz grains are visible on the surface. In the break, voids representing burnt-out vegetable temper are evident, being frequent to abundant in occurrence and ranging up to $c$. 8-10mm in length. Other rare irregular voids ranging up to $c$. 1mm in size are also present.

Amongst the mineral inclusions quartz appears predominant, with sparse, sub-rounded to rounded fine grains of medium to high sphericity and sparse very fine grains of high sphericity being present, together with very rare sub-angular grains of medium sphericity and of medium size. Secondly, there are present frequent mica grains, mainly of very fine size. Thirdly, very rare, dark brown or grey, opaque, rounded grains, very fine to fine in size, and of high sphericity occur. These inclusions may be identified as ironstone fragments. Finally, the most characteristic inclusions comprise very rare to rare opaque, whitish, sub-rounded to rounded inclusions of medium to high sphericity and ranging from medium to coarse size. These may be identified as calcitic inclusions.

**TABLE 4.2.** TERMS USED IN FABRIC DESCRIPTIONS FOR ABUNDANCE AND SIZE OF INCLUSIONS.

**Grain frequency**

| Term | Approximate frequency |
|---|---|
| Very rare | 1-5 grains |
| Rare | c. 10% |
| Sparse | c. 10-25% |
| Frequent | c. 25-50% |
| Abundant | c. 50-75% |
| Highly abundant | > 75% |

**Grain size**

| Term | Approximate size range |
|---|---|
| Very fine | < 0.5mm |
| Fine | c. 0.5mm |
| Medium | c. 1-2mm |
| Coarse | c. 2-4mm |
| Very coarse | > 4mm |

### *F1:2 (Groups 1b, 1I, 1k)*

Fabric F1:2 is closely related to F1:1, being generally the same in fired colour, hardness and texture. It has a similar high proportion of vegetable temper and most of the mineral inclusions types are also the same. However, it is distinguished by the absence of the coarse-sized calcitic inclusions.

### *F1:3 (Group 1c)*

F1:3 is an earthenware having black fired colour, occasionally with brown outer zones. The fabric is soft, with a laminated fracture and a porous texture. Several features are present on the vessel surfaces including frequent impressions from vegetable temper, frequent greyish inclusions of medium to coarse size, sparse quartz and sparse to frequent very fine mica.

In section, the fabric exhibits frequent to abundant voids, representing burnt-out vegetable temper, up to *c.* 5mm in length. The visible mineral inclusions comprise frequent, very fine quartz, rare sub-rounded to rounded quartz of medium to high sphericity and of fine to medium size, and frequent very fine mica grains. The fabric is characterised by the occurrence of rare rounded cream-coloured inclusions, probably calcitic, and the presence of sparse, greyish rock fragments having a laminar structure. These inclusions are generally sub-angular to sub-rounded, of low to medium sphericity and range from fine to very coarse size.

### *F1:4 (Group 3)*

Fabric F1:4 is an earthenware, usually of black fired colour, often with orange-brown outer zones. It is soft, with a laminated fracture and a porous texture. Frequent impressions from vegetable temper and rare rounded quartz grains are visible on the surface.

In the break, voids representing burnt-out vegetable temper are evident, being frequent in occurrence, and ranging up to *c.* 3mm in length. Amongst the mineral inclusions quartz appears predominant, with sparse, sub-rounded to rounded fine grains, of low to high sphericity being present. Secondly, there are present frequent grains, mainly of very fine size, which are likely to be quartz or micas. Thirdly, very rare, dark red, opaque, sub-rounded grains, fine in size, and of medium sphericity occur: these are likely to be ironstone fragments. Fourthly, there are very rare opaque, light grey, sub-rounded inclusions, of medium sphericity and fine size. These may be identified as polymineralic rock fragments. Finally, the most characteristic inclusions are rare, sub-angular, whitish opaque grains of low sphericity, very fine size and platy habit, which are identified as shell fragments.

### *F1:5 (Groups 2a, 2h)*

This fabric is an earthenware having a black fired colour, often with narrow orange-brown outer zones. It is soft, with a laminated fracture and a porous texture. Frequent impressions from vegetable temper and rare coarse, probably calcitic, inclusions are visible on the surface. In the break, frequent to abundant voids from burnt-out vegetable temper, ranging up to *c.* 2-3mm in length, and sparse irregular voids, *c.* 0.5mm in size, are evident.

Amongst the visible mineral inclusions quartz and feldspars predominate, with sparse, very fine quartz grains of medium to high sphericity, and sparse very fine feldspars of medium to high sphericity, being present. The third main type of inclusion consists of white, opaque, inclusions occurring as very rare to rare, sub-rounded grains of low to medium sphericity ranging from fine to coarse size, which are probably calcitic in nature. Finally, there are present rare to sparse mica grains, mainly of very fine size.

### *F1:6 [TE] (Groups 16a, 16b)*

Fabric F1:6 is an earthenware having a black core, with buff and orange outer zones, which can be very thin. It is soft, with a laminated fracture and a porous texture. Impressions from vegetable temper and rare black inclusions, likely to be a form of iron oxide, are visible on the surface. In the break are apparent abundant voids from burnt-out vegetable temper, of which the majority are of medium size but which can range up to *c.* 3-4mm in length.

Amongst the visible mineral inclusions quartz predominates, with frequent fine quartz of high sphericity and sparse quartz or quartzite grains of low to medium sphericity and of medium size being present. Secondly, there are present sparse opaque grains of a brownish-grey colour, mainly sub-angular, and of low to medium sphericity, and of medium to coarse size. This type of inclusion is considered to be either a form of iron oxide or grog. Opaque grains having a reddish-brown colour, occurring rarely, mostly sub-angular and of very low to medium sphericity, and ranging from medium to coarse size, are the third type of inclusion present. These are more likely to be ironstone than grog. The fourth type of inclusion consists of white, opaque, inclusions occurring as sparse, sub-rounded to rounded grains of low to high sphericity ranging from fine to medium size, which are probably calcitic in nature. Finally, there are present sparse mica grains, mainly of very fine size and of low to medium sphericity.

### *F1:9 (Group 4d)*

Fabric 1:9 is hard, with a dark grey or black fired colour, often with orange-brown outer zones. There can also be a third, markedly purple-tinged, zone between the core and the outer zones. In the break, the fabric exhibits a laminated fracture with a dense texture, although rare, irregular voids and frequent voids resulting from burnt-out vegetable temper are present. The latter are generally of fine size, only rarely attaining 5mm in length.

Visible inclusion types include frequent fine, sub-rounded to rounded quartz of medium to high sphericity, together with rare, sub-angular quartz, of medium to coarse size and of high sphericity. Rare to sparse calcitic grains occur, sub-angular to sub-rounded, of medium size and of medium to high sphericity. Red-brown opaque inclusions, probably of ironstone, occur in rare, sub-rounded to rounded, medium-sized grains of high sphericity. Certain opaque grains, greyish in colour, sub-angular, of medium size and of low to medium sphericity, occur rarely. These are likely to be either calcitic inclusions, or polymineralic rock fragments. Finally, a scattering of frequent very fine mica can be observed.

### *F1:10 (Group 4e)*

F1:10 is hard, with a dark grey fired colour, often with orange-brown outer zones. Sub-rounded to rounded fine quartz and sub-rounded, fine to coarse rock fragments are visible on the surfaces.

In the break, the fabric exhibits a hackly fracture with a dense texture, although rare, irregular voids and rare to sparse voids resulting from burnt-out vegetable temper are present. The latter are generally of fine size, not greater than *c.* 2mm in length. Visible inclusion types include sparse fine to medium, sub-angular to sub-rounded quartz of medium to high sphericity, together with frequent grains of very fine size which are likely to be quartz or micas. Rare black opaque sub-angular grains occur, of fine size and of medium sphericity. Red opaque inclusions, probably of ironstone, occur in very rare, rounded, fine grains of low sphericity. Certain opaque grains, greyish to brownish grey in colour, sub-angular to sub-rounded, of fine to coarse size and of low to medium sphericity, occur rarely. These are likely to be either calcitic inclusions or polymineralic rock fragments.

### *F1:12 (Group 9)*

Fabric F1:12 is an earthenware, usually of mid-dark grey fired colour. It is soft, with a hackly fracture and a dense texture, although there are present rare to frequent voids, representing burnt-out vegetable temper, ranging up to *c.* 2.5mm in length. In places where the surface is clearly eroded, sparse to frequent quartz grains and rare black grains, possibly of ironstone, are visible.

In the break, the fabric exhibits a number of visible inclusions. Quartz appears predominant with rare, sub-angular to sub-rounded medium-sized grains of medium to high sphericity, and sparse sub-angular to sub-rounded fine grains of medium to high sphericity, being present. Frequent very fine quartz particles are scattered throughout the matrix. The next most apparent features are white or cream inclusions. These occur, firstly, as rare, sub-angular, grains of medium to high sphericity and fine to medium size and, secondly, as very rare, coarse, sub-angular grains of low sphericity. They are probably calcitic in nature. The third main type of inclusion consists of rare black, or very dark grey opaque, sub-rounded to rounded grains, fine to medium in size and of medium to high sphericity. These are likely to be a form of iron oxide, probably being obtained, or derived, from ironstone fragments often associated with the Nubian Sandstone. Finally, a relatively sparse occurrence of very fine mica can be observed.

### *F1:13 (Group 13)*

Fabric F1:13 is an earthenware, usually of black fired colour, with orange-brown outer zones. In some cases the fabric is more evenly-fired to a mid-brown colour. It is soft, with a laminated fracture and a porous texture. Frequent impressions from vegetable temper, frequent very fine mica and frequent fine quartz grains are often visible on the surface. In the break, voids representing burnt-out vegetable temper are evident, ranging from sparse to abundant in occurrence, and very fine to coarse in size. Other, rare, irregular voids ranging up to *c.* 1mm in size may also be present.

Amongst the mineral inclusions quartz is present in sparse, rounded fine grains of medium to high sphericity and sparse very fine grains of high sphericity. Frequent mica grains, mainly of very fine size are present, and very rare, dark red-brown, opaque, sub-rounded to rounded grains, very fine to fine in size and of medium to high sphericity, also occur. These inclusions may be identified as ironstone fragments. Finally, there are sparse opaque, greyish, sub-rounded to rounded inclusions of low to high sphericity and of fine to coarse size. These may be identified as polymineralic rock inclusions.

### *F1:15 [TE] (Group 4g)*

F1:15 is an earthenware, generally having a dark grey core with buff exterior zone, sometimes with an additional thin pinkish outermost zone. The fabric is soft, with a laminated fracture and a porous texture. On the surface, sparse voids from burnt-out vegetable temper, rare fine ironstone inclusions, rare fine quartz and very rare possible rock fragments of medium size are visible. In the break, abundant voids from burnt-out vegetable temper are evident, ranging up to *c.* 7mm in length.

Translucent reflective grains are the most prominent mineral inclusions, occurring sparsely, in very fine size. These are generally somewhat small for secure identification, but are likely, on the basis of their reflective appearance, to be micas, although quartz grains are likely to be present. Rare grains of very fine size, of reflective appearance but clearly of low sphericity may also be identified as micas. Thirdly are present rare to sparse opaque white grains, sub-angular to rounded of medium sphericity. The majority are of fine size, but they range up to medium size.

These are identified as calcitic inclusions. Rare opaque reddish-brown grains, angular to sub-rounded, of low to medium sphericity and of very fine to fine size are the fourth type present. Such grains are either iron oxide or ironstone fragments. The fifth type of mineral present is opaque and of a light grey colour, occurring rarely in sub-angular, grains of very low sphericity and fine size. The identification of these inclusions is not certain, but they have a platy habit, which could indicate that they are shell fragments. The fabric is further distinguished by the presence of very rare polymineralic opaque inclusions. These are generally white in colour, but have black and red opaque inclusions and very fine micas visible within them. The overall grains are rounded, of high sphericity and mainly occur in coarse size. Such grains are probably calcitic or fragments of siltstone.

### *F1:16 [TE] (Groups 1d, 1l, 1m)*

Fabric F1:16 is an earthenware with a black or dark-grey core and thin mid-brown exterior zones. It is hard, with a hackly fracture and a dense texture. Features visible on the surface include frequent fine to medium quartz and feldspars, sparse medium calcitic inclusions, rare coarse quartz or quartzite and frequent voids from burnt-out vegetable temper. In the break, frequent voids from vegetable tempering ranging up to *c.* 3mm in length are present.

The most prominent mineral inclusions are sparse, sub-rounded to rounded, grains of medium to high sphericity and of medium to coarse size, composed of quartz and quartzite. Rare probable feldspar grains, of similar size and shape are also present. The next most prominent inclusion type comprises sparse black sub-angular to sub-rounded opaque grains of medium sorting, of medium to high sphericity and ranging in size from fine to medium. These are identified as iron oxides. Very fine reflective golden mica grains of very fine size also occur sparsely. The fourth inclusion type consists of rare opaque grey-coloured grains, sub-rounded to rounded, of low to medium sphericity and generally of medium size, which are likely to be a form of siltstone. Probable calcitic inclusions occur as rare, whitish opaque grains, generally sub-rounded, of medium sphericity and of fine to medium size. A small proportion ranges up to coarse size. Finally, there are present very rare reddish-brown opaque grains, generally sub-rounded, of low sphericity and of medium to coarse size, identified as ironstone fragments.

### *F1:17 [TE] (Groups 1e, 1n)*

F1:17 is an earthenware having a black core, with very thin grey exterior zones. It is soft, with a hackly fracture, and a porous texture. Sparse very fine micas, rare medium to coarse rock fragments and frequent voids from burnt-out vegetable temper are visible on the surfaces. In the break, frequent voids from burnt-out vegetable temper, ranging in size up to *c.* 4mm in length, are apparent.

Relatively few mineral types are evident in the break. The most prominent are transparent, whitish reflective grains, occurring frequently, being well-sorted and of very fine size. Grains of this type are generally too small for estimates of roundness and sphericity to be made. These grains are identified as mainly micas, but may include small quartz grains. The second type of inclusion comprises sparse opaque reddish-brown ironstone grains, moderately well sorted, sub-rounded to rounded, of low to medium sphericity and mostly of fine to medium size. A small proportion range up to coarse size. Thirdly, there are present opaque polymineralic inclusions, generally greyish in colour, but themselves containing reflective inclusions. They occur rarely, are moderately well sorted, are sub-angular to sub-rounded, of low to medium sphericity and range from medium to coarse size. Inclusions of this type appear to be rock fragments composed of micas and/or quartz, together with opaque minerals.

### *F1:18 [TE] (Groups 1f, 1o, 1p)*

Fabric F1:18 has features similar to both F1:17 and F1:3, but has been distinguished because the abundance of ironstone inclusions and rock fragments is lower than in F1:17 and much lower than in F1:3. The fabric is an earthenware having a black core and brown exterior zones. It is soft, with a hackly fracture and a dense texture. On the surface, frequent very fine micas or quartz, sparse fine quartz and rare red or black fine iron oxide grains are visible. In the break, sparse voids from burnt-out vegetable temper, ranging up to *c.* 3mm in length are apparent.

The most prominent mineral types are quartz and quartzite, together with micas and probable quartz, these types occurring frequently. The former occur in grains that are moderately well sorted, are sub-angular to rounded, of medium to high sphericity and mostly range from fine to medium size. The latter are well sorted, and are all very fine in size, being too small for degree of roundness and sphericity to be determined. Probable feldspar grains, whitish translucent in colour, occur rarely. They are generally well sorted, sub-rounded to rounded, of high sphericity and of fine to medium size. Opaque, whitish inclusions occur rarely, being moderately well sorted, sub-rounded to rounded, of medium sphericity and mostly of medium size. A proportion of these grains occurs in coarse size. They are identified as calcitic nodules. Reddish-brown opaque inclusions, identified as ironstone nodules, occur very rarely, in sub-rounded grains of medium sphericity and of medium size. The final type of inclusion visible in the break comprises polymineralic grains composed of translucent, reflective and opaque minerals. Such inclusions occur very rarely, are moderately well sorted, sub-angular to sub-rounded, of low to medium sphericity and of medium size. These appear to be rock fragments similar to those evident in Fabric F1:17.

### *F1:19 [TE] (Group 1g)*

Fabric F1:19 is an earthenware. In fired colour it has a dark grey core, with very thin brown and reddish-brown exterior zones. It is hard, with a hackly fracture and a dense texture. Very rare, very fine, red opaque and sparse very fine black opaque inclusions, both probably forms of iron oxide, and sparse voids from burnt-out vegetable temper are visible at the surface. In the break are present sparse voids from burnt-out vegetable temper, most ranging up to *c.* 2mm in length, and very rarely up to 6mm in length.

The fabric is characterised by the presence of frequent polymineralic, opaque, inclusions considered to be rock fragments. One type is reddish-brown in colour, occurring frequently, moderately-well sorted, sub-rounded, of medium to high sphericity and mostly of very fine to medium size. A few of the fragments range up to medium size. The second type is greyish-brown in colour, occurring frequently, moderately-well sorted, angular to sub-rounded, of low to high sphericity, and ranging from fine to medium size. The third component comprises translucent reflective grains which are sparse, generally well-sorted, and range from very fine to fine size. These are considered to be micas and fine quartz. Calcitic inclusions occur rarely, and are well-sorted, being present in sub-angular to rounded grains of low to medium sphericity and of medium size. Rare opaque black fragments, probably ironstone, are well-sorted, occurring in sub-rounded to rounded grains of high sphericity and very fine size. Quartz and/or quartzite inclusions are rare, well-sorted, sub-angular to rounded, of medium to high sphericity, and mostly of fine size, ranging up to a maximum of medium size. Finally, there are present very rare reddish-brown opaque grains, probably of ironstone. These are very rare, are well-sorted, sub-rounded of medium sphericity and are very fine in size.

### *F1:20 [TE] (Group 1j)*

Fabric F1:20 is similar to F1:1, being an earthenware, usually of black fired colour, with narrow buff outer zones. It is soft, with a laminated fracture and a porous texture. Frequent impressions from vegetable temper, frequent very fine mica and frequent fine quartz grains are visible on the surface. In the break, voids representing burnt-out vegetable temper are evident, being frequent to abundant in occurrence and ranging up to *c.* 8-10mm in length. Other rare irregular voids ranging up to *c.* 1mm in size are also present.

Amongst the mineral inclusions quartz appears predominant, with sparse, sub-rounded to rounded fine grains of medium to high sphericity and sparse very fine grains of high sphericity being present, together with very rare sub-angular grains of medium sphericity and of medium size. Secondly, there are present frequent mica grains, mainly of very fine size. Thirdly, very rare, reddish-brown, opaque, rounded grains, very fine to fine in size, and of high sphericity occur. These inclusions may be identified as ironstone fragments. There are present very rare opaque, whitish, sub-angular to sub-rounded inclusions of medium to high sphericity and ranging from fine to medium size. These may be identified as calcitic inclusions. The fabric is characterised by the presence of very rare, sub-angular inclusions of very low sphericity of coarse size. These are white, opaque, and have a platy habit. They are considered to be shell fragments.

### *F1:21 [TE] (included in Group 1b)*

Fabric F1:21 is an earthenware. In fired colour, it has a light brown core with broad reddish-brown zone on the exterior, together with a broad dark grey interior zone. It is soft, with a hackly fracture and a porous texture. Sparse impressions from vegetable temper, rare opaque black inclusions, rare very fine micas(?) and rare fine quartz grains are visible on the surface. In the break, sparse voids representing burnt-out vegetable temper, ranging up to *c.* 6mm in length, are evident.

Amongst the mineral inclusions, quartz and quartzite grains are rare, occurring mainly in sub-angular to sub-rounded very fine to fine grains of low to medium sphericity Secondly, there are present sparse reflective grains, of very fine size and low to medium sphericity, which are likely to be mainly micas and very fine quartzes. Thirdly, very rare, black opaque, sub-rounded to rounded grains, fine in size, and of medium to high sphericity occur. These inclusions may be identified as ironstone fragments. There are present very rare opaque, whitish, sub-rounded inclusions of medium to high sphericity and ranging from very fine to coarse size. These may be identified as calcitic inclusions. Very rare opaque inclusions, sub-angular of low to medium sphericity and medium size, composed of a greyish matrix and reflective grains are present. These are rock fragments, possibly of a mudstone. The fabric is characterised by the presence of very rare, sub-angular rock fragments of medium sphericity and of medium size. These are opaque, comprising a reddish-brown matrix with bright reflective grains within it.

This fabric has similarities with Fabric F1:2, and is considered to be a rare variant of this latter fabric. Consequently, it has not been made the basis of a new separate group.

### *F1:22 [TE] (included in Group 1b)*

Fabric F1:22 is an earthenware, having a dark grey core with a thin reddish-brown zone on the exterior, together with a very thin brown interior zone. It is soft, with a laminated fracture and a porous texture. Frequent impressions from vegetable temper, rare very fine to fine opaque black inclusions, rare very fine to medium quartz grains and very rare coarse greyish opaque rock fragments are visible on the surface. In the break, frequent voids representing burnt-out vegetable temper, ranging up to *c.* 7mm in length, are evident.

Amongst the mineral inclusions, the most predominant are sparse, reflective, grains of medium to high sphericity and of very fine size, which are likely to comprise both micas and fine quartz grains. Quartz of very fine to fine size occurs rarely, in sub-rounded to rounded grains of medium to high sphericity. The third inclusion type comprises very rare grey opaque grains, sub-angular, of medium to high sphericity. These are mostly of fine size, but range up to coarse size; they are identified as calcitic inclusions. The fabric is characterised by the presence of very rare reddish-brown opaque ironstone inclusions, sub-rounded to rounded, of medium to high sphericity. Most are of medium to coarse size, but grains of very coarse size occur.

In general, this fabric is also quite similar to Fabric F1:2, differing mainly in the presence of very rare calcitic and ironstone inclusions of coarse and very coarse size. Since only a very few examples were noted of this fabric with such coarse inclusions being visible, it has been considered to be a rare extreme of the fabric more usually having the appearance of F1:2, and has not been designated as a separate group.

### *F1:23 [TE] (Group 21)*

Fabric F1:23 is an earthenware, having a dark grey to black core with thin or very thin brown outer zones. It is hard, with a hackly fracture and a moderately dense texture. Sparse very fine micas, rare fine quartz, very rare reddish-brown opaque ironstone inclusions and rare medium impressions of vegetable temper are visible on the surface. In the break, frequent voids representing burnt-out vegetable temper, ranging up to *c.* 3mm in length, are evident.

Regarding the mineral inclusions, quartz is present in rare sub-rounded to rounded grains of medium to high sphericity and largely of medium size. Quartz is probably present also amongst the frequent reflective grains of very fine size, and of medium to high sphericity, which are likely to comprise micas in addition to the quartzes. Thirdly, there are present very rare reddish-brown opaque inclusions, mainly sub-angular, of high sphericity and of very fine size, identified as ironstone fragments. Rare, yellowish-brown inclusions, sub-angular to sub-rounded, of medium to high sphericity and of very fine size are likely to be another form of ironstone. The fifth inclusion type noted comprises very rare opaque whitish grains, sub-angular to angular, of medium to high sphericity and of very fine to fine size, which are probably calcitic. This fabric is characterised by the frequent very fine micas and/or quartz grains, together with very rare, sub-angular grains of low to medium sphericity which are likely to be quartz and/or quartzite. These inclusions fall mainly in the medium to coarse size range, but can occur in very coarse size.

### *F2:2 [TE] (Groups 15a, 15b)*

F2:2 is an earthenware having a black core and reddish-brown exterior zones in fired colour. It is hard, with a laminated fracture and a porous texture. No features were visible on the surface of the sherds examined. In the break, voids resulting from burnt-out vegetable temper ranging up to *c.* 6mm in length are present.

This fabric has a relatively restricted range of mineral inclusion types visible in the break. There is abundant quartz and/or quartzite present in well-sorted rounded grains and medium to high sphericity, mostly fine in size, with a small proportion ranging up to medium size. Sparse opaque white inclusions, considered to be calcitic, occur sparsely in moderately well-sorted grains, sub-rounded to rounded, of low to high sphericity and mainly of fine to medium size. Finally, there are present sparse black opaque grains, probably of ironstone, moderately well-sorted, sub-rounded to sub-angular, of medium sphericity and of medium to coarse size.

### *F2:3 [TE] (Groups 17a, 17b)*

Fabric F2:3 is an earthenware. In fired colour, it has a core ranging from grey to brown, with reddish-brown exterior zones. It is soft, with a hackly fracture and a dense texture. No features were visible on the surface in the sherds examined. In the break sparse irregular voids were present. These appear from their shape to be more likely to be pores than evidence for the presence of vegetable temper.

The main mineral types visible comprise, firstly, frequent rounded quartz/quartzite, well-sorted and of fine size. Secondly there are present black opaque grains, occurring rarely. These are well-sorted, rounded and of high sphericity and generally fine size. They are considered to be either iron oxide or the remains of an organic component. Thirdly, there are yellowish-brown opaque inclusions, occurring very rarely, sub-angular, of medium sphericity and of medium size, which are probably a form of iron oxide. Fourthly, there are rare opaque whitish inclusions, probably calcitic, which are rounded, of medium sphericity and mostly of medium size.

### *F2:4 [TE] (Groups 18a, 18b)*

F2:4 is an earthenware, having a fired colour ranging from grey to buff at the core, with thin reddish-brown exterior zones. It is hard, with a hackly fracture and a dense texture. No features were visible on the surface of the sherds examined. In the break, sparse irregular voids of fine size are present. These appear from their shape to be more likely to be pores than evidence for the presence of vegetable temper.

Mineral inclusions comprise, firstly, fine well-sorted quartz and/or quartzite, occurring frequently. Very fine well-sorted micas also occur frequently. Thirdly, there are present sparse black opaque inclusions, moderately-well sorted, mainly sub-angular, of low to high sphericity and of fine to medium size. These are similar to the black inclusions present in F2:3 and appear to be either a form of iron oxide or the remains of an organic temper. Finally, there are very rare opaque grains of a yellowish colour, which are rounded of high sphericity and mostly of fine size, together with very rare calcitic inclusions.

### *F2:6 (Group 6)*

F2:6 is an earthenware, usually of mid- to light-grey fired colour, often with orange-brown outer zones. It is hard, with a hackly fracture and a dense texture. In places where the surface is eroded frequent sub-angular quartz grains and sub-rounded ironstone grains are visible. In the break, rare, irregular voids of fine size (*c.* 0.5-1mm) are apparent, which are likely to represent burnt-out vegetable temper.

Amongst the mineral inclusions quartz appears predominant with frequent, sub-rounded to rounded very fine to fine grains of medium to high sphericity, and rare to sparse sub-rounded fine grains of low sphericity, being present. Secondly, there are present sparse semi-translucent, whitish, sub-angular inclusions, of medium to high sphericity and very fine size. These are likely to be feldspars. Thirdly, there are white inclusions which occur as rare, sub-rounded, grains of low to medium sphericity and fine size. They are probably calcitic in nature. The fourth main type of inclusion consists of rare dark red, opaque, sub-angular to rounded grains, very fine to fine in size, and of low to high sphericity. The final type of inclusion noted comprises rare black, opaque, sub-angular to rounded grains, very fine in size, and of medium to high sphericity. The latter two types of inclusion are likely to be forms of iron oxide.

### *F2:7 [TE] (Group 20)*

Fabric F2:7 is an earthenware having a mid brown fired colour, sometimes with a dark grey zone on the exterior. It is soft, with a hackly fracture and a dense texture. Quartz, opaque white and black inclusions, together with frequent mica are visible on the surface.

In the break, there are abundant quartz and/or quartzite grains occurring in moderately well-sorted grains, sub-angular to rounded, of low to high sphericity and of very fine to medium size. Abundant translucent whitish grains are present, moderately well-sorted, sub-rounded to rounded, of medium to high sphericity and of fine to medium size. These are likely to be feldspars. Frequent golden micas occur, well-sorted, and of very fine size. White micas and quartz occur rarely, in sub-angular grains, of medium sphericity and of fine size. Sparse opaque yellowish inclusions, probably calcitic, are present in well-sorted grains, sub-rounded of low to medium sphericity and of fine to medium size. There are present sparse black opaque grains, generally well-sorted, of very fine size, which are considered to be forms of iron oxide. Finally, there are rare opaque reddish inclusions, moderately well-sorted, sub-angular, of medium sphericity and of very fine to fine size, considered to be ironstone fragments or, possibly, grog.

### *F2:8 (Group 7a)*

Fabric F2:8 is an earthenware, usually grey to mid-brown in fired colour. It is hard, with a hackly fracture and a dense texture. No voids were evident in the break. Sparse sub-angular to sub-rounded quartz grains are visible on the surface. In the break, quartz appears predominant with frequent, sub-angular to sub-rounded very fine to fine grains of medium to high sphericity, and rare, sub-rounded to rounded medium-sized grains of medium to high sphericity, being present. Sparse semi-translucent, whitish, sub-angular inclusions, of medium to high sphericity and of very fine size are evident. These inclusions may be identified as feldspars and polycrystalline quartz. The third type of inclusion noted consists of very rare, red, opaque, sub-angular fine grains of low sphericity which are likely to be ironstone fragments. Finally, there are white, opaque, probably calcitic, inclusions which occur as very rare, sub-rounded grains of low sphericity and of medium size.

### *F2:11 (Groups 8b, 8c)*

This fabric is an earthenware, usually with a black or very dark brown fired colour. Amongst the visible inclusions are abundant medium-sized rounded or angular quartz grains, and frequent to abundant opaque whitish or cream, generally sub-rounded grains, of medium to coarse size. These grains comprise feldspars, polycrystalline quartz and sandstone fragments. They are noticeable at the surfaces of the sherds as well as in the break. The remaining main inclusions comprise sparse fine mica and rare medium-sized mica.

### *F2:12 (Group 8d)*

Fabric F2:12 is an earthenware, usually of mid- to dark-grey fired colour, often with narrow brown outer zones. It is hard, with a hackly fracture and a dense texture. No voids were evident in the break. Frequent to abundant very fine to coarse mica, frequent quartz grains and sparse sub-angular ironstone grains are visible on the surface.

In the break, quartz appears predominant with frequent sub-angular to sub-rounded fine to medium grains of low to medium sphericity, and sparse sub-rounded to rounded fine grains of high sphericity, being present. Secondly, there are present sparse sub-rounded mica grains, mainly of fine to medium size, together with rare, very fine grains of the same mineral. Thirdly, rare to sparse semi-translucent, whitish, sub-angular inclusions, of medium sphericity and of fine size are evident. These are likely to be feldspars. The fourth main type of inclusion consists of rare, red-brown, opaque, sub-rounded to rounded grains, very fine to fine in size and of high sphericity, which may be identified as ironstone fragments. Finally, there are white inclusions occurring as very rare, sub-rounded grains of low sphericity, usually of coarse size, which are probably calcitic in nature.

### *F2:13 [TE] (Group 2d)*

Fabric F2:13 is an earthenware, which can be of an orange-brown fired colour throughout, but which can have a buff core in addition. It is soft, with a hackly fracture and a dense texture. Frequent linear voids, probably representing vegetable temper, together with black opaque inclusions and quartz are visible on the surface. In the break, rare irregular voids of medium size and frequent voids from burnt-out vegetable temper, of fine size, are visible.

The inclusions comprise, firstly, frequent quartz, occurring in sub-rounded to rounded grains of medium to high sphericity and of fine to medium size. Secondly, there are frequent micas, in grains of medium sphericity and fine size. Thirdly, there are rare opaque whitish grains, probably calcitic, sub-rounded to rounded, of medium to high sphericity and largely of medium size. Rare opaque black inclusions occur as rounded grains of high sphericity and of fine size. These are likely to be either a form of iron oxide or, possibly, remains from incompletely burnt-out organic material. The final inclusion type comprises very rare rounded grains of very low to low sphericity and of medium size, which are calcitic, being present either as limestone or as shell fragments.

### *F2:15 [TE] (Group 2e)*

This fabric is generally of mid grey fired colour throughout. It is soft, with a hackly fracture and a dense texture. The main features visible on the surface are sparse linear voids of fine size, probably from burnt-out vegetable temper. Sparse voids representing vegetable temper, of fine size, are visible in the break.

Only three main types of inclusion are evident in this fabric. Firstly there are present frequent rounded grains of high sphericity and of very fine size, which are likely to comprise both quartz and micas. Secondly, there are sparse rounded quartz grains of high sphericity and fine size. Thirdly, the fabric contains frequent black opaque inclusions, sub-angular to rounded, of medium to high sphericity and of medium size, which are considered to be a form of iron oxide.

### *F2:16 [TE] (Group 7c)*

Fabric F2:16 has a mid-brown fired colour throughout. It is soft, with a hackly fracture and a dense texture. The main features visible on the surface are sparse very fine quartz and sparse very fine micas. Very rare irregular voids of fine size, which appear to be pores rather than to represent burnt-out vegetable temper, are visible in the break.

Regarding the inclusions within the fabric, quartz is predominant, occurring frequently in sub-angular to rounded grains of low to high sphericity. These are mostly of very fine to fine size, but rare grains of medium size also occur. Rare dark grey to black inclusions, probably of ironstone, occur. These are sub-angular to sub-rounded, of medium to high sphericity and mostly of very fine to fine size, with a maximum of medium size. There are rare micas, occurring in angular grains, of low to high sphericity and very fine size. Rare, white to cream inclusions occur, sub-angular to sub-rounded, of low to medium sphericity and of very fine to fine size. Their generally small size makes identification uncertain: they are likely to be feldspars, but some may be calcitic inclusions. The final inclusion type comprises rare orange-brown opaque grains, sub-angular to rounded, of medium to high sphericity. Such grains are mostly of very fine size but examples of medium size occur. These are likely to be a second form of ironstone.

### *F2:17 [TE] (Group 19)*

In fired colour Fabric F2:17 has light brown core, with an orange-brown exterior zone, and a grey interior zone. It is hard, with a hackly fracture and a dense texture. Features visible on the surface include rare very fine micas(?) and rare very fine opaque black inclusions. In the break, rare, irregular linear voids ranging up to *c.* 2mm in length are present. These may represent burnt-out vegetable temper, but do not possess clearly the features exhibited by such voids in other fabrics.

Regarding the inclusions, the most predominant type consists of sparse, dark grey or black opaque grains, angular to sub-rounded, of low to high sphericity and of fine to medium size, with a maximum of coarse size. These are identified as ironstone or another form of rock fragment. Secondly, there are present rare reddish to brown opaque inclusions, sub-angular to sub-rounded, of low to medium sphericity and of fine to medium size which are likely to be ironstone. Another form of ironstone may be present in very rare opaque yellowish-brown inclusions, angular to sub-rounded of medium to high sphericity and mostly of fine size, with a maximum of medium size. Fourthly, there are rare white opaque inclusions, probably calcitic, sub-angular to sub-rounded, of medium sphericity and mostly of fine size, occasionally ranging up to coarse size.

Fifthly, there are present very rare sub-rounded quartz grains, of medium to high sphericity, ranging from fine to medium size. The final inclusion type consists of rare reflective grains, mainly of medium to high sphericity and of very fine size, which are likely to comprise micas and some very fine quartz.

### *F4:2 (Group 11)*

This fabric is light orange-brown in fired colour. It is soft, with a smooth fracture and a dense texture. Sparse irregular voids of fine size are evident in the break.

Quartz is present as frequent, rounded, grains of low to high sphericity and very fine to fine size. There are rare black, opaque, rounded grains of high sphericity and very fine size which are likely to be a form of iron oxide. Iron oxide can also occur as reddish-brown, rounded, fine grains of high sphericity. Sparse, very fine mica is evident. Finally, there are present rare, rounded, grains of high sphericity mainly of fine to medium size which are likely to be calcitic in nature. The sherds are eroded so that the original surfaces are absent. The clear presence of ribbing on the interior may indicate that they are from closed forms.

### *F4:3 [TE] (Group 4h)*

Fabric F4:3 is an earthenware. In terms of fired colour it has a pinkish core with orange-brown exterior zones. It is hard, with a hackly fracture and a dense texture. Rare fine quartz and rare void resulting from burnt-out vegetable temper are the only features visible on the surface, which is red slipped. Rare voids resulting from burnt-out vegetable temper are evident in the break.

Regarding the inclusions within the fabric, quartz and/or quartzite grains are rare, occurring in sub-rounded to rounded grains of medium to high sphericity and generally of very fine size. A small proportion range up to fine size. Several appear to have a reddish-brown coloration on the surface, which may be a result of iron-staining. Reddish-brown opaque inclusions occur very rarely. They are generally sub-rounded, of medium sphericity and of fine size. They are identified as a form of ironstone. A second form of ironstone is likely to be present in rare opaque black grains, sub-rounded to rounded, of medium to high sphericity, and of very fine to fine size. Rare reflective grains of micas and/or quartz, generally of low to medium sphericity and of very fine size, are present. Very rare sub-angular opaque white inclusions are present, of very low sphericity and of fine size, which are considered to be shell fragments on account of their shape. The most prominent inclusion type consists of sparse whitish or cream opaque inclusions, generally rounded, of high sphericity and of very fine to fine size. A small number range up to medium size.

These are likely to be calcitic inclusions, however, there are also less discrete lenses of material of the same colour extending through the matrix in several areas. It is not clear from macroscopic examination whether two different types of feature are present, or if the white inclusions are discrete pellets of the same material as the lenses of white material. If the latter is the case, this may indicate that the fabric is composed of a mixture of a red-firing and a white-firing clay.

### F5:2 (TE) (Group 5c)

F5:2 is the second fabric considered to be of a marl clay recovered from the Surface Survey and Test Excavations. It is generally fine-grained and has a greenish-buff fired colour. In this case, it is possible the relatively strong greenish colour is due to the sherd being over-fired, more normal examples may have a colour more towards a yellowish tinge. It is hard, with a hackly fracture and a dense texture. In the break, rare irregular voids, ranging in size up to $c.$ 2mm are evident. In this fabric they have more the character of pores, rather than that of voids resulting from burnt-out vegetable temper. On the surface, sparse very fine to fine black opaque particles and rare fine opaque pinkish or cream particles are the main features visible.

Regarding the inclusions, the most prominent type comprises rare white opaque grains, generally rounded, of low to medium sphericity and of very fine size. These are likely to be calcitic inclusions. There are present rare pinkish opaque inclusions, sub-angular to rounded, of low to medium sphericity and of fine to a maximum of medium size, which are either a form of ironstone or grog. Rare black opaque inclusions are present, mostly sub-angular, of medium sphericity and of very fine to fine size. These are likely to be ironstone. Further inclusions which may be ironstone or grog are present as very rare opaque reddish-brown grains, angular, of low to medium sphericity and of fine size. Quartz and/or micas are present as sparse inclusions of low to high sphericity and of very fine size. The final inclusion type consists of very rare translucent white grains, sub-angular, of low sphericity and of fine size, which are considered to be feldspars.

## Form

A total of 11 main classes were distinguished amongst the material collected from the Surface Survey and from the Test Excavations. Brief descriptions are presented in the present report: fuller details of the form classes and the way in which sherds have been assigned to one or other of them are given in the first volume of this series (Smith 1996, 165-169). The various vessel classes are as follows:

### Class 1 (Jars)

Class 1 includes all enclosed and semi-enclosed forms that were not distinguished as other specific functional or formal categories, such as 'globular jars'. Hence, Class 1 comprises forms ranging from narrow-mouthed vessels, generally with everted rim, to broader-mouthed vessels having the uppermost part of the body and the rim inturned.

### Class 2 (Bowls)

Class 2 is a second broadly-defined category, including all open and some semi-enclosed forms not placed in Class 1 or assignable to another more specific category. Everted rim sherds of relatively broad diameter were clearly assignable to Class 2, whilst those with an approximately upright profile were also assigned to this class, together with a number of sherds exhibiting inturned upper body and rim.

### Class 3 (Drinking Vessels)

Class 3 comprises vessels, generally of open form, of relatively small diameter. In practice, vessels of 100mm diameter or less and approximately upright profile were placed in this Class.

### Class 4 ('Beer-Jars' or Globular Jars)

Sherds were placed in this class according to the criteria used at Soba, viz. those sherds belonging to handmade vessels characterised by narrow necks and wide, flaring rims (Welsby 1991, 165).

### Class 5 ('Braziers')

Class 5 comprises moderately open or broad open bowls of medium size, having a very thick solid foot, usually with a thick rounded and everted base.

### Class 6 ('Footed Dishes' or 'Stands')

Vessels in Class 6, of which only a few examples were noted, includes those of the type termed 'footed dish with handles' or those termed 'stands' by Dunham (1963, 118; 159). Such vessels consist of an open, usually fairly shallow, straight-sided dish with a high foot-ring, having one or more handles affixed between the bowl and the foot-ring.

### Class 7 (Potstands)

Class 7 comprises vessels of approximately cylindrical or bi-conical form, together with those of a simpler ring-like form which are considered to have been stands for other round-based vessels.

### Class 8 (Basins)

Basins of Class 8 were distinguished by a rim-diameter of 400mm or more, and a wall-thickness of 15mm or more near the rim.

### Class 9 (Doka)

A few sherds from apparently shallow vessels with relatively thick walls were identified as *doka* largely through their having a well-smoothed interior and a 'crazed' exterior. Such interior and exterior surfaces are similar to those of vessels identified as *doka* at Soba (cf. Welsby 1991, 179).

### Class 10 (Bases)

A relatively small number of base sherds have been recognised amongst the survey material. Sherds exhibiting shallow, solid foot-rings or deep, hollow foot-rings were clearly classifiable as bases. Certain body sherds were tentatively placed in Class 10 if they exhibited a distinct flattened area together with either an undecorated surface or a plain area in the midst of surrounding decoration.

## Class 11 (Handles)

Class 11 comprises handles that were no longer attached to a sherd classifiable into one of the above vessel classes. Rims, handles and bases from the Test Excavations were assigned to existing forms and types where possible, new types being added as necessary. Those types already recognised in the Surface Survey but also present amongst the Test Excavation material, together with the new types, are illustrated in Figures 4.1-4.2.

## Decoration

Decorations were classed according to the scheme developed for the Surface Survey, being divided into decorations executed in lines, whether incised or comb-impressed (D), painted decoration (P) and bosses (B). There were no decorations on sherds from the Test Excavations that were recognised as being classifiable into the class of decoration in relief as defined for the Surface Survey pottery. As with the Forms, the new decoration types were incorporated into the classification by being placed as close as possible to those motifs they most resembled, both in appearance and method of execution. Further decoration types were added to the end of the existing sequence for the new groups of uncertain date, or which could not be assigned to a single period. The decoration types present in the Test Excavation ceramics are shown in Figures 4.3 and 4.4.

## Groups designated in Test Excavations

### Groups 1a-1p

The majority of the ceramics from the Test Excavation sites, other than Gabati, was included within Groups 1a-1j. Much of the material was assignable to the Groups 1a and 1b, distinguished on the basis of Surface Survey material. The presence of a wider range of fabric types and surface treatments and different decorations on sherds with the same fabric as these existing groups led to the creation of several new ones, designated 1d-1j.

**Group 1a**

Sherds in Group 1a, Fabric F1:1, were present in the pottery from the Surface Survey in only a single existing form, 1:10.3, a jar form with an inward-sloping neck and concave exterior profile, having a relatively well-defined shoulder. The rim is thinned on the interior, to form a tapered top and is slightly everted. Four new types are distinguished in Group 1a material from the Test Excavations: three jars and one bowl. The jar forms comprise, firstly, a Type (1:4.1) with a moderately inward-sloping neck having a broad mouth. The rim is moderately rounded-off on the interior, forming a convex top to the rim. In cross-section the rim appears slightly thickened on the exterior, with an indentation just below the thickest part. This then merges into the exterior profile of the neck, which is slightly concave. The second type (1:8.2) is a large jar with distinct shoulder, and approximately upright neck. In cross-section, the rim is moderately rounded-off on the interior, having a slight facet at the junction of the neck and the rim top, which is convex. The rim is thickened on the exterior, the profile in consequence sloping inwards onto the exterior of the neck. The latter is a little concave around part of the extant vessel circumference but is virtually straight around the remainder. Only the uppermost part remains of the third jar, Type 1:9.6, so that the overall form cannot be determined. The extant portion comprises the rim of a jar apparently with sides sloping inwards. This is likely to have been of globular form, but the beginning of an outward turn in the exterior profile just above where the sherd breaks off, indicates that it may be a form with a short neck, though even in this case, a globular body is likely. Above the out-turn, the exterior profile is almost straight, with a sharp transition to the moderately convex rim top, which slopes down towards the interior. The rim is moderately rounded-off on the interior, which is thickened below the rim. It thins again, to the general width of the neck or body wall, at the point where the sherd is broken off.

The single bowl Type, 2:9.3, has a body wall sloping outwards at an angle of about 45°. A portion of the base is present, this appears to have been only moderately convex, or even flat. The body wall is thinned evenly from the base to the rim, which is tapered, with a plain rounded top. In the extant portion, a single hole through the body wall is present, approximately a quarter of the way down. This is likely to have been for suspending the vessel by a string, or for repairing the bowl. In addition to the rim sherds which could be designated as a specific type, there were two body sherds of moderate size, with a shallow curvature, indicating they could have been from relatively large vessels. They are smoothed and black-burnished on the interior, but have an untreated exterior surface. These could be from *doka*, although the presence of roughly-executed decoration on the exterior (D12.6, see below) may indicate they are from large bowls, since examples from a large collection of the former vessels found at Soba are not described or illustrated as being decorated with incised motifs (cf. Welsby 1991, 179, figs 95-96)

Decorations on Test Excavation Group 1a ceramics included variations on motifs known from the Surface Survey material, such as the zigzag (D1.8), the band of crosshatching between single horizontal lines (D3.4), and the single row of deep impressions unevenly-spaced vertically (D11.2). These three motifs all occur together on Type 1:8.2, a combination not previously encountered. Two decorations involving narrow elliptical impressions, or cuts, were noted. One (D12.3), has relatively close spacing of the impressions, whilst the other (D12.6) exhibits an irregular spacing, with the impressions more widely dispersed. A second version of a row of elliptical impressions was found, comprising impressions, often with one straight edge, beside which a ridge of paste is pushed up from the vessel surface (D13.5). Such impressions are likely to have been executed with a thumb or fingernail. An area scattered with irregularly-spaced circular to oval deep impressions, sometimes overlapping (D14.5), is present on a number of sherds in Group 1a, but it is particularly associated with Type 2:5.1 assigned to Group 1b. The final decorations on Group 1a material are the mat impressions. Two main types were distinguished, one consisting of thin horizontal impressions between

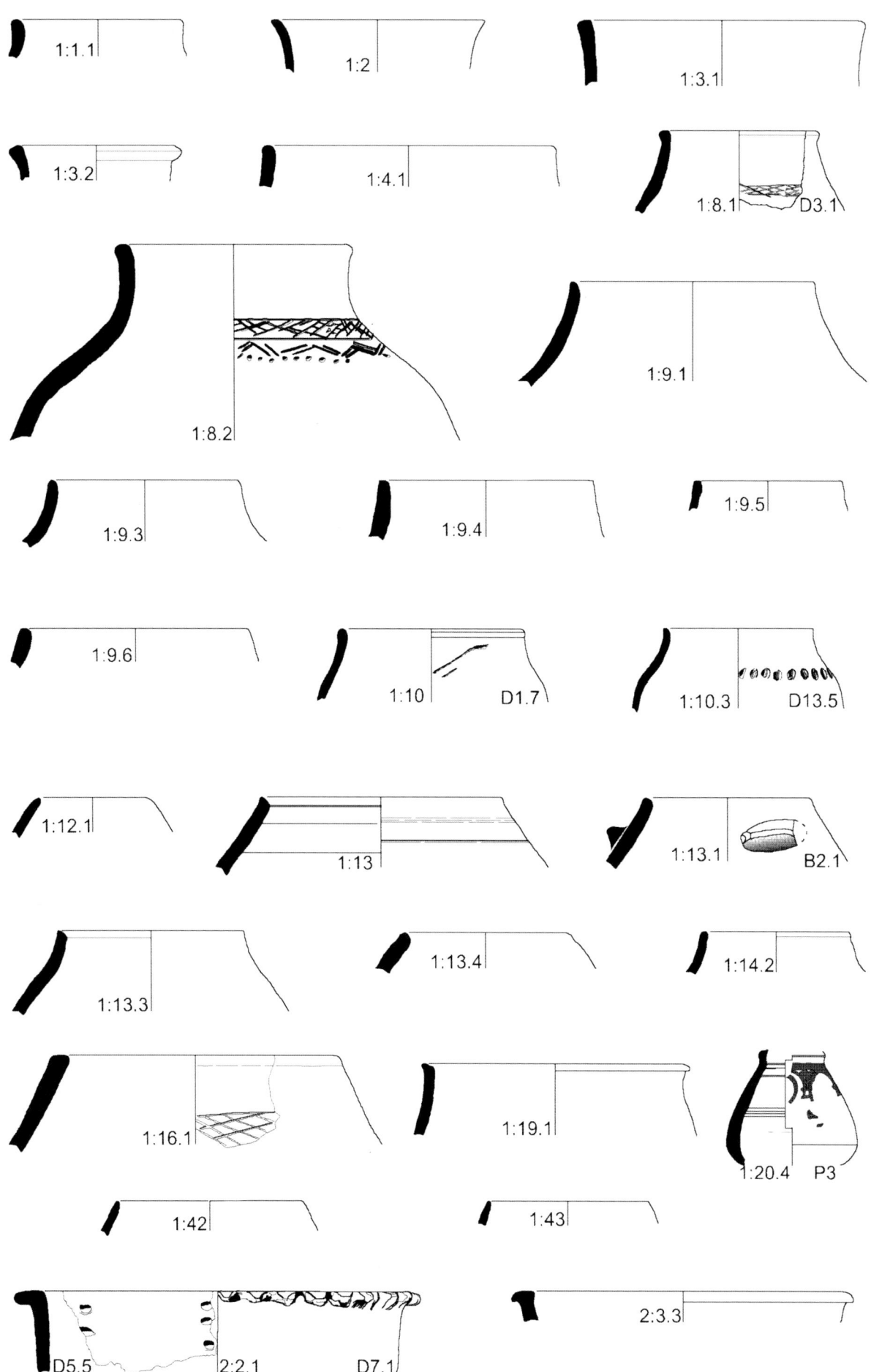

**Figure 4.1.** Forms 1:1.1 - 2:3.3 (scale 1:4).

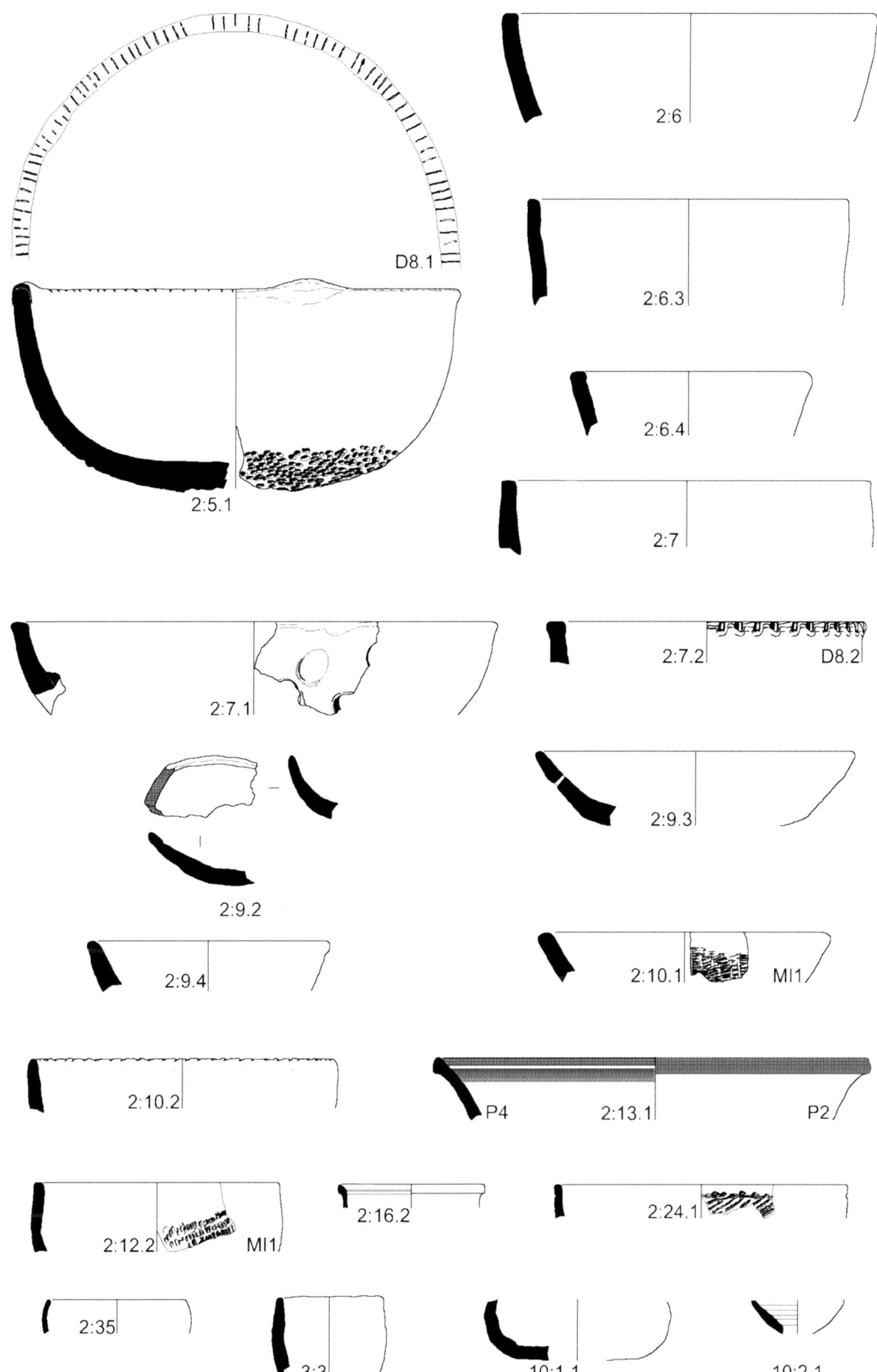

**Figure 4.2.** Forms 2:5.1 - 10:2.1 (scale 1:4).

vertical lines (MI1), the other consisting of often shallow approximately rectangular impressions in a 'herringbone' pattern (MI2). Sherds exhibiting both types are present in Group 1a.

**Group 1b**
In addition to a few sherds that could only be assigned to a general Class, viz: Jars or Bowls, and are possible *doka* sherds, 17 types were recognised within the diagnostic material placed in Group 1b. This material is in Fabric 1:2, and is characterised by smoothed but rarely burnished surfaces. Of these types, the following were already designated on the basis of the Surface Survey ceramics: 1:9.1, 1:12.1, 1:13, 2:6 and 2:10.1. Six further jar types were noted amongst the Test Excavation material. The first type (1:1.1) is similar to the existing Type 1:1, being the rim of a jar with a short, approximately upright neck, the rim thickened on the interior, but smoothly rounded-off to form a rounded top to the rim. Two additional types with a moderately out-flaring neck, similar to existing Type 1:3, were noted. The new type, 1:3.1, is wider in diameter than 1:3 but has a body-wall sloping outwards to a similar degree. In cross-section the rim of 1:3.1 is noticeably squared-off on interior lip. It has an approximately flat top, and is thickened and rounded-off on the exterior, thinning gradually to the general thickness of the body wall. New Type 1:3.2 has been placed with Types 1:3 and 1:3.1 because in cross-section the overall shape is outflared, although the exterior profile is much more strongly out-turned and modelled than in either of the former types. The rim has a flat top, sloping towards the exterior, where it is thickened and moderately squared-off, so that in profile the rim exterior is bevelled. The remaining three jar types all possess moderately inward-sloping upper parts; two, 1:8.1 and 1:9.3, having necks whilst the third, 1:13.4, appears to be of globular shape. Type 1:8.1 has a rim well rounded-off on the interior, having a flat top. It is thickened on the exterior, and slopes back into the body wall quite sharply to form a triangular cross-section. The rim of 1:9.3 is of a simple tapered form, similar in cross-section to that of 1:9.1, but less strongly rounded-off on the interior. In Type 1:13.4, the rim is slightly thickened on the interior. It is rounded-off so as to form a flat face on the interior of the rim. There is a broad shallow groove just below the top of the rim on the exterior, below which the remaining portion of the body wall is convex.

Six further bowl types were distinguished amongst the Test Excavation material in Group 1b. The first Type, 2:2.1, is a variation on the bowls considered most characteristic of the Group 1a-c ceramics from the Surface Survey. It has approximately upright sides, and is decorated with projecting bosses, in this case of roughly triangular shape, around the rim exterior. In terms of surface treatment, Type 2:2.1 is wiped on the exterior and smoothed on the interior. It is further shown to be similar to existing Types 2:1-2:3 by the presence of vertical rows of impressed dots on the interior, although in this case they are more elliptical than the dots on the former three types. The second Type, 2:5.1, is a large bowl about 330mm in diameter.

One example of Type 2:5.1 was found to be partially reconstructable (see Plates 4.3 and 4.4). This shows the type to have a virtually flat base with a smooth transition to the body wall, of which the uppermost part is approximately upright. In cross-section, the rim is plain for much of the circumference, being the same thickness as the main part of the body wall, with a moderately convex top. Only the exterior rim profile is not uniform around the vessel: a rounded slight projection being evident just below the top over part of the circumference. This type is characterised by bosses and incised decoration on the rim, together with irregular-shaped deep impressed dots on the lowest part of the body and covering most of the base.

The next two bowl types distinguished, Type 2:6.3 and 2:6.4, have a rather irregular body wall, more or less out-flared for most of the extant height. That of Type 2:6.3 is approximately vertical for the final third. There is a distinct groove on the interior at the lowest part of this upright section. The rim is plain, being the same thickness as the body wall, moderately rounded-off on both the interior and the exterior and having a moderately convex top, sloped slightly towards the interior. In Type 2:6.4 the body wall is more strongly out-flared, changing to a stance close to vertical only just below the rim. The latter is thickened and more strongly rounded-off on the interior and exterior, there being a groove below it on the interior and on the exterior around part of the circumference. The top of the rim is slightly convex. Both these types are smoothed on both surfaces but not decorated.

The next Type, 2:7.1, has been placed in Class 2, although it is unlikely to have served similar functions to most members of the Class. Rimsherds from the type give estimated rim diameters of 340-360mm. The body wall slopes out at about 30° from the vertical, and has a curving profile. The rim is thickened slightly and rounded-off on the interior, with a flat top. It is also thickened on the exterior to form a projecting lip. The unusual feature of the type is a number of large holes pierced through the body wall. The presence of a ridge of clay around the interior of these holes indicates they were made during the original manufacturing process, and are not due to the re-use of the bowl for some other function. Both surfaces have been left smooth, but are otherwise untreated.

The final Class 2 type distinguished in Group 1b is 2:12.2. The extant portion exhibits an outward-curving lower part and an upper half with relatively straight profile to the wall, which slopes inwards slightly. A distinct groove is present on the interior at the point where the direction of the wall changes. The rim is quite strongly rounded-off on the interior, and has an approximately flat top. It projects a little on the exterior, but is not otherwise modelled. Surfaces on the upper part of the vessel are left untreated but the lower part is mat-impressed (MI1).

Two further Classes of vessel are present in Group 1b. A rather roughly-made vessel, narrow in relation to its height, being only 80mm in diameter, has been placed in Class 3 ('Drinking vessels') as Type 3.3. In profile, this vessel has an outcurving lower portion, and an approximately upright upper body. The wall is thinned from about half height to the rim, which is plain and slightly flattened on the top. The surfaces are left untreated. A curved sherd in Group 1b has been tentatively identified as a base, (Type 10:1.1) since it has a flattened area, and

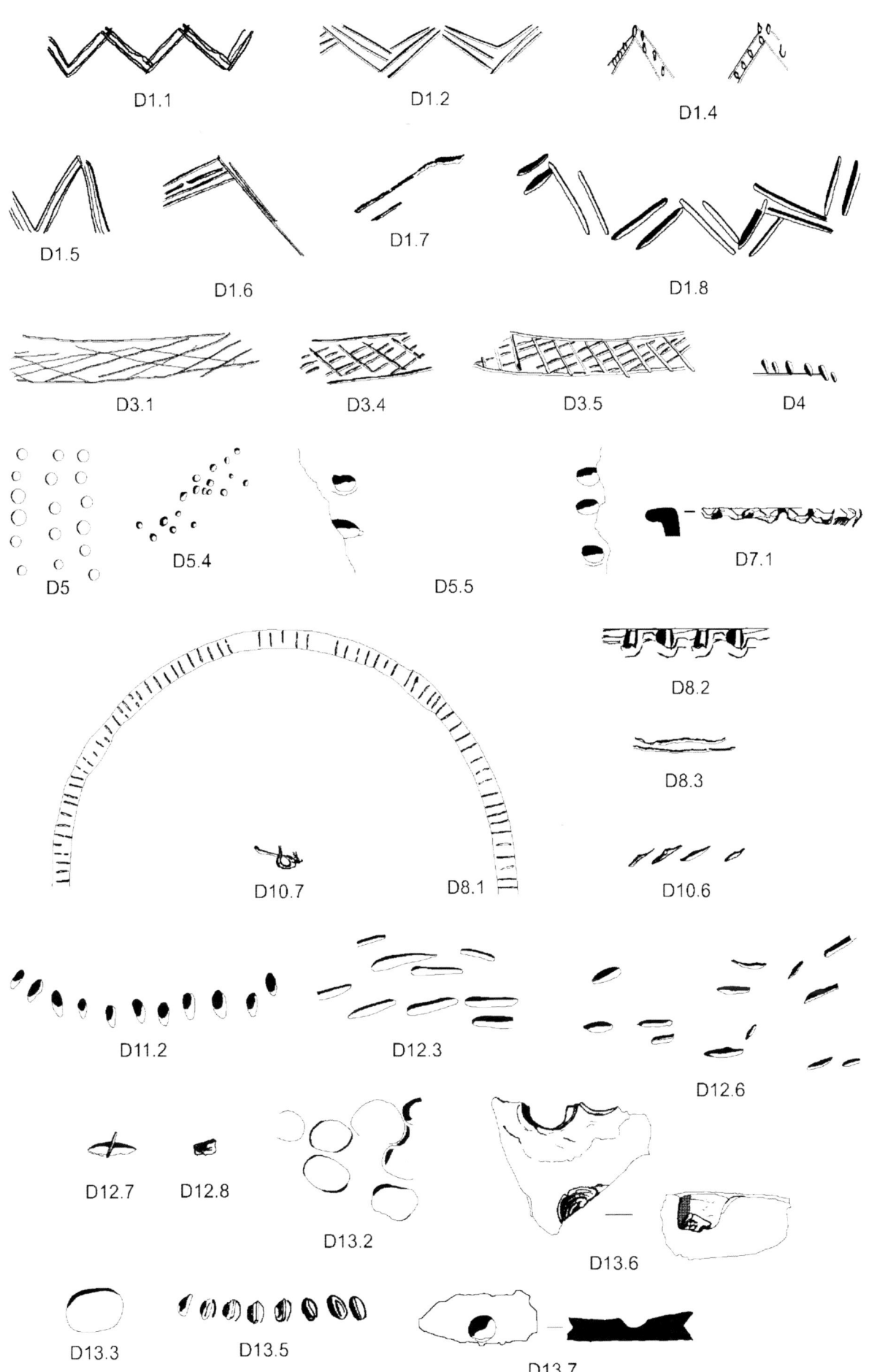

**Figure 4.3.** Decoration D1:1 - D13.7 (scale 1:2).

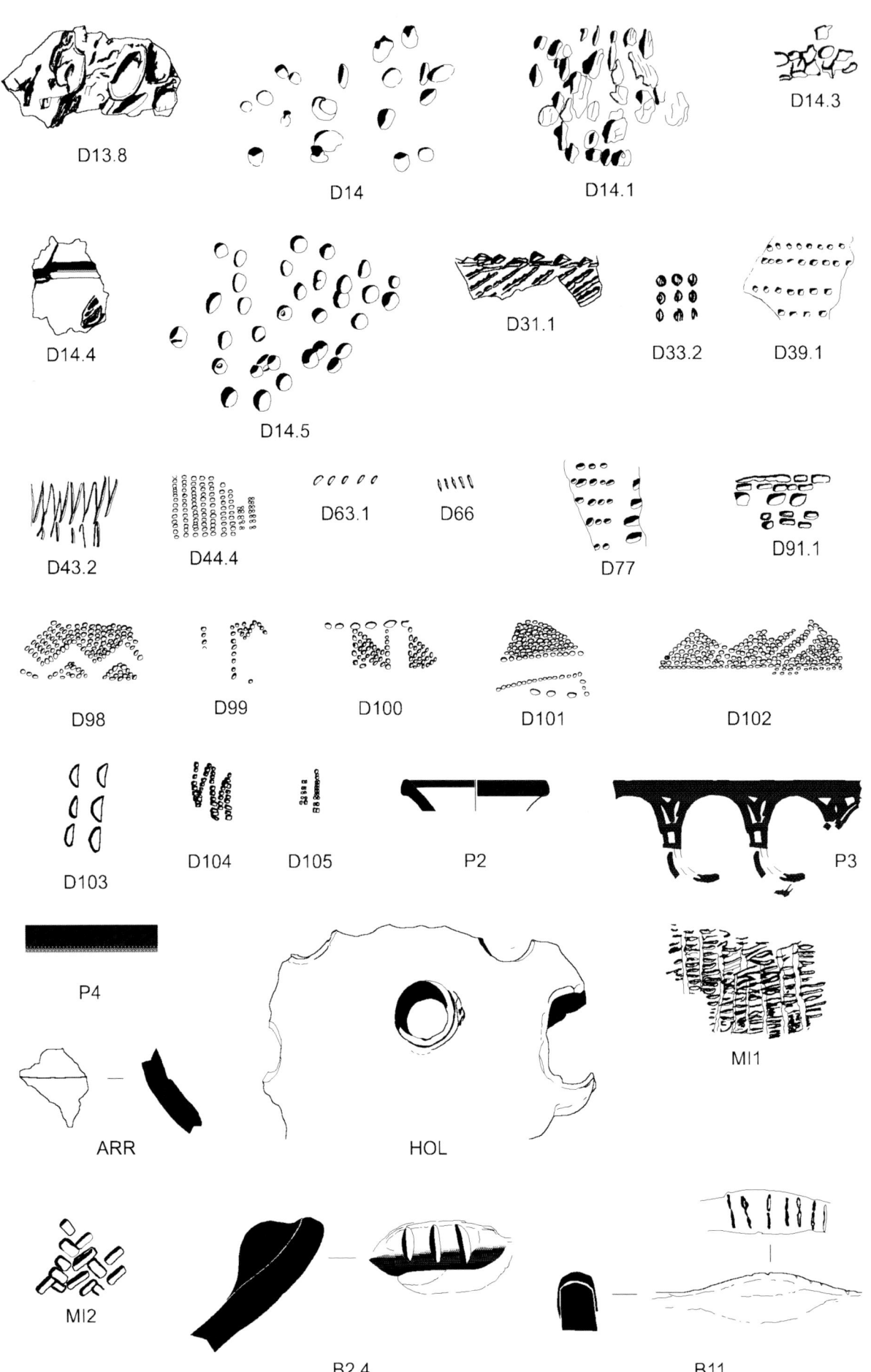

**Figure 4.4.** Decoration D13.8 - B11 (scale 1:2, P2 scale 1:4).

appears to be insufficiently regular in its curvature to be part of a body wall. Both surfaces are untreated.

As in Group 1a material, several varieties of the zigzag motif are evident in Group 1b. These include zigzags executed in thin single lines with elliptical impressions between them, zigzags in double lines slightly incurving at the top of the motif, and a version executed with quadruple lines on one side and closely spaced double lines on the other (D1.4, D1.5, D1.6). Only one version of the band of cross-hatching is present, which could be assigned to the existing Type D3.1. Interior vertical rows of impressions occur on one vessel Type (2:2.1); these are made of impressions of a more elliptical shape than those already noted, and have been designated D.5.5. The same vessel type has decoration D7.1, this being a series of strongly-projecting somewhat wedge-shaped bosses below the rim. There are two further versions of linear decoration associated with the rim: D8.1, on Type 2:5.1 comprising radial lines all round the rim top and D8.3, consisting of broken horizontal lines below the rim, on the exterior. The latter decoration is combined with a loop motif, designated D10.7.

Group 1b material exhibits a number of variations on the motif of approximately circular to elliptical impressions. These can be multiple broad, relatively shallow, sometimes overlapping impressions (D13.2), single impressions either broad and shallow or deep and narrow, (D13.3, D13.6), or 'thumbnail' impressions in rows (D13.5). The remaining version of this decoration type distinguished, comprises large irregular, mostly elliptical, impressions often with pronounced ridges between them (D 13.8). Areas of irregularly-spaced impressions of smaller size occur, these being composed of circular to oval impressions, either relatively sparse or distributed more densely (D14, D14.5). Other types are composed of irregular, often approximately U-shaped impressions, having ridges within them, occurring in groups, as in D14.1 or singly, as in D14.4. The latter decoration is combined with a ridge or cordon in the type specimen.

Both types of mat-impression seen in Group 1a, viz. MI1 and MI2, are present in Group 1b. Several large body sherds have holes pierced through them, in the same manner as those on Type 2:7.1. Such features have been designated HOL. The final decoration types in Group 1b are two bosses, the first (B2.4) being of oval shape with three vertical cuts on the exterior face, whilst the second (B11) forms a series of low, approximately oval, protrusions on the rim of the Type 2:5.1 bowl.

Concerning the new groups distinguished amongst the Test Excavation material, the majority were represented by only a small number of diagnostic sherds, some of which were types already noted within Group 1a or 1b.

**Group 1d**
Group 1d comprised sherds in Fabric 1:16; on one sherd retaining evidence for surface treatment, the latter consisted of a red matte slip. The main type in Group 1d is a bowl Type 1:10.3, already known from the Surface Survey material. This has a relatively well-defined shoulder, a moderately inward-sloping short neck, and a tapered and slightly everted rim. The example in Group 1d retained no evidence for surface treatment, but exhibited decoration D1.2, a rather irregular zigzag executed in triple lines, also known from the Surface Survey ceramics. The second featured sherd in Group 1d had evidence for the matte red slip and for mat-impressions of type MI1, but no other diagnostic features.

**Group 1e**
Group 1e was represented only by two body sherds. These, in Fabric F1:17, were both smoothed on the exterior. One exhibited Decoration D12.8, a small roughly rectangular deep impression with undulating edges, perhaps made with the end of a bone. Smudging on the exterior, together with wiping on the interior are features exhibited by the other sherd placed in this group.

**Group 1f**
Group 1f comprises material in Fabric F1:18. Characteristic sherds are either smoothed or have a brownish self-slip on the exterior, together with a light black burnish on the interior. The only type present in Group 1f is the bowl Type 2:5.1, described above under the Group 1b material. Most decorated sherds in Group 1f are probably from one or more bowls of Type 2:5.1, having the decoration designated D14.5, consisting of irregular deep circular to elliptical impressions, sometimes overlapping, which occurs on the lower part of the largely complete example of the type. Other decoration in the group consists of bosses, one of indeterminate type, the other classed as B11, again of a form present on the type specimen of 2:5.1.

**Group 1g**
Group 1g comprises sherds in Fabric 1.19, and is represented only by plain body sherds. The Group 1g sherds are of moderate wall-thickness (9mm) and have vegetable temper. It is possible that this fabric represents a variation on another fabric, such as 2:17, which are from somewhat thicker-walled vessels (c. 13mm) and have a very low content of vegetable temper. The type-specimen for Fabric F1:19, and hence for the group, has a reddish-brown thin slip or wash on the exterior, and is smoothed on the interior.

**Groups 1h and 1i**
Ceramics in Groups 1h and 1i comprised those sherds in Fabrics F1:1 and F1:2 with a brownish to greyish self-slip. It is possible some of the material recovered on the Surface Survey in these fabrics could have originally possessed this surface treatment, before being eroded. In addition to two jar sherds and one possible base sherd that were of insufficient size to assign a specific type, ten different types were recognised within these groups. Three of the jar Types, 1.9.1, 1.10 and 1:13.1, are existing ones defined on the basis of the Surface Survey collection, whilst one, 1:9.3, occurred amongst the Test Excavation Group 1b material, described above.

The first new type distinguished amongst Group 1h and 1i is jar Type 1:9.5, of which only a small proportion of the profile is extant. This is sufficient to show that it is a small jar with inward-sloping walls. The rim is quite strongly squared-off on the exterior, to form a virtually flat rim. The paste has been turned over into the interior, forming a thickened inner face to the rim, which terminates in a distinct ridge. The vessel is self-slipped on the exterior, but untreated on the interior. Type 1:13.3, the second new type in Groups 1h-1i, has a moderately distinct shoulder and an inward-sloping neck. It is distinguished by the form of the rim, which is undercut on the interior and has a plain bevelled face towards the interior. This has a sharp transition onto the exterior, with no projection beyond the profile of the wall below the rim. Surface treatment is the same as in the previous type. The third new type in Group 1h and 1i is 1:19.1. This is rather asymmetrical: one section of the rim profile, in particular, differing from another further around the circumference. In general, the extant portion of the vessel comprises a jar with inward-sloping walls, probably having a shoulder, and with a short, upright neck. The rim around most of the surviving circumference is rounded-off on the interior, and has a bevel facing towards the interior, with a slight groove along the centre. The rim projects upwards and outwards to the exterior, forming a small lip. Along part of the circumference, the exterior projection is downturned, whilst the vessel wall below is more concave than that below the upturned portion of the rim. Apart from the exterior self-slip characteristic of both groups, Type 1:19.1 is unadorned.

Bowl Type 2:3.3 is distinctive in having a rim moderately thickened on the interior, with a flat top and a strong, slightly down-turned, projection onto the exterior, which is undercut. This example has a thin reddish-brown wash on both surfaces. The second bowl type, 2:7.2, is thick-walled, with nearly vertical walls in the extant portion. The rim is moderately thickened and rounded-off on both the exterior and interior. This type is distinguished by plastic decoration in the form of impressions around the rim exterior, causing the remaining sections of the thickened portion of the rim to form shallow bosses (D8.2).

The final type distinguished in these groups is 2:9.4, a bowl with walls sloping out at about 30° from the vertical. In cross-section, the vessel wall is quite thick towards the base, but tapers strongly towards the rim, which is smoothly rounded-off on the interior to form a moderately convex top. The rim is squared-off on the exterior, with a shallow groove below. The rim projects slightly on the exterior, with a more pronounced groove below, around part of the circumference. The type has the self-slipped exterior of the groups, but is not decorated.

Remaining decoration types in Groups 1h and 1i include variations on the zigzag motif (D1.1, D1.2 and D1.7) in single and double lines, and variants of the band of cross-hatching bounded by single lines (D3.1, D3.4, D.3.5). There are present examples of a row of 'thumbnail' impressions, approximately semi-circular impressions, often with a ridge of paste pushed up on one side (D13.5), whilst there is one example of a single, shallow circular impression (D13.7), possibly on a flat base interior. Group 1i includes a new boss type (B2.4), being a form of the oval bosses with vertical cuts on its outer face, noted amongst the Surface Survey Group 1a-1c material. One sherd exhibited a feature counted as a new decoration: a shallow step or ledge in the exterior profile which has been designated as an arris (ARR). The only other decoration noted on Group 1h and 1i sherds is mat-impressions, mainly of type MI1.

**Group 1j**

Group 1j comprises material in Fabric F1:20. The type specimen is a jar of Type 1:19.1 also present within the Group 1h and 1i sherds, described above. This vessel is self-slipped on the exterior and untreated on the interior. It is not decorated. The few other sherds in Group 1j are plain body sherds.

**Groups 1k-1p**

The remaining groups contain only non-featured material. The sherds concerned have fabrics in which the Groups 1a-1j occur but were distinguished on the basis of variations in surface treatments from those exhibited by the former groups. They can be summarised as follows: Group 1k, F1:2 with light black burnish on exterior; Group 1l, F1:16 with smoothed and 'smudged' exterior; Group 1m, F1:16 with self-slipped exterior; Group 1n, F1:17 with self-slipped exterior; Group 1o, F1:18 with brown lightly-burnished slip, Group 1n, F1:18 with dark slip, lightly burnished. Most of these only occurred in a small number of sherds, for which no definite forms could be identified. They can only be dated generally to the same period as the featured sherds in the equivalent fabrics.

**Comparanda**

In addition to basic similarities in fabric and surface treatment, the presence of several vessel forms and decorations in the Test Excavation material closely comparable to a number of those seen in the Surface Survey collections has been the main reason for classing such Test Excavation material into the Groups 1a-1b. The further groups designated, 1d-1j, have included material associated with that assignable to 1a and 1b, which also exhibits a number of forms and decorations recognised in the Surface Survey material placed in Groups 1a and 1b, albeit in somewhat different fabrics and with different surface treatments. Amongst the main features seen in this Test Excavation material which are considered to link it to the Groups 1a-1c are the presence of upright-sided bowls with an indented lip or bosses on the exterior of the rim, such as Types 2:2.1 and 2:7.2, comparable to existing Types 2:1-2:3. The presence of other bowl forms amongst the Test Excavation ceramics, those classifiable as existing Types 2:6 and 2:10.1, further indicates that this material may be regarded as assignable to Groups 1a and 1b. A larger number of jar forms has been recognised in the Test Excavation collections as being the same as those already assigned to Groups 1a and 1b. These forms include those assignable to 1:9.1, 1:10.3, 1:12.1 and 1:13 in the Test Excavation Groups 1a and 1b, together with Types 1:9.1, 1:10.1. 1:10.3 and 1:13.1 in the new Groups 1d, 1h, and 1i.

Decorations that relate the Test Excavation pottery under consideration to the existing Groups 1a-c, include the use of irregular zigzags in double or triple incised lines, such as D1.6, and D1.8, and the use of bands of more or less irregular incised cross-hatching between single incised lines. Exterior rim decoration involving regular deep impressions or bosses forms a further link, the latter particularly when associated with vertical rows of impressions on the vessel interior, as in the case of D7.1 and D5.5. Decoration of areas of the vessel with irregularly-placed impressions or cut-marks, either round to oval in shape or long and narrow, constitutes a further similarity between the Test Excavation and existing Group 1a-c pottery. For instance, D12.6 in the Test Excavation material is similar to existing decoration D12.3, whilst D14.5 on new Type 2:5.1 is quite similar to existing decorations D14 and D14.1. Finally, the existence of a variety of oval bosses with vertical cuts on the outer face, designated B2.4, similar to existing boss types B2.2 and B2.3, indicates a link between the two sets of material.

The main comparanda and reasons for the dating assigned to the Group 1a-1c pottery are given in Vol. 1, where it is noted that the period to which these groups are considered to date can only be relatively broadly defined to between 13th and 16th centuries AD. (Smith 1996, 187) and that there may be some overlap between the dating of Group 1 ('Terminal Christian/Islamic') and that of Group 2 ('Christian'). The Group 1a-j pottery under consideration indicates that that the earlier part of this temporal span may be valid for at least part of these groups, since there is evidence for some overlap, in stylistic terms, with ceramics from Soba, although many of the Test Excavation types do not have exact parallels at this site. However, dating evidence (see below) indicates that the continuity of similar forms over long time-spans, especially within handmade wares, cannot be excluded.

It can be seen that Type 1:2 from the Test Excavations, lacks the boss characteristic of the type specimen from the Surface Survey. It is generally similar in the exterior profile and in the main features of the rim in cross-section, though not identical, to jar type 30.1L from Soba, with an outflared rim, dated to the Post-Meroitic to early Christian Period (Welsby 1998b, 172; 1998c, 271). However, it has some similarities with rims of Soba type 48L and 52.2L, which are from collections datable generally to the later Christian times. Test Excavation Type 1:2 also has a reasonable degree of similarity to an outflaring rim of a jar from Dar el-Mek, a form which extends in range into the latest period of occupation. This is considered to date as late as 16th century AD (Crawford and Addison 1951, 164, 179, fig. 40, 7).

Type 1:8.1 has some similarities with Soba type 24L in its exterior profile, although the Test Excavation example has a somewhat flatter top to the rim and a more vertical profile on the interior face of the rim than in 24L, where the rim is thickened to have a strongly convex profile at this point. Test Excavation Type 1:9.3 is quite close to 35L in cross-section, particularly in the tapered form of the rim. The only difference is that 1:9.3 is of similar wall-thickness from the upper part of the shoulder to the neck, whilst 35L has a thinner wall at the shoulder. The rim of 1:9.1 is similar in cross-section to that of 40L, although less sharply sloped-off on the interior. Insufficient remains of the sherds assignable to 1:9.1 to be certain whether this type has a distinct shoulder, as does the Soba type. There is no exact parallel to 1:9.4 at Soba, but in general the form of the rim is similar to that of 77L, whilst the overall form of the extant portion is closer to 79L. The rim of 1:9.5 is thickened on the interior in a very similar manner to that of Soba types 96L and 99L. It has a flat top to the rim, being in this respect closest to 99L. However, the overall form of the Test Excavation type may not be similar to these Soba vessels, since the latter are more clearly globular or gourd-shaped than 1:9.5, the extant portion of which could belong to a necked form. Definite parallels for Type 1:3.2 are difficult to establish, since the extant portion is asymmetrical. One portion of the circumference has a cross-section similar to those of 22L and 22.1L, but the stance of the latter is more upright than the former appear to be, so the parallel is not exact. In stance, it is closer to 23.1L, although the outer edge of the rim is less rounded than in this type from Soba.

Types 1:13 and 1:13.3 are distinctive through having a bevelled face on the rim interior. Type 1:13 is reasonably close to 59L in rim form and overall stance, though the exterior profile below the rim is less concave. The bevelled rim on 1:13.3 is most closely paralleled at Soba by types 107L and 107.1L, although only the uppermost part of Type 1:13.3 is close to the Soba forms, since it has a more distinct shoulder than the latter. Whilst the shape of the rim of 1:13.3 in cross-section is closest to that on Soba type 107.1L, the overall size of the vessel is most comparable to 107L. Soba type 107.1L also has a design on the flat interior face of the rim, which is absent from the Test Excavation specimen. Test Excavation Type 1:14.2 is another with an asymmetrical profile. There is no exact parallel at Soba to the form of the rim in cross-section, but the exterior profile over part of the circumference is similar to that of 74L and fairly similar to that of 71.1L. The general size and stance of the extant portion of the two types is also similar. The wall thickness of Type 1:16.1 approaches that of some examples of Class I at Soba, but it is not of so large a diameter as the majority of types in this Class, and its closest parallel appears to be with 120L, which has a similar stance and flat top to the rim. The Test Excavation type has a faint groove below the rim on the exterior, but lacks the fully developed band of ribbing present on the Soba type specimen of 120L (Welsby 1991, figs 97-103; 1998, figs 43, 45,46).

Amongst the bowl types, relatively few parallels were found with Soba forms. Type 2:5.1 is similar to 91N in rim form, but differs in the proportion of width to height, in being thicker in lower body wall and in having bosses and incised decoration on the rim. Although Type 2:6 is a larger vessel than 129N, it is similar in rim form and in the upper part of the body wall. The lower part of 129N appears to be thinned to a greater extent than in 2:6, whilst the latter may be deeper in relation to height if complete. Vessels with a similar rim form to 2:9.4 occur at Soba, in particular, in types 123.5N, 123.6N and 145N. However, the Test Excavation type is thinned at the rim, to a greater extent than 123.5N and has a slightly different stance from 145N and 123.6N. These Soba types are both relatively shallower forms and 145N is much wider, so that overall the parallels are not very close. Soba types 143N-145N have some similarities in profile with Types 2:9.2 and 2:9.3, though the latter are more irregular in profile than the Soba examples. It is not clear that these Soba types have the relatively sharp transition from wall to base exhibited by 2:9.2 and they are, again, generally forms with wider diameters. One Soba type, 89N, does have a sharp transition from base to wall but differs from the Test Excavation example in both its proportions and rim form

(Welsby 1991, figs 108, 110, 111; 1998b, figs 48, 49).

Certain decoration types evident at Soba are similar to Test Excavation examples, such as incised double- and triple-line zigzags, though at the former site, these are more often between framing lines (Soba decorations 92, 93.3, 93.4 and 94.1). Further similarities occur in the presence of cross-hatching between single horizontal lines, and areas of round to elliptical impressions (Soba decoration types 89-90, 68.1 and 70). However, it may be noted that the series of decoration types present at Soba involving cross-shapes as motifs, (e.g. nos 234-261.5) are completely absent from the Test Excavation collections (cf. Welsby 1991, Figs 125, 141, 132; 1998b, figs 60, 62, 70, 71).

A few general similarities in form are present in ceramics from Hambukol where pottery of Late to Terminal Christian type occurs, as well as Classic to Post-Classic material. For instance, 2:12.2 is of a reasonably similar form to a handmade bowl from C-6C-2 (Phillips 1991, fig. 14) though details of rim form differ. A rim of the same general form as that on 1:19.1 is present on a vase (Phillips 1991, fig. 34) though this is a wheel-made vessel, and more regular in execution than the Test Excavation specimen. A bowl with a wall tapering in thickness from near the base to the rim, and of similar stance to 2:9.3, is present at Hambukol, although the overall form is not the same since the Hambukol vessel has a thick, almost footed, base lacking the sharp transition from wall to base characteristic of 2:9.3 (Phillips 1991, Fig. 48). The moderately inturned jar Type 1:2.1 is reasonably closely paralleled by a second handmade pot from Hambukol, although the latter has decoration which the Test Excavation specimen lacks (Phillips 1991, Fig. 83). Plastic decoration involving a row of bosses around the rim exterior is exhibited by jar HBK 87-11-4 (Phillips 1991, fig. 11), currently dated to the Islamic period (J. Phillips, pers. comm., 1998). In this case the decoration includes incised oblique lines between the bosses, a combination not present amongst the Test Excavation material. Jars with moderately out-flaring rims similar to that of Type 1:2 were recovered from Kulb Fortress. One example of these has decoration on the base resembling that of D13.8. Settlement at the site lasted until 13th or 14th century (Rodziewicz and Dinkler 1972, abb. 86, 87, 706).

At Dar el-Mek there is quite a close parallel to Type 1:2, (Crawford and Addison 1951, fig. 40, 7), to the rim forms of Type 1:13 and a fairly similar rim form to that of Type 1:10 and a parallel, mainly in terms of the exterior profile of the rim, to Type 1:3.2. The overall form of Type 2:9.3, with a sharp transition between base and wall, the latter tapering to the rim in cross-section, can be paralleled in vessels characteristic of the Fung graves at Abu Geli although the Test Excavation example is undecorated, and shallower in proportion than those illustrated from this site. Concerning decoration, examples were found at Dar el-Mek of bands of incised cross-hatching framed by single lines with a band of zigzags below. One type illustrated has double-line zigzags, thus forming a design close to the combination of D1.8 and D3.4 on Type 1:8.2, though lacking the oval dots also present on the latter type. Pottery with this decoration is characteristic of the latest period of occupation at Dar el-Mek, considered to extend from the 14th century and to end not earlier than the 16th century (Crawford and Addison 1951, 179, 181, figs 45a, 44,a,d, pl. XXXII).

The above shows that the dating of the Test Excavation Group 1a-j material may be reasonably considered to include the same time-span as has been suggested for the groups on the basis of the Surface Survey pottery, viz: 13th-16th centuries. The extent to which there are similarities with vessel types from Soba, noted above, seems to confirm that it is not necessarily possible to distinguish individual vessel types from the general 'later' Christian ceramics in Upper Nubia. That this dating on the basis of stylistic comparisons is not applicable to the whole of Group1a-j is indicated by the $^{14}$C dating of the house at site 155.4, from which much material of these groups and some of Group 16a-b was recovered.

Two samples from within the structure were dated and calibrated using the OxCal program.[1] Taking the 2o range, one sample has a 91% probability of being between AD 1480-1680. The highest probability of the other (49%) is to lie between AD 1720-1820. The former sample is associated in the same excavation unit with pottery of Groups 1a and 16a, thus supporting the idea that the date for this material could cover 15th-16th centuries. The fact that the upper end of the date range is 1680, together with the second date most probably lying within 18th century, means that the pottery in these groups could also date to a later time than indicated by comparison with stylistically similar ceramics from other sites, and so could have continued into the later Islamic period. A similar situation has been noted for some of the handmade pottery of Lower Nubia, in the case of Adams' Wares H4 and H5, for which the total time-span assigned is AD 1000 - c. 1800 (cf W. Adams 1986, 427, 433-434).

## Group 2

### Groups 2a and 2b

Material assignable to Groups 2a and 2b, as defined on the basis of the Surface Survey collections, formed a relatively small proportion of the diagnostic ceramics from the Test Excavations. Only one body sherd was recovered that could be assigned to Group 2a as originally defined, being Fabric F1.5 and an unslipped surface. This sherd did not possess a rim, so that it could not be allocated to a specific type. It did exhibit a surface treatment and decoration sufficient to indicate that it must have come from a vessel of closed form, most likely from a jar (Class1). The exterior had black areas, which could have been either from the effects of firing, or have been due to deliberate 'smudging'. The sherd had areas of mat impressions (MI2), which are likely to be the result of forming the lower parts of a vessel on a mat, rather than solely a decorative effect. (cf. Rice 1987, 125, 132). The more elaborate surface treatment on the exterior in relation to the interior indicates that it is probably a closed form, whilst the sherd was sufficiently large to exhibit a degree of curvature that would be expected for a some-

---

[1] Information on the radiocarbon dating is from communication by Dr J. Ambers, Department of Scientific Research, British Museum.

what globular form, of which a jar would be the most probable Class of vessel.

A second sherd assigned to Group 2b, being in Fabric F1:1, was recovered from the same context. On this sherd the greater part of the rim was extant, but the original lip was missing. It was clear that it was from an inturned jar (Class 1), but it could not form the basis of a specific type. Sufficient of the rim remained to show that it was strongly folded over into the interior, thus forming a somewhat club-shaped cross-section, with the thickening at the rim occurring on the interior. The interior surface appeared to have been originally unslippped. The exterior surface also appears to be unslipped but it is slightly eroded, such that it could have originally had some degree of burnishing.

In the Surface Survey, Group 2a and 2b material was dated to the Christian period. A dating of the Test Excavation sherds as well to this period is supported by two main circumstances. Firstly, both are from the same context as a portion of a jar decorated in the Classic Christian style as defined by Adams (see below). Secondly, the form of the jar in Group 2b can be determined sufficiently to indicate that it was similar to types known from Soba, particularly 96L-99L, and possibly 89L or 91L (Welsby and Daniels 1991, figs 101-102). The mat-impressed sherd can only be assigned a Christian date by similarities in fabric and by context. The presence of mat-impressions on pottery has continued for long periods in Sudan, and is well known on handmade vessels from, for example, the late Meroitic period in Upper Nubia (Geus *et al.* 1986, 82, fig. 2) as well as being known on certain wares of the Christian period in both Upper and Lower Nubia (Welsby 1991, 206; W. Adams 1986, 423). While the mat-impressed sherd assigned to Group 2a would be consistent with a Christian dating, the possibility that it represents residual material from an earlier period, most likely the Post-Meroitic, cannot be excluded.

## Group 2d

Group 2d comprises material in Fabric F2:13. Only two featured sherds were recovered assignable to this group, both from Site 112.4, although several body sherds with a red slip from 112.3 were also assigned to this group on the basis of fabric and surface treatment. It was found that the two featured sherds joined although they were incorporated in the fill of two different tumuli at Site 112.4. Despite the small number of sherds involved, they were considered to be examples of a characteristic ceramic that has been designated a specific ware when recovered from a number of other sites and were, on this basis, separable into a new group.

The sherds placed in Group 2d form a bowl *c.* 330mm in diameter, with sides sloping out at an angle of *c.* 45° to the horizontal. The rim of the bowl is modelled to form a rim of the same general type as those designated as 'Samian' by W. Adams (1986, 102, fig. 39, nos 67, 68, 71). In terms of decoration, the bowl has a matte red slip on the exterior below the rim. A rim-band in dark brown, extending from the interior, covers the exterior face of the rim, but otherwise the exterior has no other decoration. It is clear that the main decoration was originally on the interior, but the motifs can no longer be determined because the surface is eroded. The general design appears to have involved decoration in dark brown on a white slip, comprising a brown band just below the rim, with another band of dark brown *c.* 3mm below this, with the white slip showing between. In its present state, it is not clear how far down the vessel the second band extended or if it might have been a background colour for further individual motifs. If the brown slip did not extend over the remainder of the vessel interior, it is probable that it formed a second band, with individual motifs below it on the white slip.

### Comparanda

Despite the eroded state of the interiors of the sherds forming the bowl, they are sufficiently distinctive to indicate that they belong to the ware designated 'Soba Ware', first identified by Shinnie (1955, 35-37). The great majority of the known examples of this ware have been recovered from the site of Soba, where a considerable range of forms and sizes of vessel have been found. Type 2:13.1 from the Test Excavations is closest in overall stance to 51N and 50.1N in the classification of Welsby, although the specific form of the rim is closest to that of 54.5N in the same classification (Welsby 1991, figs 106, 115; 1998b, figs 47, 48). In general terms, the form of 2:3.1 and its rim is similar to that of the sherd of Class 2a ('Soba Ware') illustrated by Shinnie (1955, fig. 16, 6, pl. XIIIb, centre left) although the uppermost part of 2:13.1 is more everted than in Shinnie's example.

The decoration of the sherd illustrated by Shinnie differs somewhat from that of 2:13.1, since it consists of a black slip on the exterior with a design on this in white. The exterior slip is continued over the rim onto the interior to form a band below the rim, similar to that on 2:13.1, but otherwise the decoration appears to differ in its specific motifs and positioning. From the visible remains of the decoration on 2:13.1, it seems likely to have been of a general design similar to those illustrated by Welsby as decorations 428-430 having a broad band at the top, or possibly those shown as decorations 421, 423, 425-427, which also have a broad band uppermost, but which in these decoration types, have rectangular or circular motifs painted on the bands. Such motifs are no longer evident on the Test Excavation specimen.

Whilst Soba itself has yielded the greatest quantities of Soba Ware so far known, small numbers of sherds of the ware have been found at other locations. Within Sudan, for example, Soba Ware has been recovered from as far north as Berber (D. Welsby, pers. comm. 1996) and as far south as Saqadi. Other sherds have been recovered from Burri and from quite near the Test Excavation site, at Gadu (Shinnie, 1955, 36). Sherds of Soba Ware have recently been identified from outside the present-day Sudan, at Axum in Ethiopia (Phillips 1997, 455), indicating that it was traded over considerable distances. Given the known distribution, albeit in small quantities, its presence at site 112.4 would not be unexpected.

## Group 2e

Only one partial vessel was recovered from the Test Excavations that could be assigned to Group 2e, being recovered from site 112.3. It is sufficiently distinctive in both fabric and decoration to warrant its being placed into a separate group. This sherd is the only example assigned to Fabric F2:15. The portion of the vessel available has no rim. However, sufficient of the remaining profile remained to give a good indication of the likely shape of the body and part of the base. Since the vessel was wheel-made and had a band of decoration around the upper part of the body, the approximate stance of the sherd could be determined by assuming that the wheel-marks and the upper framing lines of the decoration were horizontal when the vessel had been complete. It was considered, on this basis, that the sherd could be assigned a Form designation, viz., 1:20.4.

As reconstructed on this basis, Type 1:20:4 is a small jar, of which the small portion remaining comprises an inward sloping body, with a slight carination marking the transition between this and the base, which has a convex profile and which is relatively thin in comparison with the main part of the body wall. Towards the rim the profile of the vessel wall changes direction so as to be essentially vertical. At this point it is thickened to form a collar or cordon on the exterior, whilst the interior is concave. The cordon appears to be either the base of the neck, or a separate cordon just below the neck. The decoration on the vessel (P3) consists of a relatively broad stripe below the cordon. This dark brown stripe forms the upper border of a frieze of approximately oval elements also outlined in dark brown. The outlines of the ovals are separated by a short distance, but are linked on each side by a pair of short horizontal lines. Spaces between the upper border, the upper-most horizontal line, and the ovals are infilled by a small incompletely-closed 'loop' motif linked to the circles on each side by curved lines. The lower part of the design is insufficiently preserved to determine its form, but it is likely to mirror the upper part

**Comparanda**
Allowing for uncertainty in the exact form of the vessel, the closest general parallels are to be found in Class G, small pots and bottles, in Adams' classification, the most similar types to 1:20.4 being G31-G35 (W. Adams 1986, fig. 280). The closest individual type illustrated is G35. Form 1:20.4 appears somewhat less squat than G35, having slightly more steeply-sloping sides. Although there is no rim present, there is sufficient remains of the circumference of the vessel at the base of the neck to indicate the approximate original vessel diameter at this point: the resulting reconstruction being a vessel with a smaller maximum diameter relative to height than exhibited in Adams' G35. Despite these differences, G35 is closest in overall form, since it is the only one in the Class G vessels lacking a shoulder and it shows a thinning of the base relative to the body wall thickness, though in the illustration of the cross-section of this vessel form (W. Adams 1986, fig. 48), this thinning appears less pronounced than is apparently the case for 1:20.4, as judged from the portion of the base remaining.

Since the central part of the base is missing, it is uncertain as to whether Type 1:20.4 possessed a footring. If a footring were present, the original form is likely to have been more similar to G31, to which the overall slope of the vessel wall of 1:20.4 is closer than to G35. However,

G31 possesses a slight shoulder, so it is most probable that 1:20.4 represents a vessel most similar in overall form to G35, but with a body wall sloping less steeply, at an angle closer to that of G31.

In terms of decoration, again allowing for the abraded state of the sherd surface, 1:20.4 is closest to types 16-4 and 16-7 in the class of 'continuous friezes' distinguished by W. Adams (1986, fig. 166). The design on 1:20.4, designated P3, is closest to Adams' 16-4 in general terms, particularly in having the 'arch' motif above the circular motifs with single small loops between the circles (cf. W. Adams 1986, fig. 161). The main differences from 16-4 as illustrated by Adams are, firstly, that in P3 the small loops are relatively narrower and tend to be slanted to one side, rather than upright and, secondly, that pairs of horizontal lines linking the circles are relatively more broadly-spaced vertically. In these characteristics, P3 is closer to 16-7, in which the horizontal lines are placed closer together vertically and which is more similar in the shape of the individual small arches. However, 16-7 differs in that the arches are double rather than single. P3 appears to lack a dot within the circles as in Adams' 16-2, but it could have been similar to this design originally, although it has horizontal lines linking the circles which are absent from 16-2. These motifs characteristic of Adams' Style NIVA, 'Classic Christian Fancy Style' (W. Adams, 1986, 245-247; fig. 161). According to Adams' dating, based on Lower Nubian material, the broadest dates for Decoration P3 are between AD 800-1100 and, on the basis that the 'angle-filling lines with single projecting loops' seen in P3 are characteristic of Style NIVA only and not NIVB, most probably not later than $c$. AD 1000 (cf. W. Adams 1986, 246-247).

Material from el-Ghazali is classified by Adams into Style NIVA (W. Adams 1986, 246); however, only one motif similar to decoration P3 was noted amongst the main decorations illustrated from this site (Shinnie and Chittick 1961, fig. 28, 2). The latter motif has single solid lines outlining circular panels, together with 'angle-filling' elements. In these respects this Ghazali decoration is similar to P3, but the 'angle-filling' motifs are executed in solid, single lines, rather than being open 'loops' as in P3, so that the parallel is not exact.

Some material in Style NIVA is known to have been traded, in relatively small quantities only, to the south of the Fourth Cataract. Classic Christian Ware W5 is known from Soba (W. Adams 1986, 246: Shinnie 1955, 28, 35, pl. X; Welsby 1991, 243). The Test Excavation example, in form and surface colour, is probably closest to Ware W6 or W10 in Adams' classification. It is not possible to determine whether the sherd is from one or other of these Wares, since it is no longer possible to be certain as to whether the surface was originally matte or polished; this feature of the surface treatment being the criterion for distinguishing between them. Both W6 and W10 have been mainly known to be distributed between the First and Second Cataracts (W. Adams 1986, 494).

**Group 2g**
The remaining group of material assigned to the Christian period contains only a single diagnostic sherd. This is in Fabric F1:5 as is material in Group 2a, but has been

separated from the latter because of the presence of different surface treatment. In Group 2g sherds, the exterior surface is red slipped, and in the case of the sherd under discussion, the slip is burnished. The interior of the sherd is untreated. Below the zone of red slip on the exterior is an area of mat-impression of the same type as that in the Group 2a material (MI2). Although the Group 2g sherd has no rim, the lack of surface treatment on the interior in contrast with the exterior, indicates that it is from a closed form. On the basis of the types of vessels most commonly exhibiting a red slipped zone and mat-impressed areas, it is most likely to be from a jar.

As in the case of the mat-impressed material in Group 2a, the dating of Group 2g to the Christian period is made partly on the basis of fabric and context, the former being similar to other material stylistically quite specific to the Christian period, and the latter providing other material, including that in Group 2e, also clearly Christian on the basis of style of decoration. Vessels with red slip and mat impressions are known from the Post-Meroitic, but are present in Christian contexts in both Upper and Lower Nubia (W. Adams 1986, 423-424). Hence, Group 2g is most probably of Christian date, but it is possible that it is residual from Post-Meroitic times or earlier.

**Group 2h**

A few non-featured body sherds were assigned to Group 2h. They are in Fabric F1:5 and have a better-preserved surface than much of the other material in this fabric, such that the original surface could be clearly seen. This exhibited a self-slip, differing from Group 2a, which is unslipped. It is considered to date to the same period as other material in F1:5, and occurred on Site 112.3, from which other Christian period material was recovered.

## *Group 3*

A number of diagnostic sherds from the Test Excavations were assigned to Group 3 largely on the basis of similarities in fabric, all being identified as F1:4. The majority of sherds in this fabric are body sherds, mostly without evidence for slipping or smoothing on either surface. One example was untreated on the exterior, but had a light black burnish on the interior. Two had traces of mat impressions on the exterior, but the surface was too eroded to determine whether the type was more similar to MI1 or MI2. Of the remaining body sherds, two had incised decoration in the form of a horizontal band of cross-hatching, somewhat irregular in execution, assignable to D3.1 in the decoration classification. One sherd exhibited a number of impressions which have been assigned to D14.3, although they are somewhat irregular, and it is not certain that they represent a deliberate decorative effect. Most of the impressions are quite shallow, and vary in shape from roughly rectangular, often with a projection from one corner or side, to approximately triangular to oval. They appear to be arranged in largely contiguous rows, of which two, and a portion of a third, remain. Impressions of all the main shapes occur within each of the rows.

Only a single example of this type of impressions was found, so that the overall design is not clear, and nor is the type of implement used to make the impressions. The decoration D14.3 occurs on the sherd with light black-burnished interior, which is quite thick in cross-section. It is possible that it is from a vessel such as the *doka*s which require a smooth interior and are characterised by burnished interiors (cf. Welsby 1991, 179). However, vessels of this type at Soba are not described as being decorated on the surface, although decoration on the rim has been noted on specimens from Lower Nubia (W. Adams 1986, 223).

Concerning the only rimsherds in Group 3, one is an everted form considered to be the neck of a jar, designated Type1:2. In profile the rim of this type is slightly thickened on the interior in relation to the vessel wall, and is rounded off on the interior, forming a slightly convex surface of the rim. The vessel wall curves in sharply on the exterior, to form a sharp lip on the outer edge of the rim. Type 1:2 is smoothed on the exterior, but eroded on the interior. There is no decoration.

The second form in this group is that classified as Type 1:9.4, although it is possible that it is from the upper portion of an inturned bowl. Type 1:9.4 has a cross-section exhibiting a noticeably convex profile on the interior, with a rim rounded off on the interior, and leading in its present form to a flat rim top. This is quite sharply squared off on the exterior edge, at the junction with the exterior which is relatively straight in profile. The rim is slightly eroded, so that it is probable that the top of the rim and the exterior edge were more rounded originally. One example of the type has no evidence remaining for slip or burnish or decoration, whilst the other example has a red burnish on the exterior together with possible evidence of mat-impressions similar to MI2.

**Comparanda**

The relatively fragmentary nature of the sherds available makes the establishing of definite parallels difficult. Sherds from the Surface Survey in Fabric 1.4 and with a similar surface treatment, assigned to Group 3, were tentatively dated to the Post-Meroitic to early Christian Period (Smith 1996, 187). On the basis of the Surface Survey material, it is possible that this dating was too restricted in time. Whilst there is no clear parallel for D14.3, a decoration type comprising similar irregular shallow impressions is known from a Christian context at Soba (Welsby 1998b, fig. 60 type 68.2). Decoration D3.1 is also known in a Christian context from Soba (Welsby 1991, fig. 125, type 90) and has been considered a characteristic of Surface Survey Group 1a-1b material, considered to be Terminal Christian-early Islamic in date. As noted above, mat-impressions on jars are common in the Post-Meroitic but were sufficiently common in the Meroitic to be designated a specific variety amongst the ceramics from Meroe (Shinnie and Bradley 1980, 152) and continue through into the Christian Period (cf. Welsby 1991, 206).

Generally similar forms of everted jar neck to Type 1:2 are present amongst ceramics of the Post-Meroitic period, including material from sites such as el-Kadada, esh-Sheitab, Shaqalu and Meroe (Lenoble 1987a, pl. XII; Geus *et al.* 1986, fig. 2; Lenoble 1992, pl. I and II; Dunham 1963, fig. 188, F), although none are of exactly the same form. As concluded in the consideration of Types in Group 13 of the Surface Survey material (Smith 1996, 194), it is not possible to assign such sherds as Type 1:2 indubitably to the Post-Meroitic, because of uncertainty as to the form of the whole vessel of which they form only a part. In the case of Type 1:2, it is similar to types 30.1 L from Soba although the latter is thickened more on the interior just below the top surface of the rim, and is fairly similar to 54.1L although less strongly thinned on the interior of the lip (cf. Welsby 1998b, figs 43, 45).

It is possible that Type 1:9.4 could be from a bottle or jar similar to QM0020 illustrated from Qos Nasra, Marangan, by Edwards (1991, pl. 1X) although the estimated rim diameter for the type specimen of Type 1:9.4 appears rather large. A reasonably similar cross-section and stance is exhibited by a bowl 4012 from Gabati (Smith 1998c, fig. 6.27) although, again, the diameter is less than that for the Test Excavation samples. Both these parallels are considered to be Post-Meroitic in date. A parallel in a Christian period context is present at Soba, in type 79L, which is of more closely comparable diameter, and of very similar stance, but which has a more tapered rim than that in 1:9.4 in its present state.

On this basis, it is more appropriate to assign Group 3 to the period from the Post-Meroitic to later Christian. Such dating would appear reasonable for the site at 118FS2, considered likely to be Christian on the basis of the Surface Survey, and reasonable for 155.4 at which the majority of material is in groups 1a-j, dated to Late or Terminal Christian to Islamic periods, and where a small amount of residual earlier material might be expected. It is not so appropriate for the context at site 101.4, in the fill of Tumulus 1, which also includes material of Napatan date. It can only be suggested that if the above dating of Group 3 material is correct, the sherds in the latter burial were introduced at a relatively late period of disturbance or robbing.

## *Group 4*

**Groups 4d, 4e, 4g**
A small number of body sherds were tentatively assigned to these groups, mainly on the basis of similarities in fabric to existing groups considered to be Meroitic in date, and variations in surface treatment. They comprise: Group 4d, F1:9 with red slipped exterior; Group 4e, F1:10 unslipped; Group 4g, F1:15 with red burnished exterior.

**Group 4h**
Only one sherd that could be classed as the separate Group 4h was recovered during the Test Excavation season. Despite the single occurrence, it was sufficiently distinctive in fabric and form to warrant placing it in a new group. The sherd is in Fabric F4.3, characterised by a reddish-pink fired colour and by the presence of opaque white round or lenticular inclusions. In method of manufacture it is wheel-made. In form the sherd has been classed as a small bowl, Type 2:16.2, because the diameter is too large for it to be considered a cup in Class 3, but it could also be the top of a necked jar. The extant portion exhibits an essentially vertical body wall just below the rim. On the exterior, the profile has a slight outward curve, which could indicate that the maximum diameter of the vessel body was at least as great as the maximum external diameter at the rim.

Type 2:16.2 is distinguished by a modelled rim of approximately triangular cross-section. On the interior there is a shallow groove at the junction of the body wall and rim, above which the latter slopes up and outwards at an angle of about 45° to form a tapered top to the rim. It is folded over so as to project slightly beyond the vessel wall, forming a moderately convex, nearly vertical, exterior face to the rim, which is then merged back into the body wall. The sherd has a matte red slip on the exterior and the interior, but has no painted or incised decoration.

**Comparanda**
Type 2:16.2 has a fabric, general form and surface treatment that are closest to Meroitic ceramics. The fabric has some similarities with fabric G2, recognised amongst the Meroitic material at Gabati, particularly in the presence of the rounded opaque white inclusions.[2] Whilst there is no exact parallel to the rim form known from Gabati, a fairly similar rim, having overall a triangular cross-section, is present on vessel 511, a moderately deep bowl (Rose 1998, fig. 6.14). This differs from Type 2:16.2 in having sides sloping inwards below the rim, rather than being nearly vertical, in having a strongly convex interior face to the rim, and a slightly more concave face on the exterior. However, there are sufficient similarities to indicate that Type 2:16.2 is in a similar tradition to the Gabati example.

No exact parallel to rim form found amongst published whole vessels from Meroe (Shinnie and Bradley 1980). Reasonably similar types of rim are evident in a series of goblets from Tumulus HDN 70/24 at Kadada, where several samples have the similar characteristic of more or less triangular cross-section, with the interior profile of the rim steeply sloping up and out from the top of the body wall, a tapered top and approximately vertical, slightly convex, face to the rim on the exterior. Those specimens closest to the Test Excavation sherd are numbers B, 10, 18 and, in particular, 13, as illustrated by Lenoble (1987a, Pl. V and V1). The cross-section of the upper-most part of the body and rim of the latter vessel is almost identical to that of the Test Excavation sherd, except that the inner face of the rim comprises two facets at slightly different angles, rather than a single face. Burials of the type of Tumulus HDN 70/24 are most common in the 'Meroitique Recent' outside the Meroe cemeteries (Lenoble 1987a, 92).

## *Group 5*

**Group 5c**
Group 5c contains only a base sherd (Type 10:2.1) amongst the diagnostic material. The sherd is in a fabric (F5:2) similar to F5:1 from the surface survey, which clearly differs from the majority of the fabrics distinguished, being considered to be a marl clay of Egyptian origin. Fabric F5:2 differs from F5:1 in fired colour, being a more greenish-buff than the latter, which tends towards a yellowish-green, although this may be due to overfiring.

---

[2] I am indebted to P. J. Rose for this information.

The two fabrics are similar in hardness, type of fracture and in texture, but differ in the range of inclusions each exhibits. F5:2 has more opaque inclusions that may be forms of iron oxide, or possibly grog; has inclusions identified as probable calcite, and feldspars which are not evident in F5:1, whilst lacking the medium and coarse sized quartz grains present in F5:1. For these reasons it has been assigned to a different fabric and group.

The base in Group 5c represents only a fragment of the vessel. There are slight traces of wheelmarks on the interior and, on the assumption that the original stance can be estimated by assuming the wheelmarks to be horizontal, together with the shape of the sherd, it is likely to be from a vessel with a more or less rounded lower body. Both the relative thinness of the vessel wall and its degree of curvature indicate that the original vessel was of small to moderate size, and is likely to have been a closed vessel of the general type exemplified by a small flask or a number of globular juglets known from a later burial in the tomb of Paser at Saqqara (Bourriau and Aston 1985, 54-55, pl. 123-127). The sherd appears slightly eroded on the exterior, and exhibits no evidence for slip, or incised or painted decoration.

**Comparanda**
Given the small size of the sherd and lack of decoration, it can only be dated rather broadly. On the basis of the fabric it appears to be of Upper Egyptian origin, being made of 'Marl K5' (cf. Bourriau and Aston 1985, 52). Vessels in the same fabric as F5.2, having the forms from which the basal and body sherds in this fabric are most likely to belong, from the burial in the tomb of Paser are dated to *c*. 4$^{th}$ century BC. (Bourriau and Aston 1985, 55). Given the fragmentary nature of the Test Excavation material, it cannot be assumed to be dateable only to the same period as the Saqqara examples. Type 10:2.1 and the body sherds in the same fabric are most likely to date to the Late Period more generally, but probably from 25$^{th}$ Dynasty to the Persian Period (P. French, pers. comm., 1997). From this, Group 5c would be datable to essentially the same time range as Group 11, viz Napatan to early Meroitic.

## Group 6

**Group 6b**
A number of diagnostic body sherds, all from site 112.3, were considered to be in a fabric closest to the existing fabric designated F2:6, this being the fabric of the Group 6 material from the Surface Survey collections. The sherds from the Surface Survey are all of small size, indicating that they could have been residual. However, the surfaces were well-preserved, indicating that they had not been subjected to significant erosion. The exterior surfaces exhibited a brownish to buff colour, most likely due to the presence of a self-slip. Interior surfaces are black with a light burnish. All sherds of this type have the same decoration, designated D33.2. This consists of small elliptical to almost diamond-shaped impressions. Paste is often pushed up from the surface of the sherd, to one side of the impression, indicating that it was executed by pressing into the paste with a lateral movement. This could have been done with an implement with a curved edge, such as a shell, or with a fingernail. The similarities to the other decorations in Group 6 (D33 and D33.1, also likely to have been executed with a fingernail), led to the decoration on the Test Excavation sherds being classed with them, although a difference in size required that they be designated as a separate decoration type.

The similarity between the sherds in surface treatment and in decoration indicates that they could have been from a small number of vessels, possibly even from the same vessel. Careful finishing of the interior as well as decoration on the exterior indicates that they are likely to be from an open form, most probably a small bowl, but no conjoining rim sherds were recovered which could enable reconstruction of the original vessel form. The small size of the sherds of this type found means that it is uncertain as to whether the original decorative scheme consisted of multiple rows of the impressions, although there is evidence on some of the sherds for at least two rows.

**Comparanda**
The closest parallels for the decoration on the Surface Survey Group 6 sherds were in the 'Late Neolithic' ceramics from Shaqadud. Most examples of the Late Neolithic (or 'III$^{rd}$ Millennium') types at Shaqadud decorated with fingernail impressions, show multiple rows on the body, although one example is illustrated of a sherd with a single remaining row, the rest of the sherd being blank (Robertson 1991, fig. 7-13). In general, the decoration on the sherds from 112.3 would be consistent with their being a type similar to the 'Nabag Fingernail Impressed: Nabag Variety' at Shaqadud, although the type description mentions only that surfaces were but smoothed before being decorated, without a self slip or burnishing (Robertson 1991, 160-61).

Despite the small size of the sherds their relatively well-preserved surfaces, in contrast to the majority of the existing Group 6 material, may indicate that they are later than the Late Neolithic. Given the context of the sherds, a Christian date would be the most likely. Some decorations comprising elliptical impressions, likely to have been executed with a fingernail, occur amongst the decoration from a Christian context at Soba (Welsby 1991, fig. 125, nos 68-68.4, 70), but these do not have quite the same shape and character as the impressions on the sherds from site 112.4, and are somewhat larger in size.

Rather irregular oval impressed motifs, in a horizontal line, are illustrated by Adams in Style DIII, on the late Christian handmade wares in Lower Nubia (W. Adams 1986, fig. 121, element U-1) but, as illustrated, these impressions do not have the same characteristics as those on the sherds from 112.3. Such a design was not noted amongst the Post-Meroitic whole or reconstructable material from Gabati which may indicate that the Test Excavation sherds are not likely to be of this date. The only comparable motif is on the rim interior of the in-turned bowl associated with Tumulus 119, but not from the burial chamber, which is thought as likely to be of Christian date as Post-Meroitic. Such a motif is not present amongst the impressed decoration so far published from Meroe itself (Shinnie and Bradley 1980, fig. 58), and was not noted amongst the impressed decoration on the

Meroitic ceramics from Gabati. Given that for the sherds under consideration the main parallels for the decoration appear to be either Late Neolithic ('III$^{rd}$ Millennium') or Christian period, and that in terms of fabric, they have greatest similarities with material of the former period, assignment to the 'III$^{rd}$ Millennium/Late Neolithic Group 6b is the most appropriate, given the available sample of material.

## *Group 7*

### Group 7a

The material in Group 7a was amongst a small number of apparently residual prehistoric sherds considered to date from the Mesolithic and Neolithic. Diagnostic sherds from the sites 112.4 and the village below BM 165 were in a similar fabric, which appeared closest to Fabric F2:8, the fabric of Group 7a as defined on the basis of the Surface Survey collections.

The sherd from 112.4 is a sherd from close to the rim of what appears to be a vessel of closed form although insufficient remains to assign it to a specific form. No evidence for slip or burnish remains on either interior or exterior, but the latter exhibits impressed decoration consisting of rows of small approximately square dots, classed as decoration D39.1. The second sherd, from near BM 165, is a body sherd in Fabric 2:8, eroded on the exterior but retaining evidence of light black-burnish on the interior. This exhibits rockerstamp or zigzag decoration, composed of linear impressions rather than comb-impressions. Two rows, slightly overlapping, are preserved on the sherd. The design can be assigned to the Decoration Type D43.2.

In terms of comparative material, D39.1 and D43.2 can be paralleled by material of Neolithic date, both being known from esh-Shaheinab for example (Arkell 1953, pl. 33, fig. 8 and pl. 32, fig. 5, top right, no. 4), the latter motif being particularly closely paralleled by one of the 'curved continuous line zigzag' motifs (Arkell, 1953, pl. 32, fig. 6, right). A similar decoration to D39.1 is known from the earlier Neolithic at Geili, being very similar to the designs executed by the 'alternative pivoting stamp' technique, in the classification of Caneva (1988b, 94-96: fig. 11, no. 8), whilst the same site has designs comparable to D43.2 in some of the sherds decorated by 'rocker stamp, with plain edge' in the same classification. The closest parallel at Geili to the Test Excavation specimen is illustrated by an example with slightly overlapping rows of rocker stamp (Caneva 1988b, 86-88, fig. 5, no. 13). In view of these parallels, in fabric and decoration, this Test Excavation material has been placed in Group 7a, and hence assigned a Neolithic date.

### Group 7c

The material in Group 7c was assigned to this group on the basis of similarities of fabric with those of 7a and 7b. In general, the fabric of Group 7c, F2:16, exhibits a similar range of inclusions to Fabric 2:9 of Group 7b, particularly in the rare opaque black and brown coloured inclusions, probably forms of iron oxide, together with very fine to rounded white inclusions, possibly calcitic material, or feldspars. The overall texture is slightly finer, and the fabric lacks the fine to medium sized calcitic inclusions of Fabric 2:9 and in these respects is more similar to Fabric 2:8 of Group 7a. In fired colour, F2:16 differs from both existing fabrics, being brown rather than grey, with only a thin dark grey zone on the interior. Fine rounded quartz grains are not evident in F2:16 as they are in both Fabrics 2:8 and 2:9, so for these reasons, it was considered to constitute a new fabric type.

The diagnostic sherd in Group 7c appears to be from very close to the rim of the vessel from which it came. The lip of the rim is not present, so that a specific form type cannot be assigned. From the extant profile, although the stance can only be approximately estimated, it is likely that it is from a slightly inturned vessel, more probably of Class 2 (bowls) than of Class 1 (jars). The estimated original form of the profile is closest to Types 1:36 or 2:27 from the Surface Survey material. The surfaces on both the exterior and the interior are a little eroded, but retain evidence of burnishing. Decoration (D63.1) consists of a single remaining row of oblique semi-circular to 'tear-shaped' impressions, somewhat irregularly positioned both horizontally and vertically, perhaps indicating that they were executed individually.

**Comparanda**

The probable rim form in its original state is likely to have been reasonably close to the Group 1, and possibly Group 6, rim profiles in the classification of Caneva. It is also possible the vessel could have had a stance and rim cross-section similar to the right-handmost example of the Group 8 rim profiles from Geili (Caneva 1988b, fig. 4), and to rim type GG at esh-Shaheinab (Arkell 1953, pl. 36). In terms of decoration, the design D63.1 is not so strongly comparable to Neolithic material as is the fabric and profile of the sherd. Two types of elliptical impressions, in relatively broadly-spaced rows, are illustrated from esh-Shaheinab by Arkell (1953, pl. 35, 2 and 3) but the individual impressions are much coarser than those of D63.1, whilst those shown in Figure 4.3 are oriented with the longest diameter horizontal, rather than oblique to the vertical axis of the vessel. The decoration illustrated from Shaheinab closest to the Test Excavation example is that on a rim top, rather than on the body just below the rim, consisting of a row of semicircular impressions oriented obliquely to the sides of the vessel (Arkell 1953, pl. 37, no. 15). A somewhat similar design is illustrated from Kadero, again on a vessel rim top rather than on the body sherd (Chłodnicki 1984, fig. 2, type C4b).

On the basis of fabric and rim form, as far as it can be ascertained, together with presence of similar designs in the Neolithic of the above-mentioned sites, a Neolithic date is proposed for the Group 7c material.

## *Group 8*

### Groups 8b, 8c and 8d

A small number of sherds was recovered during Test Excavation season that could be assigned, mainly on the basis of fabric, to the existing Groups 8c and 8d, both dated from the Surface Survey collection, to the Khartoum Mesolithic. Sherds from the village below BM165 are in the fabric of Groups 8b/c, F2:11, being close in the main characteristics of fired colour, little or no vegetable temper, the presence of medium rounded or angular quartzes and feldspars, together with sparse fine to medium micas. Both surfaces are eroded, but the exterior appeared to have been smoothed. Concerning the decoration, parts of two zigzags executed with a rockerstamp making approximately square impressions remain, together with a portion of each of two rather deeper rectangular to oval impressions placed near the ends of the rockerstamp motifs. This decoration is considered sufficiently similar to D77 from the Surface Survey, to warrant classing it under the same designation.

Diagnostic material from the village below site 166.2 was in a similar fabric to F2:11, but had a moderate proportion of medium-coarse mica flakes, evident in the break and on the surface. This was, on this basis, assigned to F2:12, the fabric of existing Group 8d. Both the interior and exterior are largely eroded, but the exterior retains some areas that have been smoothed. Only a small portion remains of the decoration, which consists of about five sets of two narrow rectangular to elliptical impressions, set one above the other, above which are three irregular broad shallow impressions with another row of three sets of impressions similar to the first row above them. This decoration has impressions similar to D91 in the Surface Survey material and has, therefore, been designated D91.1. Although the impressions in the central row of D91.1 are rather ill-defined, at least one appears to be roughly V-shaped, which would suggest the design has been produced by a rocker-technique.

**Comparanda**
No exact parallel has been found for D77, partly because both examples are in a very fragmentary state. Examples with similar types of rockerstamp decoration are illustrated from the 'Early Khartoum' Mesolithic by Arkell (1949, pl. 73, fig. 1, bottom and top right). The zigzags are quite widely spaced and, as such, would fall within the class of 'rocker technique, spaced zigzags', in the classification of Caneva. Such decorative motifs are known from Mesolithic contexts at a number of sites to the north of Khartoum, such as Kabbashi, et-Temeyim and el-Qala'a (Caneva et al. 1993, 191-194, 231) and also in Mesolithic levels of Shaqadud Midden although in smaller numbers than in the Neolithic levels (Caneva and Marks 1990, fig. 2). Taking into account the fabric and the decoration, albeit fragmentary, it seemed most reasonable to assign this material to a group of Mesolithic date.

Examples of the rocker technique, producing a design similar to D91.1 are apparent also in the Mesolithic assemblages of the sites investigated by Caneva. Such designs can occur as zones of linear decoration within 'wavy-line' patterns, for example at Kabbashi (Caneva et al. 1993, fig. 8, 1 and 5), and at the Khartoum Hospital site (Arkell 1949, pl. 64, 1). As in the case of material from near BM165, the parallels found for the decoration, but mainly the similarities in fabric, warranted placing this diagnostic sherd from 166.2 in Mesolithic Group 8d. The remaining portion of the motif and the shape of the impressions is similar to varieties of spaced zigzags illustrated from Shaqadud (Caneva and Marks 1990, pl. III,

nos. 3, 5 and 8). One example, no. 5, shows broad shallow approximately V-shaped impressions at the pivoting point of each zigzag, as in the Test Excavation specimen. As noted above, the spaced zigzags would be consistent with Mesolithic dating, since they are present in these levels at Shaqadud, although they occur in greatest abundance in the Neolithic levels.

## Group 9

A single body sherd was assigned to this group on the basis of similarity in fabric to F1:12, the fabric of Group 9 as defined through the Surface Survey collection.

## Group 11

Material in Group 11 comprises only a small number of body sherds. These are sufficiently distinctive to allow identification of their general region of origin and date. The sherds are in Fabric F4:2, characterised by a light orange-brown fired colour, fine voids representing vegetable temper and rare black iron oxide fragments. Examples were recovered in small numbers from the Surface Survey and were placed in Group 11 largely on the basis of the fabric, since the sherds were eroded such that the original exterior surfaces were no longer present; only the traces of ribbing on the interior indicated that they were from wheel-made vessels and likely to be from closed forms. Group 11 sherds recovered during the Test Excavations, although all of rather small size, were not greatly eroded, retaining both surfaces. This material revealed that the exterior was generally of a greenish-buff colour, and that most sherds exhibited a form of ribbing on the exterior, consisting of shallow grooves and somewhat flattened ridges, generally distinctly 'squared-off' at the margins. Several sherds also retained evidence of wheelmarks on the interior, similar to the Surface Survey material. In terms of wall thickness, the Group 11 sherds are relatively thin-walled, attaining a maximum general thickness of 5.2mm.

**Comparanda**
The fabric, a marl clay, and method of manufacture of Group 11 sherds, together with the type of ribbing, allows identification of these sherds as coming from jars of Egyptian origin. The body sherds are probably from vessels such as storage jars, made in 'Marl A4 Variant 2' in the terminology of Aston. One likely form from which the Group 11 sherds could have come is a storage jar common in deposits of the Egyptian Third Intermediate Period. Certain small ovoid jars of the same period also have some ribbing on the body and could have been the type of vessel from which the sherds came (Aston 1996, 77, fig. 224e, b). Other probable forms include the jars from the South Tombs at Amarna, in particular types MJ1.1.1 through to MJ4.1.3 in the classification of French. This material is dated to 25$^{th}$ Dynasty (French 1986, fig. 9.17; Aston 1996, 43). Jars in the same tradition were found in Late Period caches of ceramics in the Mortuary Temple of Seti I at Thebes, dating to a later time period than the foregoing examples, being assigned to the Saite Period (Aston 1996, 48; Mysliwiec 1987, 24-26, 54-62).

Although such jars were commonly manufactured during the 25th and 26th Dynasties, it is clear that production of vessels in the same tradition continued through to the 30th Dynasty, giving an overall time-range for the Group 11 sherds of 8th to 4th centuries BC, with the most likely period being 8th to 6th centuries BC (P. J. Rose, pers. comm. 1997; P. G. French pers. comm. 1999). Therefore, the Group 11 material can be dated essentially to the Napatan Period, taking the traditional date of the division between the Napatan and Meroitic periods as about 300 BC (Shinnie 1996, 103).

## Group 13

A number of sherds from sites 118FS2, 112.4 and 153.8 were classed as Fabric F1:13, though having frequent fine to coarse vegetable temper, rare black or dark reddish-brown iron oxides and, in particular, medium to coarse greyish brown rock fragments. On this basis the Test Excavation material was assigned to Group 13. One rim sherd from 112.4 included within Group 13 is extremely eroded, such that neither original surface is present. Sufficient of the vessel remains to indicate that it is most probably a jar with an inward-tapering neck. Although the precise original rim profile and stance could not be determined, it seems to be of a form comparable to type 1:9.6, also from the Test Excavation collection, since a portion of the interior with a surface appearing to be closest to the original indicates that the rim was thickened on the interior in a similar manner to that of 1:9.6, though it is likely that the original wall-thickness was less than for the type specimen of the type. No evidence for the original surface treatment or decoration survives on the sherd.

Body sherds considered to be in Fabric 1:13 included three from site 118FS2 with the same decoration, D13.2, already noted amongst the Surface Survey material. This decoration consists of irregularly placed, broad shallow impressions, probably executed with a fingertip or thumb. The impressions often overlap so that the effect is of an area of oval or irregular depressions separated by relatively sharply-defined ridges. No evidence for slip or burnishing remained on the sherds.

**Comparanda**

Group 13 was considered likely to be of Post-Meroitic date, although it could not be assigned indubitably to this period stylistically because of uncertainty as to the forms of the complete vessels from which the sherds came. One type distinguished could have exhibited characteristics more comparable to late Christian forms than to Post-Meroitic ones, if a greater proportion of the vessel profile had been present (Smith 1996, 193-194). Examination of the main published reports of Post-Meroitic material, especially from Shendi reach (Geus *et al.* 1986; Jaquet-Gordon and Bonnet 1972; Lenoble 1987a, 1987b, 1989, 1991, 1992; Lenoble *et al.* 1994; Edwards 1989) did not provide any clear parallels for the Test Excavation material assigned to Group 13.

The eroded rim sherd considered to have been originally similar to Type 1:9.6, although classed as a jar, could have been from a bowl similar to that from the Post-Meroitic tomb near Berber, illustrated by Lenoble (1991, fig. 4, 10). This exhibits a similar manner of thickening on the interior on the uppermost portion of the vessel and has a relatively straight exterior profile, possessing a stance approximating that estimated for the Test Excavation sherd. Otherwise, despite the uncertainty as to the precise original form of the Test Excavation specimen, enough remains to indicate that it does not have profile of the majority of the characteristic Post-Meroitic bowls, such as those illustrated by Lenoble in the same article or by Edwards (1991, pl. I-III; IX-X) nor is it like the necks and rims of characteristic Post-Meroitic globular jars. On none of the examples illustrated is there a rusticated area of decoration comparable to D31.2.

In comparison with the Post-Meroitic material from Gabati, the Test Excavation sherd could be similar to bowl 2116 (Smith 1998c, fig. 6.27), which is also relatively thick on the interior and is of similar diameter, though possibly the stance of the two vessels is not quite the same. Gabati does provide a parallel to D31.2 on the lower body and base of vessel 643/1, which has a similar type of rustication, though somewhat heavier than the Test Excavation examples of the decoration (Smith 1998c, fig. 6.29). The Gabati vessel is not from a closed context and may be of Post-Meroitic or medieval date. Other likely parallels for the original form of the sherd comparable to Type 1:9.6 occur in a Christian context at Soba; these include types 84L, 85L and 97L. Overlapping impressions of a type similar to D21.2 also occur as Decoration 71 at Soba, although the type example illustrated indicates a single row rather than an area of such impress (Welsby 1991, fig. 125). Such parallels as have been found for the material in Fabric F1:13 together with the relatively fragmentary state of the material from both Surface Survey and Test Excavation means that Group 13 as presently constituted may only be dated rather broadly from the Post-Meroitic to Christian periods.

Three further body sherds have been classed as Fabric F1:13 and have, on this basis, been included in Group 13. None retains evidence for slip or burnish, but two are mat-impressed, each having one of the main types distinguished, MI1 and MI2 and one has remains of decoration, in its present state similar to D4. The third exhibits only a couple of thin incised lines, presumably a fragment of a larger motif. These sherds are not clearly diagnostic in terms of period and so do not aid the dating of the group. As noted above under Groups 2g and 3, mat impressions are present on ceramics made over a long period, including the Meroitic, Post-Meroitic and Christian. However, the remains of a motif on the third F1:13 sherd is too fragmentary to be diagnostic.

## Group 15

**Group 15a/15b**

Groups 15a and 15b both contain material in Fabric F2:2 and will be considered together. The two groups have been distinguished on the basis of evidence for vessel surface treatment and decoration. One sherd placed in Group 15a was recovered from site 101.4. This sherd did not have sufficient rim remaining to warrant assigning it to a specific type, but it was considered to be probably a sherd from a bowl (Class 2). An approximate diameter of 18 cm was estimated for the original vessel. In terms of surface treatment, the exterior was clearly smoothed, but the interior was too eroded for the surface treatment to be determined. The sherd retained no evidence of slip or any form of decoration.

Two further sherds in Fabric F2:2 were found at site 112.3. These comprise body sherds with a somewhat eroded exterior surface, but probably originally having a smooth brown self-slip. The interiors of the sherds are better preserved, exhibiting a well-smoothed or lightly-burnished black surface. Decoration on these sherds consists of small elliptical to almost diamond-shaped impressions, often with paste pushed up to one side of the impression, which appear to have been made with an implement such as a shell, or with a fingernail. This decoration is essentially the same as that on sherds in Group 6b and is also designated D33.2.

**Comparanda**
The closest parallels for the Group 15 sherds are amongst the same material as that adduced for Group 6b, viz the decorations composed of elliptical impressions, types 68-68.4 and 70, at Soba (Welsby 1991, fig. 125), Adams' Style DIII, on the late Christian handmade wares in Lower Nubia (W. Adams 1986, fig. 121, element U-1) and the decoration on 'Nabag Fingernail Impressed: Nabag Variety' ceramics at Shaqadud (Robertson 1991, 160-61). The small size of the sherds concerned indicates that Group 15 is most likely to represent residual pottery from a period earlier than that of the main occupation at Site 112.3 from which ceramics are present. The fabric of Group 15 is closely related to that of Group 6b and to that of Group 22a-c (see below). Although the decoration cannot be attributed solely to one period, the presence of D33.2 taken in conjunction with the fabric, indicates that Group 15a and 15b can most probably be assigned to the later Neolithic period.

*Group 16*

**Groups 16a-c**
All of the groups designated 16a-c comprise material in Fabric F1:6 and so will be treated together. Group 16a contains sherds having an unslipped but sometimes smoothed surface on either exterior, interior or, in some cases, on both surfaces. Sherds assigned to Group 16b have a mid-dark brown to reddish-brown exterior surface colour, considered most likely to be a self-slip rather than a deliberately applied and coloured slip, whilst sherds in Group 16c, of which only a very few were found, have a red exterior slip and no decoration.

A number of new types were distinguished within the Test Excavation material assigned to Groups 16a-c, the majority in Group 16a. The first is Type 1:8.1, a jar with an inward-sloping neck, having an irregular body wall cross-section which, in the portion extant, is relatively thin at the part furthest from the rim, but which then thickens to a maximum finally becoming thinner again as it approaches the rim. The thickest part is probably at the junction of the neck with the vessel body. The rim itself is rounded on the interior margin and approximately flat on the top. It projects on the exterior to form a slight ledge, being again moderately rounded on the exterior margin and then sloping at an acute angle to merge with the exterior of the neck.

Type 1:14.2 is the second form recognised within Group 16a, comprising another jar form with inward-sloping sides, having a slightly concave exterior profile. The extant part of the vessel is somewhat asymmetrical, such that one section of the circumference is slightly convex just below the rim, whilst another section has a distinct step in the exterior profile a short distance below the rim, forming a shallow groove around part of the vessel. On the interior, the vessel wall has a broad shallow groove at about the same level as the step on the exterior. Above this groove, the vessel wall continues almost vertically to the flat rim, slightly rounded of on both interior and exterior, but otherwise not modelled. The vessel is smoothed on the exterior only and is undecorated.

The third form recognised within Group 16a is Type 1:16.1, classed as a jar because it is inturned, although it has a diameter at the mouth somewhat greater than the majority of the other types in Class 1. Type 1:16.1 has a form occurring very rarely amongst the Survey and Test Excavation material, in that the cross-section of the vessel wall increases uniformly in the portion extant, from the point furthest from the rim, up to the rim itself. The latter is further thickened on the interior and quite distinctly rounded-off, there being a broad flat top to the rim. The exterior of the rim is also slightly thickened and similarly rounded-off, to form a shallow groove just below the rim similar to the exterior of Type 1:14.2, although on a larger scale.

The final new type within these groups is 2:10.2, recognised only in Group 16b. This type comprises another broad open bowl, the walls of the extant portion being very slightly everted. The vessel wall is of moderate thickness, and is of approximately equal width from the break until just below the rim, where it thins out slightly and continues into a tapered rim. The vessel is self-slipped on the interior and exterior surfaces.

Group 16a also includes an example of Type 2:7, noted previously amongst the Surface Survey material in Groups 1a-c. This type is a broad open bowl, relatively thick-walled, with a flat rim-top, rounded on the interior and squared off on the exterior. It is distinctive in that in the cross-section the overall stance of the extant portion is approximately vertical, and the vessel wall increases in thickness quite noticeably from a short distance below the rim. In the Group 16a example the exterior surface is untreated whilst the interior is smoothed. There is no decoration evident on the extant portion of the vessel.

In terms of decoration, the types in Groups 16a and 16b exhibit a fairly restricted range of motifs. Both Types 1:8.1 and 1:16.1 are decorated with incised motifs, being variations on the horizontal bands of cross-hatching designated D3.1-D3.5. There is some variation in the style of execution of the motifs; that on Type 1:8.1 being more irregular than that on 1:16.1. The overall visual impression of the cross-hatching was thought to be sufficiently similar to warrant assigning the decoration on both types to D3.1. The rim of Type 2:10.2 has decoration consisting of approximately elliptical impressions of unequal length, placed at an oblique angle to the diameter of the vessel (see D10.6). Body sherds exhibit mat impressions, mainly close to MI1, and several types of zigzag motif, including D1.1 and a new variation, D1.5. An elliptical impression with a line across it forms D12.7.

**Comparanda**
The types in Groups 16a-c are generally comparable to those in Groups 1a-c, Type 1:8.1 fitting in with the range of types from the Surface Survey, having moderately inward-sloping necks and slight projections of the rim on the exterior. In particular, they are similar to Types 1:5 and 1:8, and, to a lesser extent, to 1:7 (Smith 1996, fig. 1). Type 2:10 has similarities with both Types 2:9.1 and 2:10; the tapering rim is closest to the former, but the overall stance is closest to 2:10 and, for this reason, the Test Excavation specimen was placed close to this type in the classification. As noted, both the other bowl type and two of the decorations in Groups 16a-c can be assigned to form type and decoration types first recognised amongst the Group 1a-c material. Only Type 1:14.2 and Decoration D10.6 do not fit closely to the existing types. Together with the fabric, which tends to be somewhat harder-fired than those of Groups 1a-c and to be characterised by sub-angular grains of brownish-grey colour, the Type 1:14.2 and Decoration 10.6 have led to the placing of this material in a series of groups separate from 1a-c. However, it is considered that the similarities in form and decoration indicate that the groups can be dated to the same period- i.e. Late or Terminal Christian to early Islamic.

A dating to approximately this time-span is partly supported by the $^{14}$C dating of the structure at site 155.4. Sherds assigned to Group 16a were recovered from the same excavation unit as a charcoal sample for which the most probable calibrated date at 2o is AD 1480-1680. This indicates a date for at least part of Group 16a-c in the later part of the range suggested by comparisons on stylistic grounds, probably extending into the later Islamic period.

## *Group 17*

**Groups 17a/b and 18a/b**
Groups 17a and 17b and 18a and 18b will be dealt with together for although they are separated on the basis of some differences in fabric, they are similar in surface treatment and exhibit similarities in method and style of decoration. Group 17a/b comprises material in Fabric F2:3 whereas Group 18a/b contains material in Fabric F2:4. Sherds in Groups 17a and 18a may be commented on but briefly. Only a small number of unslipped and undecorated sherds have been found in the two fabrics. These appear to be from vessels of moderate wall-thickness, generally c. 7.5mm wide. Examples of Groups 17a and 18a are often well smoothed on the exterior. This contrasts with the interior, which has a rough texture. It is not clear to what extent this represents the original surface texture, since in some cases, the interior appears to be flaking off, so that the roughness is likely to be due to erosion of the surface. However, not all examples are clearly flaking, so that the texture indicates that the interior was probably less-well finished that the exterior and, consequently, that the sherds are from vessels of closed form.

Material in Group 17b (Fabric F2:3) and in Group 18b (Fabric F2:4) has similar surface treatment, consisting of a reddish-brown burnished slip on the exterior. A few plain sherds were found, but most assigned to these groups were decorated with impressed motifs. Areas of decoration within the motifs were probably executed with a comb, whilst other impressions, including those of larger size and more widely and unevenly spaced than the former, appear more likely to have been executed with an instrument having a single point such as the end of a stick or a bone.

As in the case of the unslipped sherds, those in Groups 17a and 18a have a somewhat rough texture on the interior, which may be due to erosion of the surface, although no examples were found with a clearly flaking surface. It is possible that such a texture could have been originally formed by scraping the vessel interior when in a partly-dried state, to complete the shaping process. If the present surface is indicative of the original, it would, in conjunction with the more elaborately finished exterior, imply that these sherds were also from vessels having closed forms.

Whilst no rim sherds have been found in Groups 17b and 18b, so that no more specific indications of vessel form can be given, several decoration types, all executed in a similar manner, have been noted on sherds assigned to these groups. The motifs, although fragmentary in some cases, together with method of execution are distinctive of these two groups. The first decoration comprises a single zigzag approximately 5mm wide, which is defined by its being a plain red zone in between two registers of comb impressions. These zones consist of a broad area with a zigzag edge on one side, and on the other a similar area, of which only an intermittent portion of the zigzag edge remains (see Decoration D98). The second decoration comprises two comb impressed lines c. 9mm apart, with an area of comb impression having a zigzag edge, extending at right-angles from one end of the longest remaining of the two lines (D99).

Decoration D100 consists of a line of relatively large ovoid impressions, somewhat widely-spaced. Extending from one side of the line are the remains of two bands of comb impressed decoration, separated by c. 7 mm. One is incomplete, but appears similar to the second, which consists of a band of comb impressions c. 14mm wide, having an oblique end, touching the horizontal line at one point only. One edge of the band is formed by a line of comb impressions of diminishing diameter, which extends up to the horizontal line. The next decoration noted within these groups consists of an approximately triangular area of comb impressions of small to moderate size, bounded on the base by a line of larger ovoid impressions. Below a broad band lacking comb impressions, in which the general red background colour of the surface shows through, a second, slightly oblique comb impressed line extends for the width of the triangular area, then turns downwards to form a third line nearly at right angles. Three large, irregularly-spaced, oval impressions occur within the space defined by the second and third lines (D101).

The final decoration in the two Groups, D102, comprises an area of comb-impressions applied in a rather irregular manner, using a rocker-technique, but leaving some gaps between adjacent pairs of rocker-stamped lines. The original form of the motif is uncertain, but it appears to have formed one or more triangular areas, perhaps to create a zigzag edge as in D98. The longest remaining edge of the area of comb impressions is bounded by a continuous comb-impressed line, with two rows of smaller comb-impressed dots below and parallel to it for approximately one third of its length

**Comparanda**
It has not proved possible to find exact parallels for the decorative motifs on the sherds in Groups 17b and 18b, partly due to the rather fragmentary nature of the sherds assigned to these groups. Material noted having similarities in some aspects to the Test Excavation sherds has been dated to more than a single period, so that only a date range can be suggested for the two groups.

Decoration executed in a similar manner to Decorations D98-102 can be related to a general tradition of pottery decoration in Nubia extending back to C-Group times, in Lower Nubia, but also being well-known from the Meroitic and Post-Meroitic Periods through into the early Christian Period. Such decoration occurs in Adams' 'Meroitic Domestic Style' (Style DI) being executed in comb impressions rather than the more common incised lines (W. Adams 1986, 238).

Zigzags consisting of a plain surface defined by surrounding areas of comb impressions, as in D98, are known from the Meroitic, for example on a cup from the Faras Cemetery,[3] dated to 1$^{st}$ to 2$^{nd}$ century AD (Griffith 1925, 145, 157, pl. XLI, 21). A similar motif is present, on a somewhat larger scale, on a jar or 'petite bouteille' from a Meroitic grave at Kadada (Lenoble 1995, pl. III, 21). In this example, in addition to the difference in size of the decoration, the method of decoration is not quite the same as in D98, since the triangles defining the plain zigzag are more regularly executed, with the lines of comb impressions forming the triangle all originating at the apex, in contrast to the lines of comb impressions in D98, which are sometimes oblique.

There are some similarities in terms of style and method of execution of Decorations D98-102 with material from the habitation site at Abu Geili. Decoration on much of the pottery at this site consists of geometric shapes composed of areas of comb impressions, sometimes irregular in execution. Some decoration types exhibit arched motifs formed by a plain surface between areas of comb impressions. No exact parallels for the specific motifs comprising D98-102 could be recognised amongst the published Abu Geili or Saqadi material (Crawford and Addison 1951, pls XXXVA, 8, 9, XXXVIB, 7, LXV-LXVI), so that these sites can only act as an example of a generally similar tradition of decoration. They may give some support to the dating of the Begrawiya-Atbara sherds, since the material from Abu Geili in this tradition is associated with some pottery and other objects clearly of later Meroitic date (Crawford and Addison 1951, 40, 81-84), although it cannot necessarily be assumed that particular ceramic traditions were contemporaneous in the Shendi Reach and the southern Gezira.

Sherds exhibiting a similar method of decoration have been recovered from Soba. There are a number of decoration types at this site comprising triangular panels made up of small impressed dots, similar to the areas of comb impressions in D101 and D102. The examples at Soba closest to the Test Excavation Decorations are 135.1, 135.3, 135.9, 138.6, 138.8, 138.85, 138.9 and 141.2 (Welsby 1998b, figs 64, 65). However, several of these decorations at Soba have one or more sides of the triangle bounded by incised lines, which do not occur on the sherds from the Test Excavations. In terms of the execution of the motif in D102 there is a further parallel with some material at Soba: both have the comb-impressions applied somewhat unevenly, so that there are areas within a general zone of comb-impressions in which the background surface shows through. This is the case, for example, with Soba decoration 167.4 (Welsby 1998b, fig. 67), although the motif is not the same as D102.

The zigzag motif D98 is, in general form, paralleled by decoration 135.8 at Soba. There is a difference in that in D98 the zigzag is defined by the absence of impressions, being formed by the plain background slip colour showing between two zones of impressed dots. In Soba decoration 135.8, the zigzag line is itself formed of impressed dots, although there is some similarity in that the latter decoration has an intermediate zone between two zigzag comb-impressed lines such that there is also a zigzag defined by the absence of impressed dots, in a similar manner to that in D98 (Welsby 1998b, fig. 65). No close parallels for the Decorations D99, D100, or D101 could be found amongst the Soba material in terms of motif, although both D100 and D101 are similar in method of execution particularly to Soba type 141.2, with the use of irregular areas of small dots associated with areas or lines of larger dots which tend to be approximately oval in shape.

Given that the decoration D98 can be paralleled in terms of motif and a similar method of execution on pottery of later Meroitic date and that a similar technique of decoration existed at Abu Geili in association with later Meroitic material, such a date could be suggested for Groups 17b and 18b. However, the decoration types from Soba with similarities to D98-D102 are considered to belong to a Post-Meroitic decorative style (Welsby 1998b, 119). The Group 17b and 18b sherds are in the fill of Tumulus 1 at Site 101.4, near the foot of Jebel Ardeb, and are associated with sherds of Groups 5 and 11, which are considered to be of Napatan date. Since all the pottery in this tomb was very fragmentary, it seems probable that it represents residual material incorporated into the fill. From the foregoing, a date for Groups 17b and 18b in the late Meroitic to Post-Meroitic periods is the most likely.

## Group 19

Only a small number of sherds has been placed in Group 19, which is one that has been defined largely on the basis of a distinctive fabric (F2:17) and to which no forms could be assigned. One sherd is unslipped, whilst the other

---
[3] The cup is in the Ashmolean Museum, Oxford; Registration no. 1912.366. I am indebted to Dr H Whitehouse for this information.

has a reddish-brown burnish on the exterior. The sherds come from a large vessel, and from a small, relatively thin-walled one. Both of these appear to be closed forms, as indicated by the finer finish of the exterior surfaces. However, the interior surface of the sherd from the thin-walled vessel is somewhat eroded, so that the original surface treatment cannot be determined with certainty. It is possible, therefore, that this sherd is from an open bowl, rather than an inturned bowl or a jar.

The sherds assigned to the group do not exhibit sufficient distinctive features to allow it to be dated through its own attributes. Material assigned to the group comes only from site 112.3, where it is associated with the portion of painted jar dateable to the Classic Christian period. On these grounds, Group 19 may also be dated to the Christian period, unless it represents residual material from an earlier period. The latter is a possibility because its presence in such small quantities, although the condition of the sherd surface does not indicate that the material was subject to much weathering before incorporation into the deposits.

## Group 20

Only two examples assignable to Group 20 were found in the Test Excavations, both from site 101.4. These are sherds from handmade vessels in Fabric 2:7, which is characterised by a dense texture, the lack of identifiable voids from burnt-out vegetable temper, and the presence of abundant fine to medium probable feldspar grains. In addition to the fabric type specimen, which is a body sherd with an eroded exterior and a smooth interior, Group 20 contains Type 2:24.1, decorated with D31.1. This type is from a broad open bowl, with an approximately upright stance. The vessel wall of the extant portion is of quite even thickness, widening on the interior slightly at the point furthest from the rim. The latter is moderately rounded-off on both the interior and exterior, to form an almost flat top, having a slight ridge at the midline. Concerning the surface treatment, that on the exterior is uncertain because the surface is eroded. On the least eroded portion, just below rim, there is definite evidence for smoothing and a few patches which appear to be burnished. On the interior, which is better preserved, the surface is black in colour. It exhibits parallel oblique striae, presumably from scraping, and still retains some areas of burnishing. This may indicate that the interior of the vessel was first scraped to carry out final shaping of the bowl, subsequently being burnished relatively lightly, the which process resulted only in the partial removal of the scraping marks.

It is in the decoration that Type 2:24.1 is distinctive amongst the Test Excavation material. The design consists of a horizontal row of triangular elements, virtually all in alignment, but with an occasional one dropping below the majority. Below the line of triangles, there is a zone of oblique parallel lines, made up of semi-circular impressions, which decrease in size from the body of the vessel towards the rim (D31.1). These impressions may have been made with a form of comb, or a cord.

## Comparanda

Fabrics lacking in vegetable temper are characteristic of the Mesolithic and Neolithic periods, and have also been noted in some Medieval wares recovered during the Surface Survey and in excavations at Gabati. This material is stylistically similar to sherds from Soba that have been compared to ceramics of the 'Funj' Period at Abu Geili and Dar el-Mek, although the Gabati sherds also appear similar to ceramics from the Eastern Desert associated with tumuli of which two have been dated to $7^{th}$-$8^{th}$ centuries AD (Edwards 1998e, 183; Welsby 1991, 206, 213; Sadr et al. 1995, 212-221).

The Test Excavation sherds seem unlikely to be as late as the Fung Period, in terms of their context in Tumulus 1 of Site 101.4, which otherwise contains material considered to date to the Napatan, Meroitic and possibly Post-Meroitic Periods. Furthermore, the decoration seems quite different in method of execution and motifs from the 'Funj' wares, such as those illustrated from Abu Geili, Gabati or the Eastern Desert (Crawford and Addison 1951, pl. XXXII; Smith 1998c, fig. 6.32; Sadr et al. 1995, figs 25-26).

Comparison with material of Neolithic date shows that sherds with quite similar rim profiles and stance are known from both Geili and Kadero, the Test Excavation example being closest to rim form Type 2a, and being perhaps from a vessel of shape type III or IV, illustrated from the latter site (Caneva 1988b, figs 4, 5; Chłodnicki 1984, fig. 1). Both the Geili and Kadero rim profiles have the characteristics of moderately rounded-off flat rim top, differing little in width from thickness of the vessel wall below. Despite the similarity in rim profile, no clear parallel for the decoration was found. The shapes of the impressions are reasonably well paralleled in an example of Late Neolithic pottery from Geili, but forming panels of horizontal rather than oblique lines (Caneva 1988c, fig. 14, 2). Individual, apparently rectangular, elements forming oblique lines are illustrated by Chłodnicki in several motifs (Chłodicki 1984, fig. 2, B-02, G1, G2, B3-02), although the lines are generally illustrated as being less oblique than in the decoration of the Test Excavation example. Overall, the latter is most similar to Chłodnicki's motif B-02. However, the single row of triangular impressions is absent from this, so that the parallel is not exact.

An example of comb-impressed oblique lines on the vessel body, extending to a comparable distance below the rim as in D33.1, occurs in the Neolithic of Shaqadud. The design as a whole in this case differs from D33.1 in that above the zone of oblique lines there are two horizontal comb-impressed lines rather than single line of relatively large triangular impressions (Caneva and Marks 1990, pl. IX, 3). It is not clear whether the Shaqadud decoration and D33.1 are executed in the same way, since in the former, the individual impressions appear to be all the same size, rather than decreasing in size towards the rim, as in the latter. The closest similarity to D33.1 found occasionally in the Late Neolithic or 'III$^{rd}$ Millennium' ceramics from Shaqadud, in the 'Wisal Comb-Impressed, Wisal variety' in the classification of Robertson (1991, 157, 160). One type has approximately parallel oblique comb-impressed lines, the impressions forming which decrease in size at the ends of the lines. Above the zone of oblique lines are triangular impressions of similar relative size to those in D33.1. The Shaqadud specimen does differ in that there are at least four rows of triangles, and they are inverted, in contrast to the single row on D33.1 (Robertson 1991, fig. 7-12, b). Type 2:24.1 would be consistent with Shaqadud material according to descriptions of the types of paste, and the forms of vessels occurring in Wisal Comb-Impressed, and in terms of surface treatment, since Wisal Comb-Impressed can have a burnished appearance. However, no mention is made of prominent wiping or scraping marks on the interior of Wisal Comb-Impressed, and it is likely that the method of execution of motifs does differ, given that in Wisal Comb-Impressed the comb impressions are usually rectangular in shape and closely spaced, rather than semi-circular and somewhat widely spaced as in D33.1. In general, although an exact parallel for the Group 20 type and its decoration has not been identified, the closest similarities to vessel form and Decoration appear to be in the Neolithic, as broadly defined, including the Late Neolithic or 'IIIrd Millennium' time period, as defined on basis of material from Shaqadud.

## *Group 21*

Group 21 comprises sherds in Fabric 1:23. Only two types from the Test Excavation collection have been placed in this group, comprising two rim sherds of different form. A third sherd was identified as being likely to have come from an inturned vessel, probably a jar (Class 1), but it is too small to be assignable to a specific type. The remainder of the material assigned to the group consists of undecorated body sherds. The two types in Group 21 are handmade, with a brown self-slip or wash on both exterior and interior. One rim sherd appears to be lightly burnished on both sides.

One rim sherd is only slightly inturned, and could be from either a bowl or a jar. The lowest part of the body present appears to curve outwards at a slightly more acute angle to the horizontal than does the uppermost part of the wall immediately below the rim. This, together with the more finely-finished exterior surface and the presence of striae, probably resulting from wiping the surface, on the interior implies that the sherd is more likely to be from the neck of a jar than to be from a bowl. It has, accordingly, been assigned to Class 1, Form 42. Type 1:42 is relatively thin-walled, having a maximum thickness of *c.* 6.5 mm. The wall progresses without a marked transition into a rim thinned on the interior to form a tapered-off rim. This is not formed so as to exhibit a sharply-pointed cross-section, but one slightly flattened at the lip of the rim. The vessel, as represented by the sherd, appears to have been somewhat irregular in execution, so than on point on the circumference of the lip appears more rounded-off than at another, where it is more squared-off.

Type 1:43 has a relatively similar form to that of 1:42, both being inturned jars of comparable diameter, and having quite a thin body wall. In comparison with Type 1:42, Type 1:43 appears likely to have been of a more globular shape, since the exterior profile is noticeably convex, and a projection of the profile on the basis of the extant portion would produce an approximately semi-circular form. In Type 1:42, the profile of the extant part is slightly out-turned and appears more appropriate for a form possessing a more-or-less distinct shoulder. The rim of Type 1:43 is quite distinctive. The vessel wall is thinned at the rim to produce a tapered top, although one more asymmetric than the type example of a tapered rim illustrated by W. Adams (1986, Fig. 95). The inner face of the rim is smoothed off almost vertically, then passes into a slightly concave section as the vessel wall thickens out. On the corresponding part of the exterior, the vessel wall is thinned to give a concave section immediately below the rim top, which then changes to a convex profile as it merges into the main body wall. This type has no decoration, but both interior and exterior surfaces are smooth and have a matte brown colour, which is probably a self-slip, rather than due to deliberate coating with a separate brown-coloured slip.

It has not been possible to find close parallels for the two types in Group 21, largely because of the small size of the sherds concerned. The majority of the material assigned to the group was found at site 101.4, just to the north of Meroe town-site, in association with sherds of amphorae of Napatan date. In view of this, Group 21 may be considered to date to the Napatan or Meroitic periods, probably to the latter of the two periods.

## *Group 22*

### Group 22a/b/c/d

Groups 22a, 22b and 22c contain material in Fabric F2:6. The groups are distinguished by the evidence available for surface treatment. Group 22a is unslipped, but may be smoothed, 22b has a lightly burnished red surface, 22c is smooth, with the remains of a brown burnish, whilst 22d has traces of white material, possibly a slip, remaining. Sherds in Group 22a, exhibit two types of decoration. The first, D103, consists of vertical or horizontal rows of semi-circular impressions. In some cases a portion of the surface adjacent to the impression is in relief, a result of clay being pushed up when the impression was formed. Such examples are likely to have been made with a fingernail rather than an implement. This decorative motif is very similar to D33.2, but has been distinguished from it because the impressions appear more angular than those of D33.2, so that it is not definite that it was produced by the same technique. The second type comprises D104 and D105 exhibiting moderately-spaced rocker-stamp impressions, which form curved lines in the former and straight lines in the latter. The single sherd in Group 22b exhibits the semi-circular impressions of D103.

The sherds in Group 22c and 22d have been assigned to Type 2:35. Only the difference in apparent surface treatment has led to their being subdivided into Groups 22c and 22d. The Type 2.35 comprises a bowl of *c.* 10 cm diameter at the mouth. From the small portion extant, it appears likely to be a vessel of approximately globular form. It is characterised by very thin walls, of only about 3-3.5mm thickness. The vessel wall, with a convex profile, is thinned only slightly from the maximum wall thickness to the rim, which is essentially rounded, but which has a slight ridge at its mid-line; it is not otherwise modelled. Neither sherd bears any trace of painted or incised decoration.

**Comparanda**

The dating of Groups 22a-d has been uncertain, since the fabric resembles that of material in Group 6a and 6b, considered likely Late Neolithic in date. Decoration generally similar to D103 is present amongst the Late Neolithic ceramics at Shaqadud, in the 'Nabag Fingernail Impressed' type and comb-impressed decoration also occurs (Robertson 1991, fig. 7-13, b-d). However, the rockerstamp decoration D104 is more characteristic of the earlier Neolithic Period, and could be included within D42 in the Surface Survey material (Smith 1996, pl. 13, 2); none of the comb-impressed motifs illustrated from Shaqadud are clearly the same as D103 and D104 (Robertson 1991, figs 7-4, 7-5, 7-9). Semi-circular impressions of exactly the same type as D103 were not encountered amongst the Surface Survey collections, but decoration presenting a similar appearance is present on a bowl from el-Ghaba considered to be contemporary with the 'Khartoum' Neolithic (Geus 1986, 67, fig. 3). The decoration classed as D105 is not exactly paralleled amongst the material illustrated from Geili, but appears similar to some portions of designs executed in 'rockerstamp', with evenly-spaced packed dots (Caneva 1988b, 88-91; fig. 8, nos 1 and 9).

It has not proved possible to locate exact parallels for Type 2:35. The sherd in Group 22c with apparent traces of a brown burnish would be quite unusual for the majority of Meroitic to Christian period material if brown were the original colour intended, since most wares of these periods are either red or black-burnished. Thin-walled bowls are known from the Christian period, having been recovered from Soba (Welsby 1991, 193). Of the forms published, only 3M (Welsby 1991, fig. 103) has a rim form reasonably similar to Type 2:35. However, the Soba example is thickened slightly on the interior just below the rim, thus presenting a different cross-section to the former. The overall form of the Soba vessels is more inturned than the Test Excavation form, and their exterior profile is less convex. The surface treatments of the Soba Class M vessels are usually red or black burnished, slipped or painted, whilst incised decoration is known on a few examples (Welsby 1991, 193). Although the Group 22d specimen could be comparable to a slipped Class M vessel, overall there is insufficient similarity between the Test Excavation and the Soba sherds to allow the former to be identified as examples of the latter.

Both the sites from which Group 22c and 22d material came had evidence of Napatan or Meroitic-period ceramics either from the Surface Survey or Test Excavations. The thinness of the vessel walls and the presence of a white surface would indicate that parallels might be found amongst the Meroitic 'Family M' wares. Groups 22c and 22d cannot be identified as Family M, since the fabric is completely different from that of the latter wares, being more suitably described as 'siltware'. Such a ware, having thin walls, in one case a white surface colouring, might be an imitation of the Family M wares in a siltware fabric.

However, the form is not common in the Family M wares, the most similar bowl form amongst those illustrated by Adams as being typical of Ware W30 in Family M (W. Adams 1986, Fig. 255, Form C, 29) has more upright sides, being less inturned. It appears that a form similar to Type 2:35 is not amongst those characteristic of the Post-Meroitic ceramics of the area, as known from Gabati, situated only some 6 km to the south of Site 170.1. From the above, it appears that the type is not clearly close to the thin-walled vessels known from the Christian period in the southern region, nor is it closely allied with the Meroitic material (P. J. Rose, pers. comm. 1996), so that an earlier date would be reasonable.

Given that the fabric is similar to Fabrics 2:6 and 2:2 of the existing groups assigned a Late Neolithic date, a parallel might be expected in material of this date. Form Type 2:35 is not close to any of the Surface Survey rim sherds placed in Group 6, and differs from the examples of rims illustrated from Shaqadud (Robertson and Marks 1989, Fig. 5). Whilst 'incurving walled' bowls with walls as thin as those possessed by Type 2:35 are described verbally in the main report on the Shaqadud Late Neolithic ceramics under the 'Bamia Black' and 'Samr Zoned Incised' types (Robertson 1991, 130-131, 153-154), the illustrations are of the decorations alone, so an exact comparison of forms could not be made. It is sufficient to indicate that Test Excavation Type 2:35 could be within the range of variation exhibited by the Shaqadud pottery. The type is not closely paralleled amongst the range of rim forms illustrated from Geili (Caneva 1988b, fig. 4) nor by those illustrated by Mohammed-Ali (1982, figs 36-40).

The main difficulty with Groups 22a-d is that the fabric is close to that the 'IIIrd Millennium'/'Late Neolithic' material but the form and decoration within these groups cannot be exclusively attributed to the 'Late Neolithic', even though Form 2:35 appears uncharacteristic of ceramics from the Meroitic to Christian Periods. Hence, it is possible only to suggest a relatively broad date for these two groups. On the basis of both fabric and decoration, the most probable time-range would be from 'Khartoum' Neolithic to 'Late Neolithic', although the possibility of a date in the protohistoric period cannot be definitely excluded.

## Group 23

A single sherd from site 112.3 was assigned to Group 23. It has a fabric classed as F1:5. The similarities are mainly in density, the presence of abundant relatively fine vegetable temper and in the presence of fine to medium calcitic inclusions, and in the possession of a light brown to reddish-brown exterior zone. The Group 23 sherd is thinner-walled than the fabric type specimen, so it does not have a broad dark grey zone. This fabric was first identified amongst the Surface Survey material, and is the fabric of Group 2a. The sherd is, however, different from Group 2a, and has been assigned a new group designation on the basis of surface treatment and colouring. The interior is eroded, so that the presence of any slip originally present cannot be determined. On the exterior, there is a yellow slip, probably originally matte rather than polished. There is no remaining evidence for incised or painted decoration. The sherd exhibits a further feature relatively unusual amongst the Test Excavation ceramics, in that it has a single hole through it, appearing to have been executed post-firing. This hole could have been either for the suspension of the vessel, for the attachment of a lid (possibly of basketry, as used at present in Sudan) or could have been one of a pair drilled for the repair of the vessel.

The sherd is not otherwise diagnostic, so has been tentatively assigned to a period largely through the similarity of its fabric to Group 2a and its slip colour. Yellow-slipped wares are not common amongst Meroitic ceramics, a plain white colour being more usual. At Meroe, yellow-slipped wares were not noted in the main classification of ceramics from the site and nor are yellow-slipped vessels mentioned in the register of whole pots (Shinnie and Bradley 1980, 152, 129-151). Yellow-slipped vessels were manufactured more consistently in the Christian period of Lower Nubia; in earlier periods, including the Meroitic, they are considered to have been a 'rather uncommon variant of white' (W. Adams 1986, 201-2). A hard grey fabric was produced in the former period and area, for instance being particularly associated with Classic Christian wares from the manufacturing site at Faras (cf. W. Adams 1986, 490), although also known from succeeding wares (W. Adams 1986, 497, 504). Given its characteristics and context, it is most likely to be of Christian date.

## Conclusion: Summary of featured sherds

The nature of the pottery from the Test Excavations, excluding Gabati, as indicated by the featured material can be summarised by barcharts, giving the percentages by count of the various groups. These are shown in Figures 4.5-4.11. The first barchart gives the percentages of the groups in the whole set of featured sherds, calculated over all nine sites, whereas the subsequent charts show the percentages of each group within the collection of featured sherds from each site from which more than a few such sherds were recovered.

From Figure 4.5 it is clear that the overall assemblage is dominated by material of Groups 1a-j and 16a-b, especially that of Group 1b, which reaches over 30% of the collection. This demonstrates the predominance of the material considered to be of later Christian-early Islamic date within the ceramics from the sites other than Gabati. The high percentage of Groups 1a-j and 16a-b represents mainly the contributions of two sites, 153.8 and 155.4, which can be seen from the barcharts (Figures 4.10-4.11) to have yielded material almost exclusively of these groups, with virtually no featured sherds assigned to earlier periods. These barcharts may be contrasted with that from Site 101.4 (Figure 4.6) in which material in Groups 1a-j, 2a-g and 16a-b is essentially absent. Some further pottery from these groups has come from other sites, including 112.3 and 112.4. Neither of the barcharts for these latter sites (Figures 4.7-4.8) show the concentration of material in Groups 1a-j seen in the charts for 153.8 and 155.4, the sherds being more evenly spread amongst certain of the other groups, so that the contribution of these sites to the proportion of Groups 1a-j and 16a-b in the whole assemblage of featured sherds is relatively small.

In contrast to Group 1a-j and 16a-b material, the amount identifiable as being earlier Christian in date, or as likely to be of this Period, viz Groups 2a-g and 23, is relatively small. It forms just over 3.5% of the collection *in toto*, with the most common group, 2d, reaching only 1.61%. As seen from the barcharts, most material in Groups 2a-g and in Group 23 is from Sites 112.3 and 112.4, with that from Group 2d forming the highest percentage of the small number of featured sherds at the latter site. Pottery of Groups 2a-j and 23 together forms *c.* 30% of the featured material from 112.3, although almost an equal proportion is from groups considered to be later Christian to Islamic in date. This reflects the Surface Survey collection, in which there was a large percentage of featured sherds of Groups 1a-c but also several 'beer-jar' rims similar to those from Soba (Mallinson *et al.* 1996, 48, Part II, fig. 4).

Outside Gabati, no material was assignable definitely to the Post-Meroitic Period, only Groups 3 and 13 are considered to be of Post-Meroitic to (early) Christian date and Groups 17a-18b are likely to be of Meroitic to Post-Meroitic date. The two former groups are relatively well-represented in the overall collection, reaching 3.6% and 2.82% respectively, whilst Groups 17b and 18b are less evident, amounting together to only *c.* 3.6%.

The relative prominence of Group 3 can be explained by the presence of sherds classed as Group 3, in small quantities, at several sites otherwise dominated by material of other groups. These include Site 155.4, yielding predominantly ceramics of Groups 1a-j and 16a-b and Site 101.4, where most material is from Groups 11 through to 22. It forms a large proportion of sherds from 118FS2, together with sherds of Group 13, but the total number of featured sherds from this site is small. Group 13 is similar, occurring at Site 153.8 in a small percentage, and at 112.4. This may reflect the presence of Post-Meroitic to Christian Period sherds as residual material at sites where the main occupation dates to later periods, but could also be a function of the small number of vessel forms in these groups, which are not especially distinctive chronologically.

Groups 17b and 18b, in contrast to Groups 3 and 13, were found only at one site, 101.4, where these two groups account for *c.* 15% and *c.* 12% of the featured sherds. Their attribution on stylistic ground, to the Meroitic to Post-Meroitic Periods, rather than the Post-Meroitic to Christian Periods, would be consistent with the location of the site just to the north of Meroe.

Given that Gabati was the only site having a substantial amount of Meroitic material, which it was considered vital to investigate in the Test Excavation season, Meroitic pottery from the other Test Excavation sites was very limited in occurrence. Apart from the Group 17b/18b sherds, only Group 4h, containing a single featured piece, was clearly assignable to the Meroitic Period. Consequently, Meroitic groups form only a very low proportion of the overall assemblage. The Group 4h sherd was from a village site near BM 165, from which only a small number of featured sherds was recovered; the other sherds were from Groups 7a and 8c (Neolithic and Mesolithic) and from Group 1d.

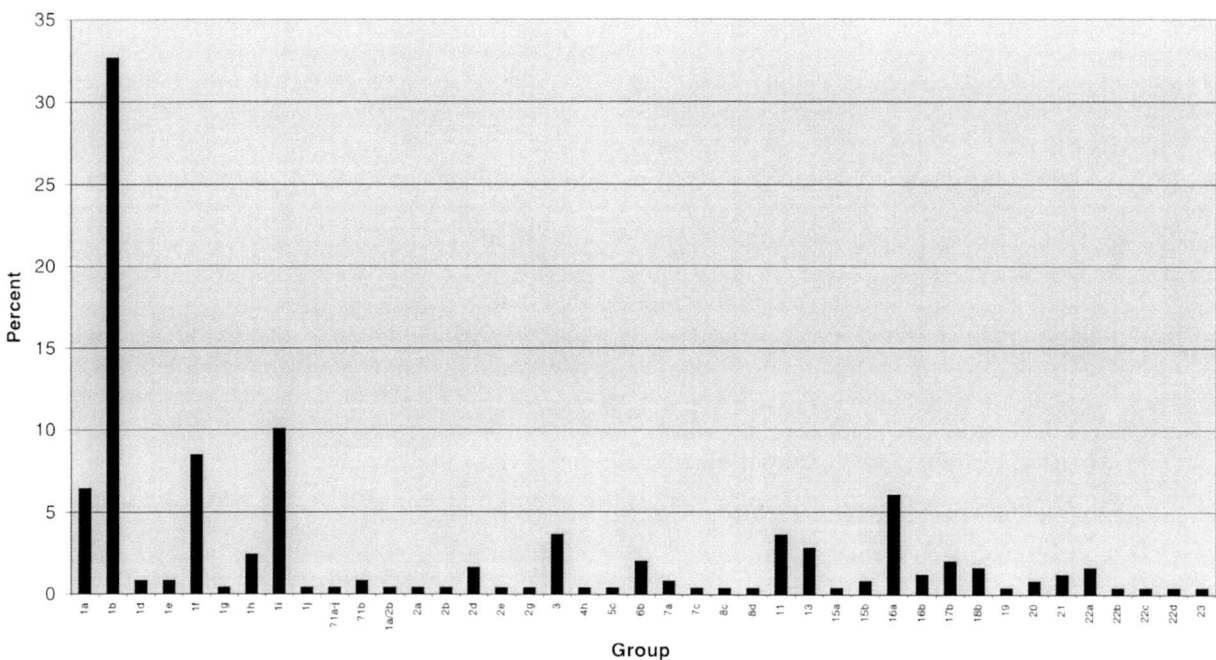

**Figure 4.5.** Percentages of the featured sherds calculated over all sites considered in the present report.

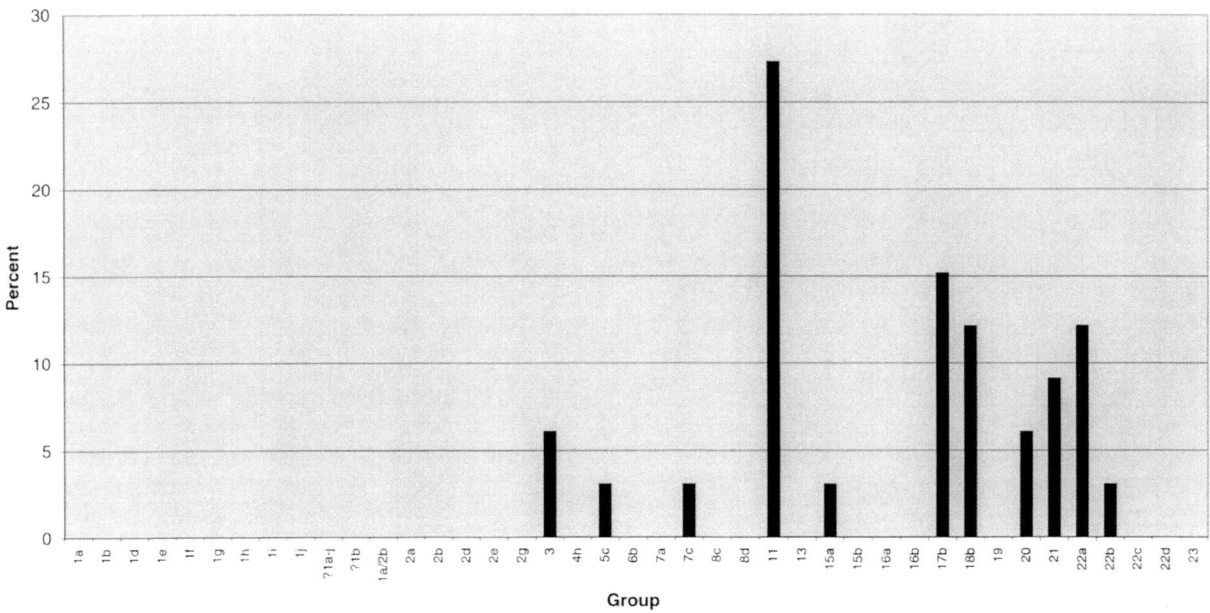

**Figure 4.6.** Percentages of the featured sherds calculated for Site 101.4.

The barchart for this site is not illustrated.

Sherds in Groups 5c and 11, assignable to the Egyptian Late Period, most probably equivalent to the Napatan Period in date, were distinctive when encountered in the field because of their marl clay fabric. However, such sherds only form a relatively small proportion of the total collection, only reaching a total of *c.* 4% of the featured sherds. Featured pottery from both these groups was, again, encountered only at one site, 101.4. Here, it forms a relatively high percentage of material from the site, amounting to over 25%. Sherds of Group 21, possibly also Napatan, formed only 1.2% of the featured sherds. Sherds in Groups 22c and 22d may date as late as the Napatan, but in terms of fabric, they have more similarity to the 'Late Neolithic' material.

Group 6b contains sherds comparable to the 'Late Neolithic' or 'IIIrd Millennium' of the Surface Survey in both fabric and decoration. It amounts to *c.* 2% of the overall collection. The majority of featured Group 6b sherds are from Site 112.3. Here, they are present as a number of very fragmentary sherds, likely to be residual. Their fragmentary state results in their forming a rather high percentage by count of sherds from that sites, which otherwise contained material predominantly of Group 1b, 2a-g 19 and 23 generally datable to the Christian through to Islamic Periods.

Material in Groups 15a, 15b and 22a-d, all forming a very low percentage of the total assemblage, appears to be related to Groups 6a and 6b in terms of fabric, although, as discussed above, the decoration is not especially characteristic of the 'Late Neolithic' ceramics encountered on the Surface Survey. Consequently, these groups have been

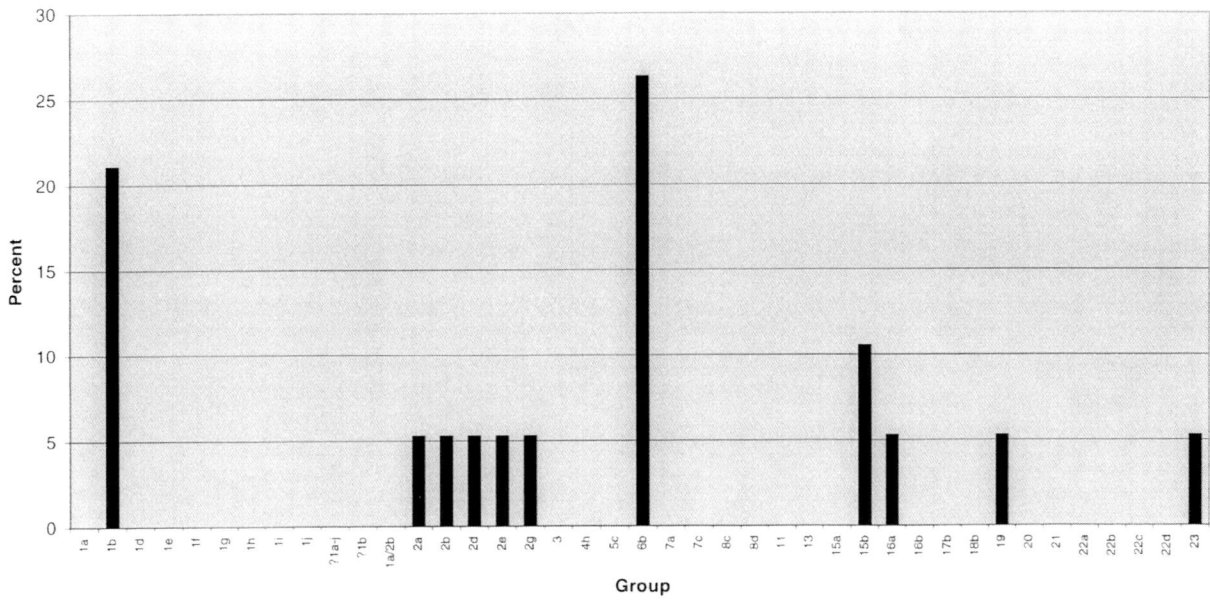

**Figure 4.7.** Percentages of the featured sherds calculated for Site 112.3.

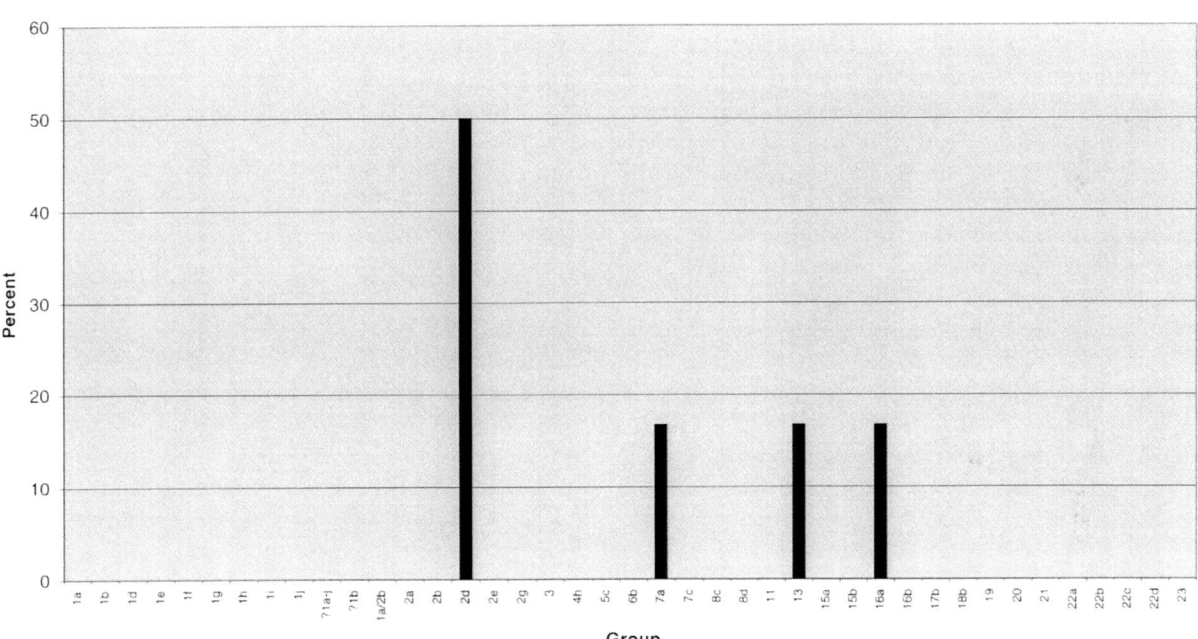

**Figure 4.8.** Percentages of the featured sherds calculated for Site 112.4.

only tentatively assigned to the 'Late Neolithic', with the caveat that a similar fabric could have been used earlier, and possibly later, since gritty fabrics lacking vegetable temper are known from the 'Funj' Period. Sherds placed in Groups 15a, 15b and 22a-d were found mainly at Sites 101.4 and 112.3 where, again, they are likely to be residual, although they form a relatively high proportion of sherds at each of the sites, owing to their fragmentary state (see Figures 4.6-4.7). The two sherds in Groups 22c and 22d were the only featured material at the sole site where they were found, 170.1. The barchart for this site is not illustrated.

## Evidence for trade contacts

The main ceramic evidence for longer-distance contacts amongst the sites investigated came from Gabati. This comprised mainly the 'oil jars' in an Aswani fabric, recovered from several tumuli including Tumulus 2 and Tumulus 4 excavated in the Test Excavation season, together with the vessels <T5/93C> and <36> with decoration stylistically similar to Early Christian ceramics known to be made at Old Dongola (Smith 1998c, 184-186, figs 6.31-6.32). Evidence from the sites other than Gabati was quite restricted, the great majority of pottery being in fabrics and styles that, at the present stage of research, can only be regarded as 'locally' produced. It may be assumed that most vessels were manufactured within the Shendi Reach itself, although the possibility cannot be excluded that a wider, yet still 'local' source area existed for the pottery at the sites

Apart from the Gabati 'oil jars', material from four groups distinguished at the other sites studied is probably from regions other than that within which the Surface Survey and Test Excavations were conducted. The body

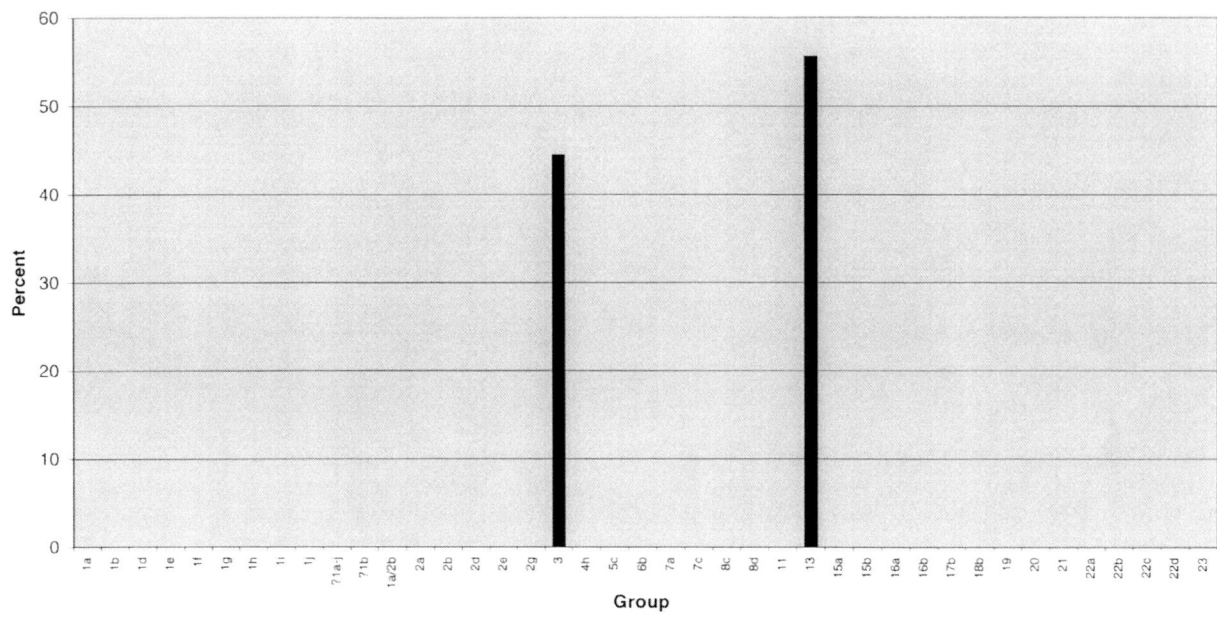

**Figure 4.9.** Percentages of the featured sherds calculated for Site 118FS.2.

and base sherds in Groups 5c and 11 are in marl fabrics, which can be ascribed to an Egyptian provenance, the fabrics of both groups being best known from Upper Egypt. Chronologically, these groups have a similar range, generally within the Napatan Period. Hence, they are most likely to represent vessels imported into the region during the period of Nubian rule in Egypt during the 25th Dynasty, or the continuation of contact and trade with Egypt subsequent to the Nubian withdrawal from that country.

The other two groups containing pottery likely to have been imported to the sites at which they were found are both of Christian Period date. The first of these groups is 2d, comprising sherds comparable to 'Soba Ware'. Whilst no definite evidence for the manufacture of 'Soba Ware' has been discovered so far at Soba itself, most of the pottery there may be assumed to have been produced in the town or close by and 'Soba Ware' is considered to be a 'local' product (Welsby 1991, 324; 1998b, 91). On this basis, Group 2d was probably imported from the vicinity of Soba.

Jar 1:20.4 in Group 2e from Site 112.3 is stylistically similar to Adams' Style NIVA, for which Faras in Lower Nubia and el-Ghazali in the Merowe area are the two known production centres. The appearance of the fabric of the jar (F1:15) has some similarities to the description of wares in Style NIVA known from Faras, as given by W. Adams (1986, 490) although it is not fully consistent with the latter. Given the distribution of wares in this Style, predominantly to the north of the Fourth Cataract, and the present lack of identified production sites to the south of el-Ghazali, Group 2e can be considered an import from at least the Merowe area, and possibly from Lower Nubia.

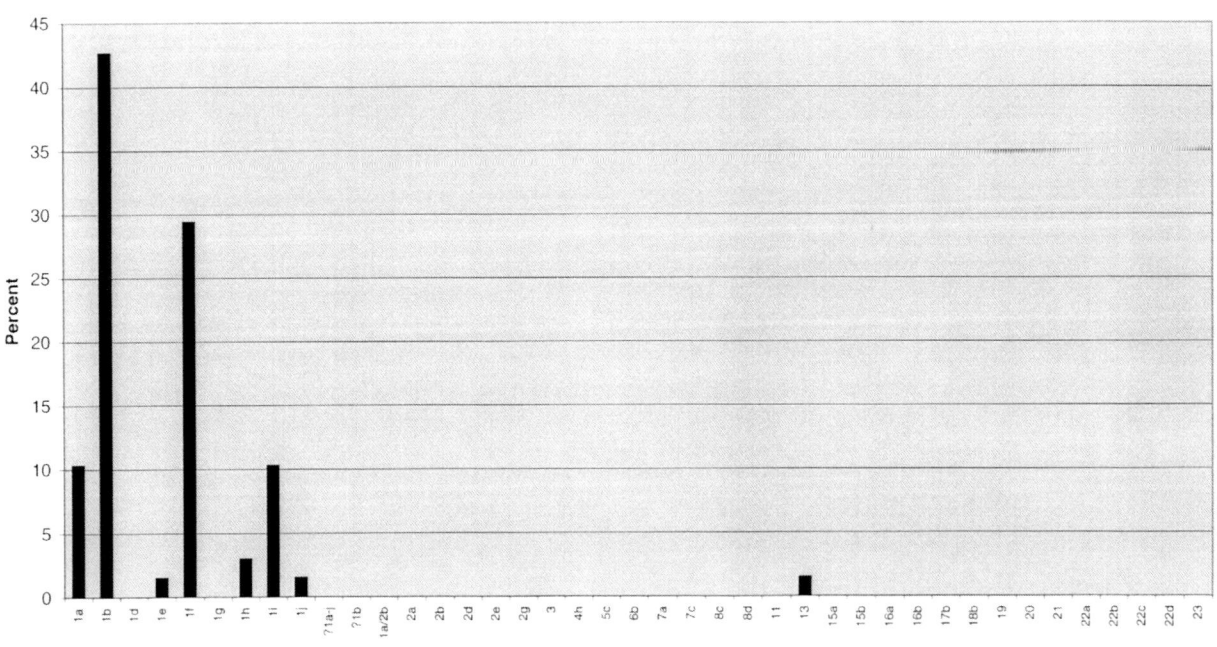

**Figure 4.10.** Percentages of the featured sherds calculated for Site 153.8.

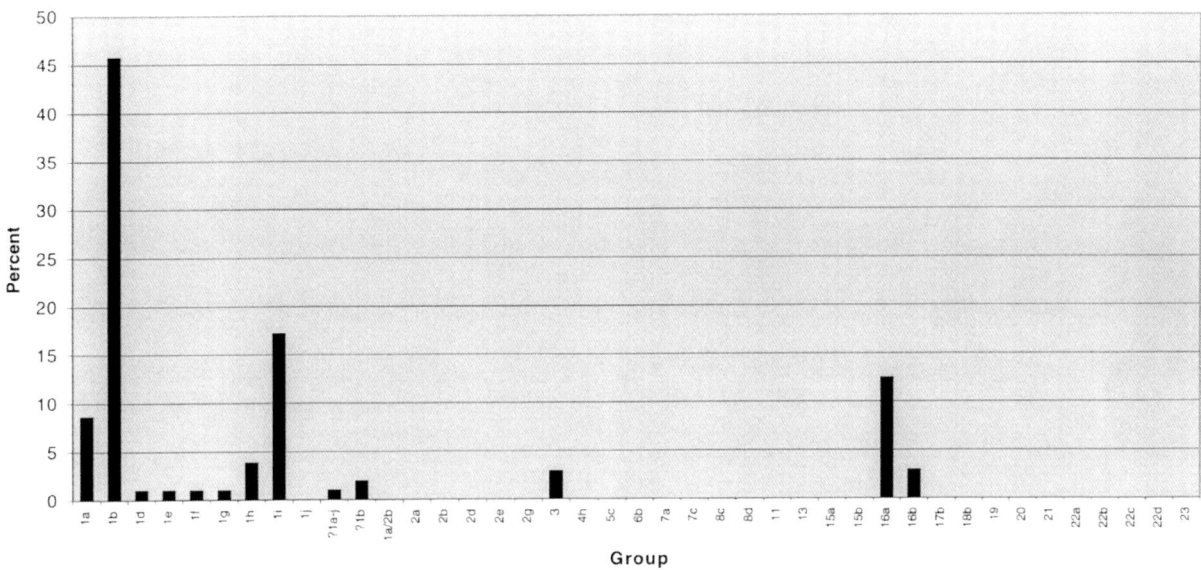

**Figure 4.11.** Percentages of the featured sherds calculated for Site 155.4.

## Comparison of Test Excavation and Surface Survey collections

Given that there are many areas for which surface collections are likely to remain the substantive source of information for the near future, it was considered of interest to see how the impression of the nature of the main sites investigated, particularly with regard to pottery dating, gained from the Surface Survey compared with that obtained from the Test Excavations. For example, site 101.4 did not have any featured sherds on the surface, so the main indications of a Post-Meroitic and Christian date for the burials investigated were given by the Test Excavation, which also produced evidence for Neolithic and Napatan Period ceramics.

Site 112.3 was considered to be earlier Christian and late Christian to Islamic on the basis of the Surface Survey (Mallinson *et al*, 1996, 48). These periods were confirmed in the Test Excavations as being the main ones for which ceramic evidence was present, with a small amount of probable Neolithic material also being found. No featured sherds were collected from the surface of Sites 112.4 and 118FS2, but on other grounds a Christian date was suggested for both (Mallinson *et al*, 1996, 49, 57). The sherds from the Test Excavation at the former site were datable to the Post-Meroitic to Christian Period and to the late Christian to early Islamic Period, with possible residual Neolithic. Test Excavation sherds from the latter sites dated from the Post-Meroitic through to Christian Period. At Site 155.4, the Surface Survey material indicated a Late/Terminal Christian to Islamic date, which was confirmed by the great majority of the Test Excavation ceramics, with only a small proportion of Post-Meroitic to Christian material being found. Only at Gabati was the predominant surface material substantially different from that excavated, since the surface featured sherds were virtually all Meroitic, with a few Christian and Late Christian to Islamic ones. The surface ceramics did not clearly indicate the presence of the substantial number of Post-Meroitic burials, although the range of date suggested initially for the site, on the basis of all available information, included this period.

From the foregoing, it appears that the Test Excavations generally confirmed the impression of the dating gained from the surface collections with regard to the material most prevalent at the sites. However, the evidence of the Test Excavation ceramics did extend the known temporal range of activity at a site. Often this was only through the presence of small numbers of featured sherds, which are likely to be residual, but in one important case, that of Gabati, the body of ceramics collected was significantly increased by the Test Excavations. In this way, the pottery from the Test Excavations resulted in an expansion of the periods to which several of the sites could be attributed, whilst furnishing further evidence for inter-regional contacts.

## Acknowledgements

I am indebted to the following for advice and help during the study of the pottery from the Test Excavations: Dr J. A. Alexander, Ms. J. Bourriau, Dr D. N. Edwards, Ms J. Filer, Mr P. G. French, Dr P. Lenoble, Dr J. Phillips, Dr J. Reinhold, Dr P. J. Rose, Dr K. Smith, Dr D. A. Welsby (who kindly allowed me to examine material from Soba in the British Museum), Ms I. Welsby Sjöström and Dr H. Whitehouse. The season of the Test Excavations and the main study of the material was undertaken whilst the author was a Visiting Scholar at the Department of Archaeology, a Junior Research Fellow at Wolfson College, and subsequently attached to the McDonald Institute, University of Cambridge. I am grateful to these institutions and for the support of the following members of the Department: Dr C. A. Shell, and Ms J. C. Rippengal and to Ms J. Boreham, then of the Department of Archaeology, University of Cambridge.

# 5. Gabati Basketery and Cordage

*Willemina Wendrich*

The rescue excavations of the tumuli on the Gabati road section of the North Challenge Road (Khartoum-Atbara) retrieved a small number of basketry and cordage samples (Table 5.1). The conditions of all fragments was extremely fragile and it was, therefore, not possible to determine the plant species of which the objects had been made. From macroscopical appearance all cordage and basketry seems to have been made of a broad sturdy leaf, probably dom palm leaf.

Cordage was found in Tumuli T1, T5 and T12. All cordage was made in the same way, the only variety occurring in the diameter of the string and spun yarns. Generally, daily use string is usually made of two z-spun yarns (zS2 string), but the Gabati string was designed to be slightly stronger and was made with three z-spun yarns which were plied in S-direction (Figure 5.1).

The string was found with dismantled fragments of *angareeb* beds. Since leather webbing was found in relation to the bed frames, it seems likely that the string was not used for making the webbing, but for maintaining the tension in the leather webbing fabric. The warp of the webbing was probably fastened to a set-up line of string, connected with zig-zagging string to the frame (Figure 5.2). When the bed started sagging the tension could be restored by pulling the string at the head of the bed.

The basketry fragments all came from the same tumulus (T4). All three fragments represent fine workmanship. The first coiled basket is represented by five fragments, one of which was the centre (Figure 5.3). The wrapping leaf is 2mm wide and wound alternately over one and over two bundles. The bundles are 2.5mm in diameter. The start of the basket is formed by a *snail centre*. For two coils the wrapping strand has been stitched through the previous bundle. This is a stronger method than the coiling over one and two bundles, and better suited to make the start of the coil. All four wall fragments are wrapped alternatively over one and two bundles. The thickness of the wall of the basket is 5mm. The diameter of the basket must have been approximately 60mm, the height or shape of the basket is unknown.

The second basket fragment sample is also a finely coiled centre. As in the previous one the snail centre is stitched through the previous bundle. The wrapping strand is 2mm wide, the bundle has a diameter of 2.7mm. After two coils the rest of the basket seems to have been made in the *lazy basketmakers' stitch*, a time saving method in which the palm leaf strip is wrapped a number of times around the bundle and fastened only now and again onto the previous bundle (Figure 5.4). This technique is used whenever basket makers figure that nobody will see the stitch. New Kingdom foundation deposits, for instance, have token baskets made in this technique. Most probably, however, the fragment found at Gabati once was covered completely with an intricate wrapping pattern or leather, which is now lost. Parallels for the combination of the *lazy basketmakers' stitch* have been found at Qasr Ibrim (Wendrich 1999, 219).

A small fragment of twill plaited matting was the third basketry find find from Tumulus 4. The plaiting strips are 3.7mm wide. The thickness of the finely plaited mat is 1.7mm. The plait pattern runs over 2/ under 2 (Figure 5.5).

**TABLE 5.1.** CATALOGUE OF GABATI BASKETRY AND CORDAGE.

| Context | Description | Diameter | No. of fragments | Drawing |
| --- | --- | --- | --- | --- |
| T1 (20) | Deteriorated palm leaf string, zS3 | 4/2.5 to 6/3.5mm | 15 | 1 |
| T1 (26) | Deteriorated palm leaf string, zS3 | 44/2.5 mm | 20 | 1 |
| T4 (13) A | Coiled basket fragment | 31.5 x 28.7 mm | 5 | 3 |
| T4 (13) B | Coiled basket fragment | 21 x 21.2mm | 1 | 4 |
| T4 (13) C | Twill plaiting | 20 x 20mm | 1 | 5 |
| T5 (23) | Deteriorated palm leaf string, zS3 | 4.5/3.4mm | 8 | 1 |
| T12 (20)+(23) | Deteriorated palm leaf string, zS3 | 5.5/3mm | 2 | 1 |

**Figure 5.1.** All string found at Gabati (in Tumuli 1, 5 and 12) was made of zS3 string (three z-spun yarns which were S-plied).

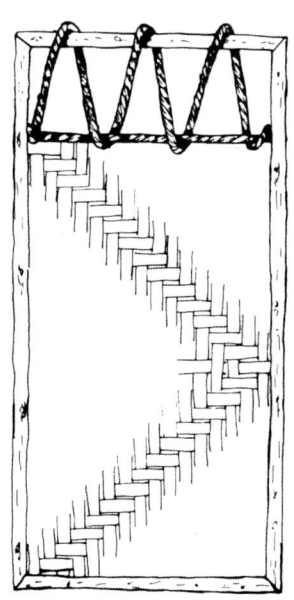

Palm leaf string
sZ3                                  T1   20 & 26

                                     T5   23

Centre: leather straps               T12  20 & 23

**Figure 5.2.** Schematic drawing of an *angerib* bed with string at the head for maintaining the tension of the plaited leather webbing. Evidence for these beds was found in Tumuli 1, 5 and 12.

↻

T4 - 13

Stitch scheme

Centre

Base and sides

**Figure 5.3.** Fragment of a coiled basket (scale 1:1) found in Tumulus 4. The two schematic drawings show respectively the technique employed near the centre (stitch through previous bundle) and the 'over1/over2' wrapping technique.

↺

T4 - 13

Stich scheme

Centre

Base

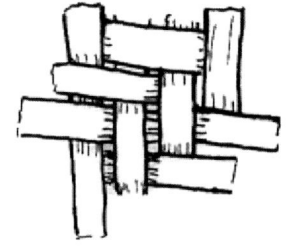

**Figure 5.4.** Fragment of a coiled basket (scale 1:1), found in Tumulus 4. The schematic drawing shows the lazy basketmakers' stitch.

**Figure 5.5.** Small fragment of a twill plaited mat found in Tumulus 4.

# 6. The leather samples from Test Excavations at Gabati (site 159.2)

*Barbara Wills*

Samples from leather items taken during the test excavations at the site of Gabati (159.2) were studied to obtain information on the species represented, methods of tanning and the manufacture of the objects from which they came. The results of this study for individual specimens concerning the samples examined and, where appropriate, the type of object from which they were taken, identification of the species from which the skins came, results of tests for tanning agents and comments concerning evidence for the ways in which the leather items were made, including types of seams and stitching, are presented in Table 6.1. and Figure 6.1. A summary of the general conclusions derived from the whole collection of samples is given below.

The leather falls into several distinct categories:

a) Thin leather, of 1mm thickness, or less. This is probably from goat, kid or hair sheep. Such material is used for pouches, covering for baskets, clothing etc.

b) Thicker strips of up to 4mm thickness. These are evidently used on the plain-woven bed webbing. Many knots (ring hitch) survive, tied over a plied leather cord. Species identification was not possible, but the thickness of the cross-section is consistent with, for example, the larger animals, such as cattle.

c) Leather threads and cords. No textile threads were found in the stitch holes; all extant stitching was done in thin leather or rawhide strips. The leather threads were used in a variety of stitching techniques. Leather cords (S-twist) were two or three-ply.

Some of the seams and edge-bindings are very sophisticated, e.g. T4/26, T5/11c,11f, with complex stitching and incorporating an extra fillet of leather (welt or rand) for additional strength. Examples of the 'herringbone' type of stitching, as in sample T6/29, were found. This type of stitching was present in the material recovered from the Rescue Excavations at the site (cf. Edwards 1998d, 125-126; Mould 1998).

Tanning tests (see following section) were of interest in showing that it was possible to obtain results from degraded leather from an archaeological context. However, these results were not always clear-cut. Broadly, the thin goat leather tended to give positive results for vegetable tan, as would be expected. The thicker leather strips or thongs, which were perhaps more likely to be rawhide or very simply tanned, gave a wide range of results.

**TABLE 6.1.** (CONT). LEATHER SAMPLE DESCRIPTIONS. EXAMPLES FROM SPECIMENS MARKED * ARE ILLUSTRATED IN FIGURE 6.1.

| Sample number | Description | Material | Type | Tannage Type Vegetable | Tannage Type Alum | Comments | Illust. |
|---|---|---|---|---|---|---|---|
| T1/21 (cont.) | Fragments of strips/thongs | Leather | Possibly goat or bovine | Possible | Negative | Many pieces of twisted and stretched thongs, probably once interwoven. Associated with 3-ply, S-spun vegetable fibre string. | |
| T1/22 | Fragments and knotted thongs | Leather | Thin, probably goat or kid. Knots thicker, possibly bovine | Negative | Positive | Some pieces have associated fibrous (woollen?) material on one side. | * |
| T1/23 | Fragments of strips/thongs | Leather | Thickness > 0.5 mm: possibly goat/kid. Species identification difficult due to obscuring of grain layer. | Positive | Positive | Many fragments of thongs/strips of skin or leather. Leather appears stretched and some pieces bent and twisted. Some pieces are overlapping or of double thickness. | |
| T1/25 | Fragments | Leather | Thin, probably goat or kid | - | - | Associated pieces of wood and twig. Fibrous material on the back of some of the pieces. | |
| T2/3 | Small number of leather thongs and a knot | Leather | Diameter > 2 mm: bovine(?) | Positive | Negative | Associated small bones and plain-woven woollen textile. Includes knots of type in samples T1/22 and T2/32a-d. | |
| T2/4 | Pieces | Skin | Appear to be human skin | Positive | Void | The positive result for vegetable tan is a surprise: the skin may in fact be tanned leather, or perhaps was once in close association with tanned leather, thus picking up the tanning agent. Associated with textile. | |
| T2/6 | Area of stretched skin and fragments | Skin | Possibly human | Possible | Positive | This result is also a surprise: again possible close association with a tanning agent. | |
| T2/10 | Broken strips/thongs and knots | Leather | Species identification not possible | Negative | Negative | S-twist (two-ply?) thongs together with knots. | * |
| T2/20 | Thin, stretched skin | Skin | Resembles human skin | Negative | Negative | - | |
| T2/21 | Fragments of skin with hairs | Skin and hair | Resembles goat skin and hairs | No test | No test | This sample untanned. | |
| T2/23 | Pieces of skin | Skin | Resembles human skin | Positive | Positive | A surprise result for possible human skin, perhaps due to association with tanned material. | |
| T2/32a | Thongs, including knots | Leather | Thick (> 4 mm): bovine(?) | Negative | Negative | Ends of some thongs knotted onto a twisted leather cord. Knots are of the same type as in sample T1/22. | * |
| T2/32b | Thongs | Leather | Thick (> 4 mm): bovine(?) | Negative | Negative | | |

**TABLE 6.1. (CONT.). LEATHER SAMPLE DESCRIPTIONS. EXAMPLES FROM SPECIMENS MARKED * ARE ILLUSTRATED IN FIGURE 6.1.**

| Sample number | Description | Material | Type | Tannage Type Vegetable | Tannage Type Alum | Comments | Illust. |
|---|---|---|---|---|---|---|---|
| T2/32c | Thongs | Leather | Thick (> 4 mm): bovine(?) | Negative | Possible | | |
| T2/32d | Thongs, including woven pieces and knots | Leather | Thick (>4 mm): bovine(?) | Positive | Negative | Woven over and under each other to form lattice of bed stringing. Includes five sets of knots, mostly paired, similar to those in sample 32a. | * |
| T4/2a | Pieces, possibly covering for basket lid | Leather | Very thin, possibly goat/kid | Negative | Possible | Striations in leather indicate that it might once have covered a coil of basketry, possibly a lid. | * |
| T4/2b | Assorted fragments | Leather | Thin, probably goat or kid | No test | - | Many pieces of leather in two or more layers. Includes numerous edge pieces, some with diagonally-placed stitch holes, and some with holes in zig-zag pattern along strip. Associated with 3-ply vegetable fibre string and piece of wood (dowel?). | * |
| T5/8 | Fragments | Leather | Thin, probably goat or kid | Negative | Positive | Includes piece fixed over a cord. Also edge stitched as in sample T5/11c, f. Associated with plain woven woollen textile. | |
| T5/11c-k | Fragments | Leather | Possibly goat: species identification difficult due to surface damage | c, d, f, k, positive | f positive; c, k negative; d possible | Includes edging(?) pieces of square cross-section, with additional folded fillet of leather inserted between the two layers (rand). A strip of leather or skin (2-3 mm wide) holds the layers together with a running stitch. Some pieces tightly folded, almost knotted. Sample c has strip of different coloured leather (reddish brown) stitched behind. | * (c) <br> * (f) |
| T5/20 | Assorted pieces | Leather | Thin, probably goat or kid | Positive | Negative | Includes pieces of seams, some retaining stitching, others only with stitch holes and remains of leather fillet on back. Seam has extra fillet of leather (rand). | * |
| T5/23 | Fragments | Leather | Thin, no species id. | No test | - | Associated with rope sample | * |
| T6/29 | Large piece of twisted leather and textile, from apron | Leather | Thin (> 1 mm), probably goat | Positive | Positive | Includes several examples of seams with leather stitching. One piece has stitching and exhibits a cut edge. | * |
| T6/56S | Thongs or strips | Leather or skin | Thin, broad thongs, probably goat | Negative | Possible | Strips are distorted, indicating that they might have been woven. Most are of thicker type of leather. Some are overlapping strips, associated with yarn. | |

# 7. Analysis of tanning agents from the excavated leather samples

*David Thickett*

Thirty-two leather samples excavated from post-Meroitic burials in the cemetery site at Gabati in Sudan have been analysed for the presence of tanning agents. Tests were carried out for aluminium salts, in this case alum, and for tannins (polyphenols derived from vegetable matter). Tannins can be subdivided into two types, catechol or condensed and pyrogallol or hydrolysable. The dark colour of the leather samples precluded the use of colorometric spot tests directly on the leather surface. Therefore, each leather sample was extracted with a 1:1 mixture of acetone and distilled water at 60° in a screw top vial for one hour. The extract was tested separately with ammonium ferric sulphate solution (2% in water) and aluminon solution (1% in water). A blue/black precipitate with the ammonium ferric sulphate solution indicated the presence of an hydrolysable tannin. A green/black precipitate would indicate the presence of a condensed tannin. All of the ammonium ferric sulphate tests gave brown or black precipitates, indicating that tannins were present but these colours do not differentiate between hydrolysable or condensed tannins. A red colour with the aluminon solution would indicate the presence of aluminium and hence alum. The results of these tests are presented in Table 7.1.

The tests indicate that several of the samples were from leathers treated with tannins and/or alum. A negative test should not be taken to mean that the leather was definitely not tanned, as the survival of a tannin cannot be assumed and alum is water soluble and therefore may have been leached out of the leather over time. The possibility of cross contamination from leathers buried together cannot be excluded.

**TABLE 7.1.** RESULTS OF TANNIN TESTS WITH AMMONIUM FERRIC SULPHATE AND ALUMINON.

| Sample | Extract Colour | Extract Reaction with: | |
|---|---|---|---|
| | | Ammonium ferric sulphate | Aluminon |
| T1/4,4a | Dark yellow | ? | ? |
| T1/12 | Pale yellow | N | N |
| T1/14a | Yellow | N | N |
| T1/14b | Yellow | ? | ? |
| T1/14c | Pale yellow | N | N |
| T1/15a | Pale yellow | N | N |
| T1/15b | Pale yellow | N | N |
| T1/16 | Pale yellow | N | N |
| T1/19a | Very pale | N | V |
| T1/19b | Yellow | N | N |
| T1/21 | Yellow | ? | N |
| T1/22 | Yellow | N | P |
| T1/23 | Dark yellow | P | P |
| | | | |
| T2/3 | Yellow | P | N |
| T2/4 | Yellow | P | V |
| T2/6 | Orange | ? | P |
| T2/10 | Yellow | N | N |
| T2/20 | Yellow | N | N |
| T2/23 | Pale Yellow | P | P |
| T2/32a | Yellow | N | N |
| T2/32b | Yellow | N | N |
| T2/32c | Yellow | N | ? |
| T2/32d | Yellow | P | N |
| | | | |
| T4/2a | Yellow | N | ? |
| | | | |
| T5/8 | Dark yellow | N | P |
| T5/11a | Orange | P | P |
| T5/11c | Orange | P | N |
| T5/11d | Orange | P | ? |
| T5/11k | Yellow | P | N |
| T5/20 | Yellow | P | N |
| | | | |
| T6/29 | Orange | P | P |
| T6/56S | Very yellow | N | ? |

Key:  N, negative test
P, positive test
?, possibly a positive test, some slight colour change
V, test void, dark colour of solution obscured any colour change

# 8. Gabati grave goods from the Test Excavations and the consideration of funerary practices

*Laurence Smith*

## Introduction

The small finds from the Post-Meroitic burials at the cemetery of Gabati have been considered in some detail in the main report on the site as a whole (Edwards 1998a). In that publication they were treated typologically, with examples of each class of artefacts being dealt with together, but separately from the other items in the same tomb. In the present report, since the Gabati tumuli excavated in the Test Excavations are described, it was considered appropriate to include a summary of and, in some cases, additions to information on parallels for the artefacts from these burials considered in the previously published report. In this case, the material from each tumulus will be treated as a grouping of the different broad classes of artefacts, viz. pottery, metal objects, beads and the various items of organic materials, rather than typologically. By examining the material from this perspective, it is intended to draw some general conclusions about the nature of the assemblage of objects placed with these burials and to put forward suggestions regarding the significance of the grave goods. It must be emphasised that such conclusions can only be of a descriptive nature, and the suggestions tentative, given the small size of the database for the Test Excavations. They should be considered in relation to the general summary of the Post-Meroitic use of the site given by Edwards (1998f). Information on the burials is derived from notes taken in the field and during post-excavation work, and previously published reports including those giving further details of pottery and finds (Mallinson 1994 and this volume; Filer 1994; 1995; Smith 1994; 1998a; 1998b; 1998c; Taylor 1998; Wills, this volume).

## The Tomb Groups

The material from Site 159.2 Tumulus 1 (Figure 1.4) placed around the burial chamber, not directly associated with the body, includes two pottery vessels. One, T1/41C, is a broad open black burnished bowl. The second, T1/42C, is a bowl with upright sides (Figure 8.1 – after Smith 1998c and Plates 8.1 and 8.2). The first vessel is closely comparable to a broad open bowl recovered from a tomb near Berber excavated by Lenoble. The parallel is close, since the bowl from the Berber tomb also exhibits incised decoration on the interior forming a 'zigzag' motif (Lenoble 1991, 170; fig. 5, no. 11). The second bowl may be related to vessels having a moderately concavo-convex profile, often with a thinned rim, and having the maximum diameter about the mid-upper body, known from Abu Geili (Crawford and Addison 1951, pl. XXVII, Types III.1-III.11). Of these Types, only Types III.1, III.5 and possibly III.7 resemble T1/42C, although the similarity is not close. Another parallel that may be tentatively drawn is with material from northern Nubia. In form, T1/42C may be related to that of cups from the cemetery at Wadi Qitna, illustrated by Strouhal (1991, figs 6 & 8). These cups are in the fine handmade ceramic termed 'H-ware' which can be dated from the second half of the $3^{rd}$ century to the end of the $5^{th}$ or early $6^{th}$ century AD. It is not likely that T1/42C is part of the tradition of the 'H-ware' cups as known from Lower Nubia, since the decoration characteristic of the latter is generally quite different, although some use is made of bands or, occasionally, panels with a slip colour differing from that of the overall surface, delimited by incised lines (Strouhal 1991, 3-5; figs 2-8, 9a-b, 10a-b, 11a-b). It may have derived from a related, more southerly, tradition.

Other items not directly associated with the body of the deceased include the bed, strung with rope and leather webbing, as is the case in Tumuli 2, 4 and 5. The legs of these beds are decorative. The upper part of each, containing the sockets for fitting the side members, pass into a section comprising either a single roughly spherical motif or a pair of more elliptical shapes, one above the other (see Figure 8.1, T1/8). The legs are terminated by a splayed-out feet, again of approximately square cross-section at the base. These feet appear to have been carved with a flat face on all four sides, rather than forming a truncated cone.

Two tools are included in this burial. One is an iron arrowhead of approximately leaf shape with long tang (T1/18). The tip is missing, and there is an area on one edge that appears likely to have been the position of a barb, also now missing. The second is an iron blade, T1/24, which could be the end of a slender spear point, or part of a knife or small dagger blade (Figure 8.1, Plates 8.3 and 8.4). The arrowhead is an example of one of the main types known widely from Post-Meroitic period, being generally of a relatively short 'leaf-shape'. These form one of the artefact types considered to show continuity from Meroitic to Post-Meroitic burial contexts (Geus and Lenoble 1985, 87) and are known from sites ranging from Lower Nubia, e.g. Aman Daud (Firth 1912, 198, pl. 38, f) to el-Hobagi (Lenoble et al. 1994, 53, pl. 11), through to Qoz Nasra near Wad Medani on the Blue Nile, also generally dated to the Post-Meroitic period (Edwards 1991, 47-49, 52-53, pl. VII, QM0010).

The second item is similar to, although smaller than, an iron knife from Firka Cemetery A, Tomb 14, dated generally to the 'X-Group' period ($5^{th}$ century to mid-$6^{th}$ century) which has a similarly relatively thin and narrow point, almost symmetrical but with one side, presumably the cutting edge, being more convex than the other (Kir-

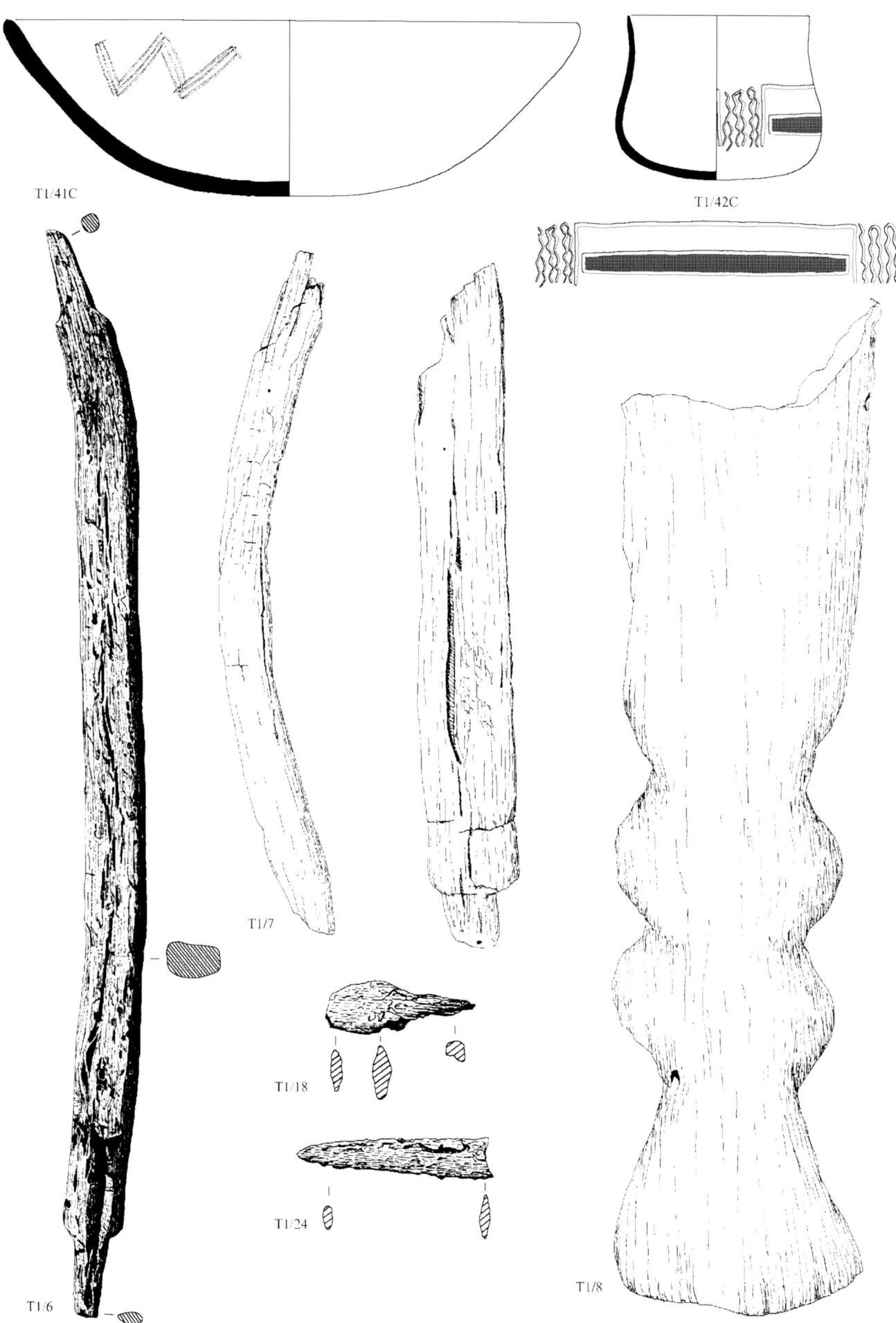

**Figure 8.1.** Items from the tomb chamber of Tumulus 1, including examples of bed frame and legs (scale 1:2; T1/6 and pottery – scale 1:4;).

wan 1939, 35, pl. X, 3). Find T1/24 is also of similar form to a small knife, 96mm long, from Abka site 416, Grave 9, although somewhat narrower, dated to the late Meroitic or 'X-Group', c. AD 100-600 (Säve-Söderbergh et al. 1981, 169, pls 64, 89). The blade was associated with a leather sheath. The remaining items with the burial comprise further items of leather, and textiles on or around the body.

In Tumulus 2, containing the burial of a young woman and an infant, the artefacts placed in the chamber comprised the bed, of generally similar form to that in Tumulus 1, but clearly dismantled. One of the main side-members of the bed, together with the plaited leather webbing, lay along one side of the chamber, beyond the bodies. The two identifiable legs were in the north and north-eastern parts of the chamber.

As in the case of the Tumulus 1 burial (Figure 1.5), only two pottery vessels were provided: one, a red wheel-made oil bottle (T2.1/31C), placed near the head, and the second a black burnished bowl with inturned sides (T2.1/32C) placed near the feet (Figure 8.2 – after Smith 1998c – Plate 8.5). The oil jar is most comparable, particularly in terms of fabric and surface treatment, to ceramics classed as 'Family A' in the system of Adams and as 'Egyptian A' ware in the classification of Hayes. Both authorities note that there is evidence for a major manufacturing centre, or the only manufacturing centre, in the vicinity of Aswan (W. Adams 1986, 525; Hayes 1972, 387; 1980, 530). These wares have been found in small numbers as far south as the Shendi Reach; the distribution of the Family A Ware U2 extends at least to the Fourth Cataract (W. Adams 1986, 545), and some specimens of Family A Ware R30 have been recovered from the townsite at Meroe (Shinnie and Bradley 1980, 158). A relatively small number of sherds having 'Aswani' fabric are known from Soba, including fragments of 'oil-jars' (Welsby 1998b, fig. 35 Types 40C, 43C, fig. 51, 21.1U). However, the closest parallels to the oil bottle from T2, together with that from Tumulus 4 (see below) occur amongst the ceramics from Elephantine illustrated by Gempeler (1992, Abb. 76, nos 10, 14-17, Abb. 77, nos 3, 5).

Although it has not been possible to locate an exact parallel to T2/32C, there are a few fairly similar forms to that of this vessel known. One occurs amongst the wares recovered from Garstang's excavations at Meroe, comprising his type ZT, although this is not strongly carinated. A parallel for the incised motif is illustrated from amongst sherds of the black-burnished handmade wares from the same site (Garstang et al. 1911, 41, pl. XLVI, no. 48; pl. LII, no. 6). A second reasonably similar form is present amongst the bowls from el-Kadada, Tomb 100/1, in particular vessel 28, whilst a third comparable vessel was recovered from a Post-Meroitic tomb at Shaqalu, near Shendi. The tomb yielded a bowl, also black burnished, of similar proportions to the Gabati vessel, but differing in the lack of a distinct carination and in the lack of decoration (Geus and Lenoble 1985, fig. 10; Geus et al. 1986, fig. 1, no. 1).

The other items recovered from the burial in Tumulus 2 were essentially coverings and personal adornments closely associated with the bodies. Several sets of beads were found, made of both inorganic and organic materials. Brown beads, apparently the husks of seeds, were near the forearms, whilst various types of beads of eggshell (probably ostrich) glass, faience and semi-precious stones (see examples, Plate 8.6) were near the feet and in the region of the upper body. Colours range from semi-translucent, white, amber, reddish-brown, through blue and green, purplish-brown and white striped, to dark grey.

One set of the eggshell beads was associated with a cowrie shell (Figure 8.2 – after Smith 1998a – Plate 8.7) from which the back had been removed. The placing of both worked and unworked cowries in burials has a very long tradition in Nubia. Examples of cowries with the back removed are known, for example, in Neolithic contexts, such as graves at el-Kadada (Geus 1984, 64, no. 14). Numerous examples, both in their natural state and lacking the back, have been recovered from Meroitic burials and tombs

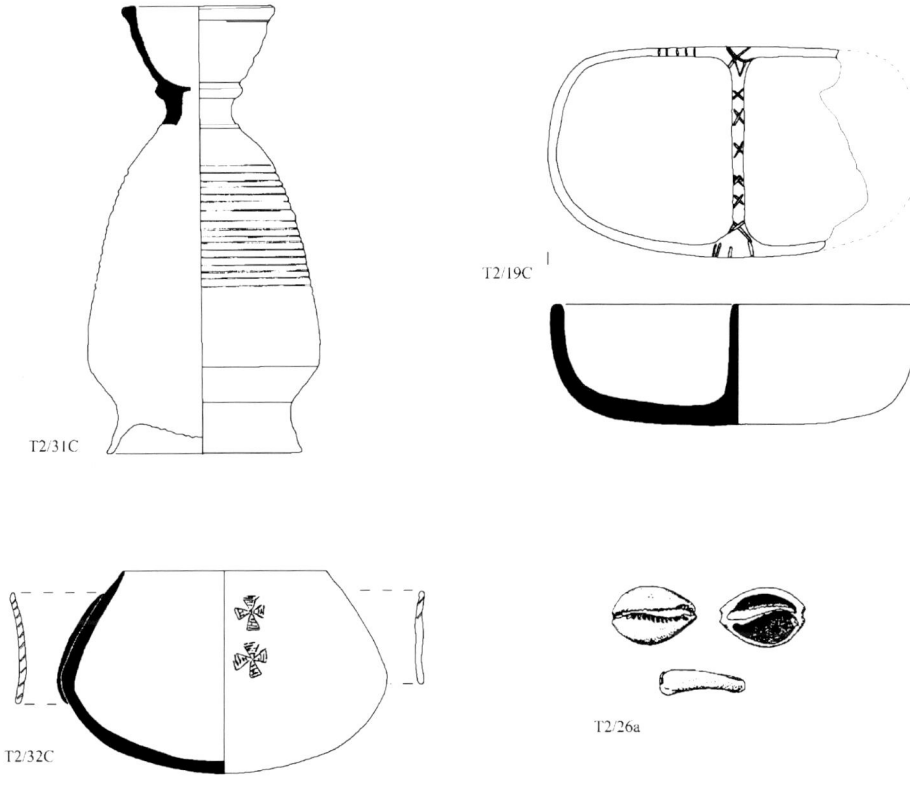

**Figure 8.2.** Items from Tumulus 2. Includes T4/19C from the tumulus fill (cowrie – scale 1:2; pottery – scale 1:4).

from 'X-Group' contexts in Lower Nubia (Dunham 1963, 108, fig. 81, g; Pellicer and Llongueras 1965, fig. 44, type 155). Such shells are considered to have been imported from the Red Sea, this being the nearest source (cf. Geus 1984, 65).

Tumulus 4 (Figure 1.6) contained the burial of a sub-adult woman. The largest item of grave furniture is a bed, generally of the same form as those in the other two Tumuli. This was apparently dismantled, part of the frame lying behind the body, and legs in the corners of the chamber.

In this interment, only a single pottery vessel was present, this being a red wheel-made oil-bottle (T4/49C). This is intact apart from the rim, which was apparently broken in antiquity (Figure 8.3 – after Smith 1998c – Plate 8.8). Despite the ancient damage, it is clearly a vessel of the

1927, 48, pl. LXVII, fig. 33). The artefact concerned is in iron, rather than copper alloy, but the basic shape is comparable to that of the Gabati examples.

The artefacts on or associated with the body include a number of pieces of metal jewellery. An iron anklet, T4/1, of which only half was found, consisted of a semi-circular element with an approximately round cross-section (Figure 8.3, after Smith 1998a). One end (presumably originally both ends of the complete object) is flattened and bent back to form a loop. Plain types of anklet are known, for example, from Nag el-Arab (Pellicer and Llongueras 1965, fig. 32, 4, 5). These are reasonably close to T4/1 in overall form, consisting of a circular rod of metal bent round to form an open, slightly flattened loop. In both these cases, the ends are more or less rounded-off to form a plain conical profile, lacking the small loops

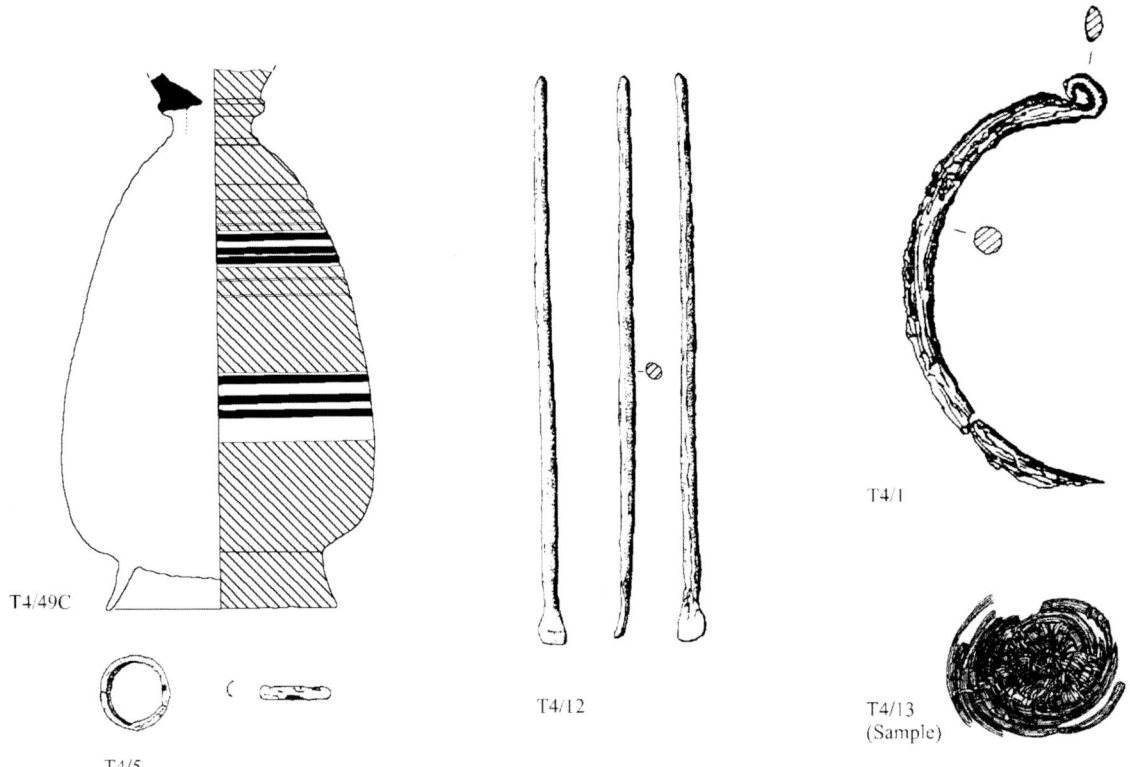

**Figure 8.3.** Items from the tomb chamber of Tumulus 4 (pottery – scale 1:4 other objects – scale 1:2).

same type as that in Tumulus 2 and would be expected to be from the same source. It differs in having painted decoration on the body, comprising two broad cream bands each with three sets of dark reddish-brown stripes. This was one of only two oil bottles recovered from the Gabati excavations in both the seasons having such decoration. Oil bottle <12405> has four sets of thin purplish-brown stripes (possibly intended to be dark red), one around the rim and three on the body (Smith 1998c, 184).

The other items placed in the burial chamber include both organic and inorganic materials. The former are a gourd, dom nuts and small baskets (see example of the basketry, Figure 8.3, T4/13). The latter consists of a copper-alloy spatula, or *kohl* spoon, T4/12 (Figure 8.3 – after Smith 1998a – Plate 8.9). The closest parallel to the form of T4/12 noted in the published reports is from Gamai, Cemetery 100, Grave 147 (Bates and Dunham

at each end. Another example of a similar type, although with some incised and relief decoration near each end, was found in burial Z4 at Gamai (Bates and Dunham 1927, pl. LXVIII, fig. 51). The closest type, although not identical, is another anklet from Gabati itself, <12411>, also from a Post-Meroitic burial (Edwards 1998c, 97; 1998d, 129, fig. 5.12).

A copper-alloy finger ring of thin metal remained in position on one hand (T4/5, Figure 8.3). The only ring of closely similar form to T4/5 noted in the published literature is one from Ballana, Tomb B.80 (Emery and Kirwan 1938, pl. 85, B). It appears to be relatively thin-walled, and although the material of which it is made is not clear from the illustration or from the grave catalogue, it appears to have a surface differing in texture and colour from the iron artefacts in the Ballana graves. A single copper-alloy ring from the Rescue Season at Gabati is more irregular

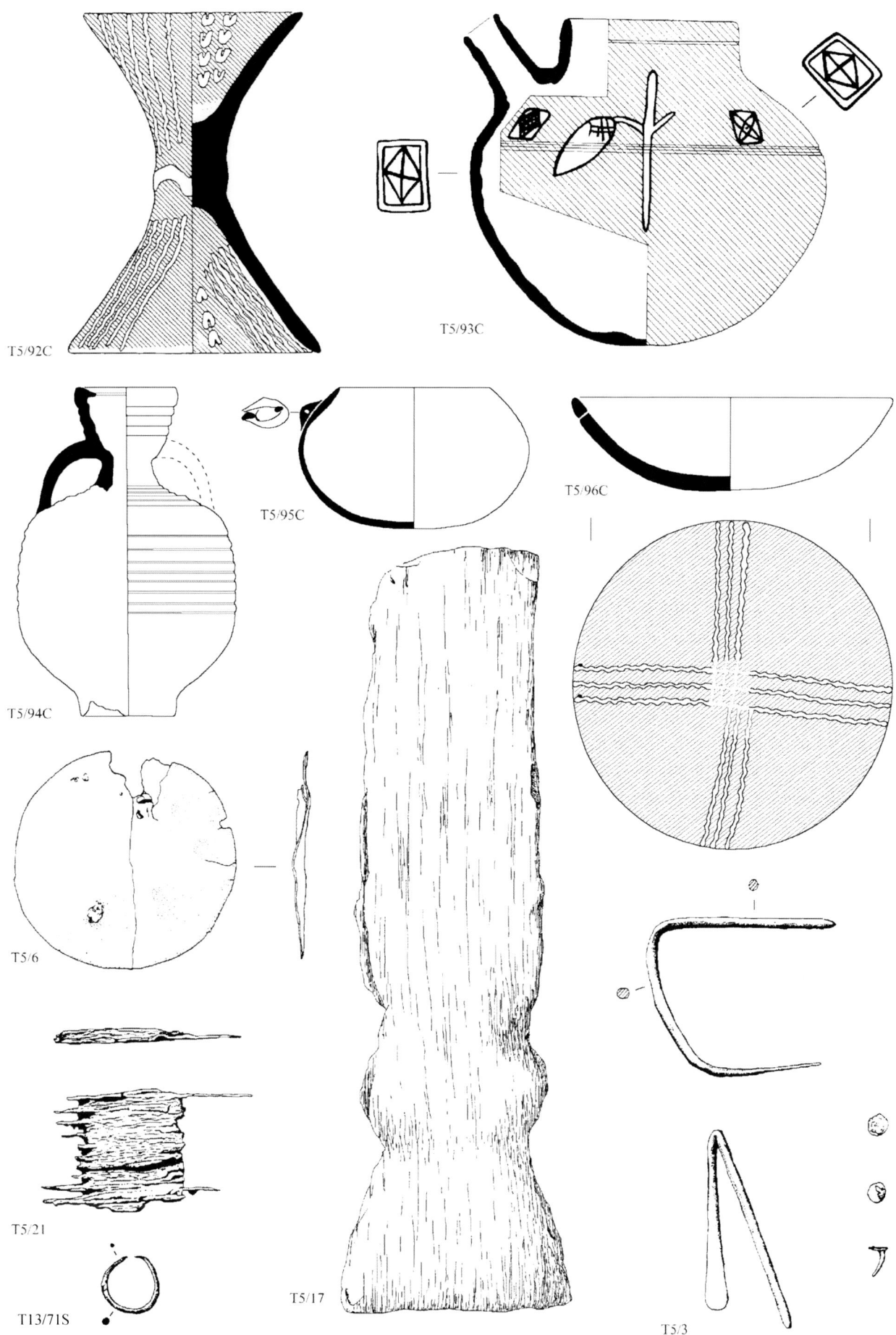

**Figure 8.4.** Items from Tumulus 5 and earring from Tumulus 13 (pottery – scale 1:4 other objects – scale 1:2).

in form and of approximately round rather than crescentic cross-section (Edwards 1998d, 129, fig. 5.12).

The remaining items comprise personal adornments and coverings associated with the body. Numerous beads were present, particularly in the region of the ankles. These may have been worn in life, or been placed in the tomb, in addition to the iron anklet. The beads range in colour from light tones, such as semi-translucent, white, yellow and orange, through amber and brown/white banded, green, blue and bluish-green, to dark shades, including dark purplish-brown and grey (see examples, Plates 8.10-8.12). Shapes include mainly spherical, barrel, cylindrical and biconical, with a few rare forms such as cuboid, ellipsoidal and lozenge-shaped.

Remains of some of the textile, dark brown in colour and of relatively coarse weave, was apparent beneath the chin and hands. This indicates that it is likely that the body had been laid upon a mat or blanket.

In Tumulus 5 (Figure 1.7), the burial of an adult woman, the usual bed is present, albeit of somewhat more massive construction than in other interments. As noted in other tumuli, the legs (see example T5/17, Figure 8.4) lay one in each corner of the chamber, but in this case part of the bed-frame lay across the neck of the body. This indicates that this bed could have been installed whole, and collapsed, rather than having been dismantled, unless part of the bed frame placed to one side of the body had later fallen across it.

This burial is exceptional amongst those excavated in the Test Excavations, and even amongst the Post-Meroitic graves in the excavated portion of the cemetery as a whole, for the range and number of objects buried with the deceased. Five pottery vessels were present, all close together near one corner of the chamber. These comprised the following: a handmade potstand of biconical form T5/92C (Figure 8.4 – after Smith 1998c, Plate 8.13), and a bowl, T5/96C, with dark purplish-red slip and decoration in white, composed of two sets of three wavy lines in white paint, extending from the centre of the base interior, approximately at right angles, also in thick white paint (Figure 8.4, design restored and corrected from Smith 1998c, Plate 8.14); another handmade bowl, T5/95C, with inturned sides, red slipped and with a single oval lug (Figure 8.4 – after Smith 1998c, Plate 8.15). These handmade vessels are associated with two wheel-made ones. The first is a red-slipped spouted jar (T5/93C). This has incised decoration in the form of grooves, and painted geometric decoration in black lines on isolated white panels, together with a vegetal motif in the centre of each side. The direction of this motif is reversed on one side of the vessel with respect to the other side (see Figure 8.4 – after Smith 1998c, with vegetal motif restored, Plate 8.16, reverse side showing best-preserved vegetal motif). The second wheelmade vessel, T5/94C, is a red ribbed 'table amphora' originally having two handles (Figure 8.4 – after Smith 1998c, Plate 8.17, showing the form of the rim and handle attachments).

The handmade wares are of types known in Post-Meroitic to 'Transitional'/early Christian contexts. The potstand is reasonably well paralleled in the excavations of the Middle Necropolis at Meroe. The closest form to the Gabati specimen amongst those illustrated by Garstang is his Type P4 from Tomb 307 (Garstang et al. 1911, 40, 42, pl. XLII, no. 3). Two further parallels are illustrated by Lenoble from a tomb near Berber, dated to c. 5$^{th}$ century AD. Neither is identical to T5/92C but one is close in form since it has a relatively short central stem, as does the vessel from Gabati (Lenoble 1991, 170, 174, fig. 3, no. 4).

In the same way, the bowl with a similar style of decoration (T5/96C) is, approximately, similar to a number of shallow open bowls that are known from contexts dated to the Post-Meroitic or early Christian periods. It is reasonably similar to examples known from Burri. This site as a whole was dated to the 7$^{th}$ century AD, within the early Christian period (Addison 1930, 286-288; fig. 1, nos. 3 and 4). Close parallels for the overall form of the Gabati vessel T5/95C are vessel nos 6 and 9 from the tomb near Berber (Lenoble 1991 fig. 4), although the latter is somewhat greater in maximum diameter relative to the height. Other examples of bowls of this form with lugs occur further south. They are known from early contexts at Soba (Welsby 1998b, 89, fig. 46, 6.2N) and from sites along the Blue Nile dated, in general terms, to the Post-Meroitic period, such as those from the cemetery at Qoz Nasra on the Blue Nile, (Edwards 1991, 53, pl. IX).

The characteristics of the 'table amphora' are comparable to Adams' 'Family A' and Hayes' 'Egyptian A' ware. In form, T5/94C is paralleled within Adams' classification of 'Table Amphorae' only by J13. This form resembles the Gabati vessel in having a rather squat cylindrical body with a footed base, two loop handles and a relatively narrow neck. J13 differs strongly in the shape of the rim in that, although the lower part is out-flared, the upper part of the rim is inward-sloping (W. Adams 1986, 144, fig. 55). As in the case of the oil bottle, vessels with similarities to the overall form of the 'table-amphorae' are also known from Elephantine. The closest example is that of Form K617 (Gempeler 1992, Abb. 118, no. 1) although the shape of the body is not identical.

The spouted jar can be quite closely paralleled amongst the forms characteristic of Adams' Ware R2 in Group NII (the 'X-Group' wares), being similar to form G42. Ware R2 is a 'Transitional' ware dated, in Lower Nubia, to AD 550-650 (W. Adams 1986, 470, fig. 269). The form is known from Old Dongola, from the fortifications and other 'Transitional' contexts (Godlewski 1991, 116-117, fig. 7, b; pers. comm. 1994). The style of decoration is similar to motifs on the Transitional or Early Christian pottery both from the Old Church and Building X and from a kiln site at Old Dongola (Pluskota 1990, 319, fig. 5; 1991, 39; 41, no. 1; 42, no. 8; 44, no. 15). A similar geometric motif is known on a sherd from Meroe (Garstang et al. 1911, pl. LI), although it appears to be part of a larger design and might not seem so similar to the Gabati decoration if a larger portion were visible. There are parallels at Soba, where several types of cross-hatched geometric decoration and some forms of vegetal motifs occur (Welsby 1998b, fig. 74, Decoration Types 700-703, fig. 77, Decoration Type 776).

In association with the bed and pottery vessels are several objects of inorganic and organic materials. The first of the former is a circular copper-alloy mirror (T5/6,

Figure 8.4, after Smith 1998a, and Plate 8.18). It is of a type known in both similar and more elaborate forms from the Meroitic period, at Meroe and Faras, and in similar relatively simple forms, with the addition of a short tang, in 25th Dynasty-Napatan contexts (Dunham 1963, 106, fig. 79, o, 358, fig. 190, F; 374, fig. 202, C; Griffith 1924, pl. LV, 21). Secondly, there is a copper-alloy spatula, bent around two right angles, and a nail (T5/3, Figure 8.4 – after Smith 1998a, Plate 8.19). The former has some similarities in form to the example from Gammai, Cemetery 100 Grave 147, in iron (Bates and Dunham 1927, 48, pl. LXVII, fig. 33) and from Karanog, albeit in a somewhat earlier context than Tumulus 5 (Woolley and Randall-MacIver 1910, pl. 36, 7384, 7386, 7397). Thirdly, a shallow stone bowl, or more probably a mortar, was found near the knees of the skeleton (T5/97S, Figure 8.5 – after Smith 1998a, Plate 8.20). Mortars have been recovered from both Meroitic and Post-Meroitic (particularly 'X-Group') burials, as at Ballana. Examples, slightly more circular in plan but with only slight or moderate 'lug'-type projections on the corners, occur in Tomb 15 at Ballana (Farid 1963, 60, 131, fig. 68, 3) and in the Faras Meroitic cemetery (Griffith 1924, pl. LVII, 22).

The two main organic artefacts comprise a *kohl* pot and lid of turned ivory and a double-sided wooden comb (T5/5, Figure 8.5; T5/21, Figure 8.4 – after Smith 1998a, Plates 8.21 and 8.22). In very general terms, the style of the former and its methods of construction, particularly the technique of turning and of drilling out the interior, are similar to some items from the Ballana and Qustul tombs, although no vessel of this particular form has been published from there. The general style of the base of T5/5 can be quite closely paralleled in a lid from Q17 (Q17-60), as can the method of manufacture and finishing, the lid being lathe-turned and polished. A second lid, Q17-59, also lathe-turned and polished, has some stylistic similarities with the stopper on T5/5, but is more elaborate (Emery and Kirwan 1938, vol. I, 343, vol. II, pl. 86, D). A small *kohl* pot lid similar in general conception to T5/5 is known from Arminna West,[1] although it is assigned a rather earlier date than the Tumulus 5 burial, in the terminal Meroitic (Trigger 1967, 42, 84, 86, fig. 27a). An object of somewhat similar form to the stopper was recovered from Soba. This is considered to be the stopper of a water bottle, rather than a *kohl* pot. The piece does not provide a close parallel, but is of an analogous form (Shinnie 1955, fig. 33, 3). Hence, there are a number of elements of shape and style exhibited by T5/5 that can be paralleled individually in several similar types of containers from the later Meroitic through to the Post-Meroitic (X-Group) burials. Despite this, a vessel with the elements combined in similar proportions and level of elaboration has not been noted to date.

In the case of the comb, whilst such objects are known, for example from the Ballana and Qustul burials, they tend to be more elaborate in decoration, but to have only a single set of teeth (Emery and Kirwan 1938, 41, pl. 86, E, Q3-83). The closest parallels to the Tumulus 5 comb are from another burial at Gabati, two examples being recovered from Tumulus 124. These appear to be of very similar type, although one does appears to lack a double row of teeth (Edwards 1998d, 128, fig. 5.10).

Items associated with the body comprised a similar assemblage of beads, leather and textile objects encountered in the other Tumuli, apart from the absence of metal jewellery such as anklets or rings. The beads, found especially near the ankles and neck, exhibit the same range of main shapes and colours as seen in the other tumuli, particularly Tumulus 4. There are a few types not seen in the other Test Excavation burials, including one gold-in-glass bead and one glass bead having a green background with yellow roundels, with dark brown radial lines, irregularly spaced around the perimeter,[2] and irregular trapezoidal shapes outlined in dark brown on the ends (see examples, Plates 8.23-8.26).

In a few cases, it has been possible to gain an indication of the order of the beads, since strung examples survived. One set from the Tumulus (T5/9) makes up a short section of 'necklace' comprising two strings of dark blue-green annular glass or faience beads on either side of a single white cylindrical bone bead. Another set (T5/2) has a couple of examples of cylindrical bone beads with small annular dark blue to blue-green glass beads stuck in each end. From this, a 'necklace' was reconstructed, with the multicoloured beads at the same findspot being strung together in a symmetrical fashion (Plates 8.25 and 8.26).

Tumulus 6 (Figure 1.8) contained the 'poorest' burial of the Post-Meroitic graves investigated during the Test Excavations. This burial, of an elderly man, contained no bed and no objects placed around the chamber, all material being associated closely with the body. This was covered by a shroud and laid on a 'blanket'. These are of dark brown, or dark brown and natural striped wool. Much leather, probably goat leather, forming an 'apron' or kilt was found over the lower limbs. The only other artefacts were fragments of worked wood, including a piece shaped into a 'finial' together with leather strips, in the fill of the Tumulus (Figure 8.5, T6/55S). This is likely to have been a bier, seemingly somewhat different from the substantial beds in the other Tumuli.

Leather and textiles were common to several of the burials. In the case of the textiles (Taylor 1998), four main types were present. One comprises a band of reddish-brown woollen textile from the head of the body, in Tumuli 1 and 2. This is likely to have been from a form of cap or other headgear. It is noticeable for the very fine quality of the weave. The second major textile items were woollen blankets, in several cases placed underneath the body. These can be plain, often dark brown, or decorated with narrow stripes, including stripes of blue and light buff (the natural colour of the cloth). Thirdly, there were the wrappings considered to be shrouds, mainly of plain dark brown wool. Fourthly, textiles coloured in brown red and blue, of wool in cases where they were identifiable, occurred in Tumuli 2 and 5.

Leather forms an important material within the burials. Leather objects, likely to have been garments comprising a form of 'apron', or kilt, were around the lower limbs of several of the bodies. In Tumulus 1 a sheath (T1/19), showing evidence of stitching, was present with the iron

---

[1] I am indebted to D. Q Fuller for drawing my attention to this find.

[2] Terms used are those of Beck (1928, 3).

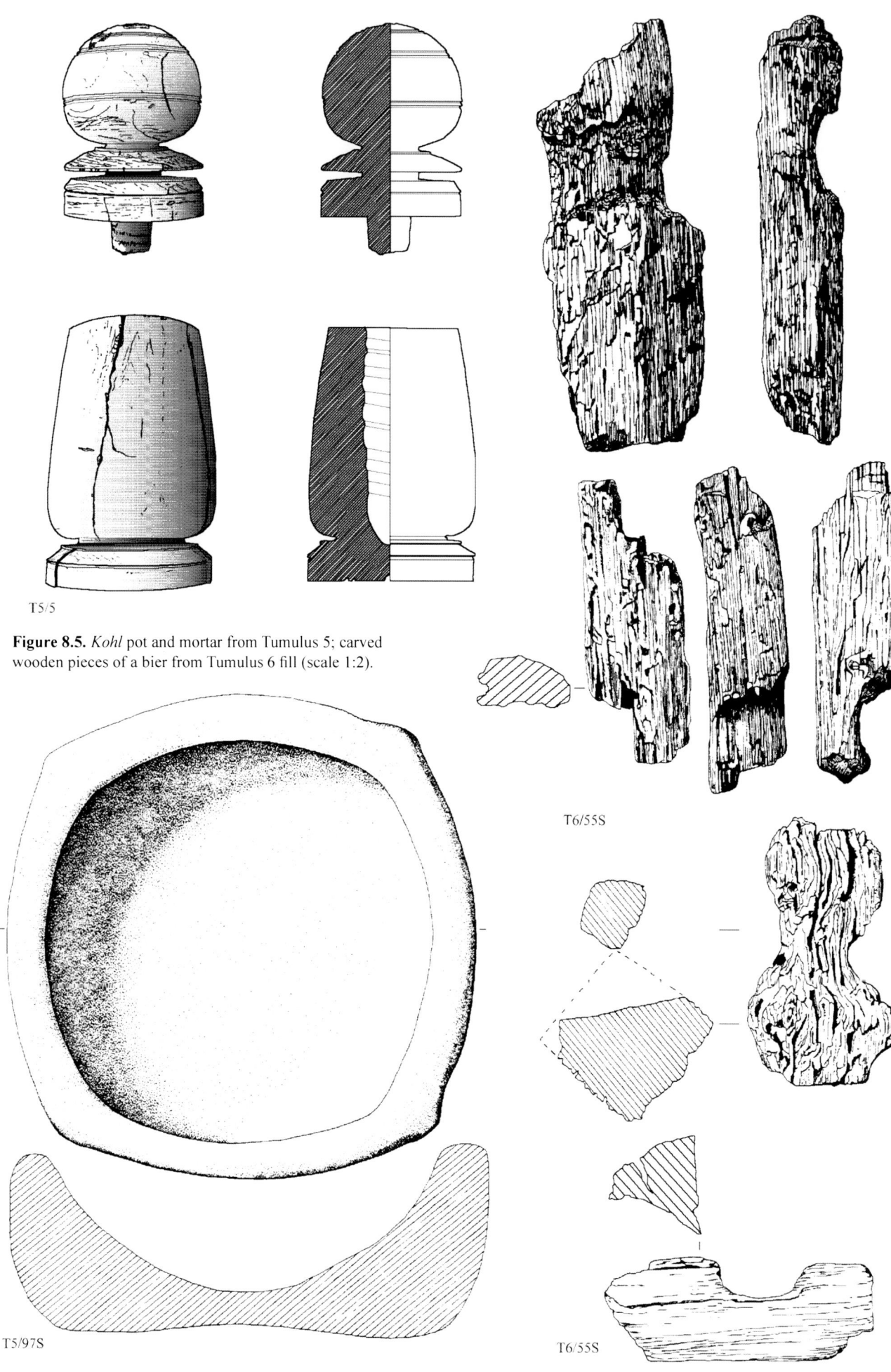

**Figure 8.5.** *Kohl* pot and mortar from Tumulus 5; carved wooden pieces of a bier from Tumulus 6 fill (scale 1:2).

knife/dagger blade. Examination of the leather (Wills, this volume) showed that the leatherwork under or on the bodies was largely of goat or kid, whilst the leather associated with the bed stringing included goat leather, but was possibly of bovine leather for some parts of the webbing.

Tumulus 13, being a Christian-period burial, did not contain any objects other than the shroud of dark greyish-brown wool wrapping the body, and some personal adornments. There was a copper-alloy earring (T13/71S, Figure 8.4), similar in shape to earrings from one of the 'Coptic' burials at Gamai, (Bates and Dunham 1927, 97, 99, pl. XXXVII, 4, W; pl. LXXII, fig. 45) and from Christian contexts at Soba and Geili (Shinnie 1955, fig. 31, 1; Caneva 1988c, fig. 37, b, top left). The only other items were beads, including banded blue and white, yellowish-brown carnelian, biconical amber and banded carnelian and a unique amber seven-sided facetted bead (Plate 8.27).

## Consideration of the Tomb Groups

Investigation of the main artefacts in the tomb groups and the degree to which they can be paralleled elsewhere indicates that they can be divided at two levels. At one level, they can be divided into those associated closely with the body, mainly covering it or under it, and those placed around the body. At another level, the grave goods can be divided into three classes according to their source: firstly, material that conforms to what is so far known of artefacts of the period within the Shendi Reach and southwards, generally assignable to local sources; secondly, those objects that, on the basis of their material and style, are reasonably certain to have been imported from outside the immediate region; thirdly, artefacts that are probably but not indubitably from outside the region. This is the case, for example, with the spouted jar from Tumulus 5, stylistically similar to ceramics known from the Dongola Reach, but for which it is not possible to carry out fabric studies to confirm this. It is the case with the Tumulus 5 *kohl* pot, which is not closely paralleled in excavations to date in the local region. As noted by Edwards (1998d, 127) ivory has been generally assumed to be a Nubian export, but evidence for local working is lacking. It is possible that the raw material could have been exported to the north, and some of the items made from it re-imported. In this case, and in the case of the other imported, or probably imported, material its significance in terms of the burials lies in the possibility that such objects were of 'value' and could have conferred 'prestige' on their owners, thus acting as some index of social status.

The nine tumuli fully excavated in the Test Excavation season exhibited almost the full range of variation in grave goods seen in the whole area excavated during both seasons, ranging from the 'poor' burial of Tumulus 6 through to the 'rich' burial of Tumulus 5. Christian-period burials can lack grave goods, as in Tumulus 7 and the Christian burial dug into T11, but may contain some personal adornments as in Tumulus 13.

It is to be expected that the Tumulus 13 burial would contain no objects deposited in the burial chamber, but could have items closely associated with the body itself. Whilst the exact significance of this is not readily apparent, it can be compared with other burials of the Christian period. It appears that in the early part of the Christian period, even high-ranking ecclesiastical dignitaries could be buried without grave goods, clothing or even shrouds, as is the case for a burial in the Cruciform Church at Old Dongola. This has been interpreted as relating either to the Christian ideal of poverty, or to the concept of appearing at the Day of Judgement in a condition 'as created by God'. It is evident that burial without grave goods was not universal, even at this period, or did not apply so strictly to lay persons. Another burial in the Cruciform Church, considered to be of a 'Queen Mother', included vestments and grave clothes (Dzierzykray-Rogalski and Promińska 1978, 91-93; Jakobielski 1975, 358-360).

At Soba, excavations of Christian burials during both campaigns similarly revealed some examples of burials with fine clothing, and several with shoes or sandals. Other items were very rare; some objects presumably being associated with Christian burials were found in the crypt of Building A, but otherwise only examples of metal crosses from two tombs were reported from investigations in the 19th century (Welsby 1998c, 278). Bishops could be buried in their vestments, with a few grave goods. For example, the late 14th century burial of Bishop Timotheos at Qasr Ibrim contained shroud, clothing, a benedictional cross and his Letters Testimonial, confirming his consecration (Adams 1977, 480-481; Plumley 1975, 3-4). It thus appears that the Tumulus 13 burial generally conforms to the norm expected for Christian burial in terms of lack of grave goods but can be seen as part of a tradition, present in Nubia at least, in which personal items or even items associated with the deceased's office were permitted.

In the case of the pre-Christian Post-Meroitic burials investigated in the Test Excavation season, it appears that there are certain common elements, which can be noted. These are as follows:

1. Isolation of at least a part, or the whole, of the body from contact with the ground. This is achieved by laying the body on a mat or blanket in some cases, and is likely to have been done by the burial on a bed in other cases. In this respect, these Gabati burials represent the latest manifestation of a long tradition of burial in Nubia, certainly extending back to the Kerma period (cf. Adams 1977, 198). It is clear that in some burials the body was not actually placed on the bed, but the latter was dismantled and the portions placed in different parts of the chamber. In a few cases, not all elements of the bed were found in the chamber; one tumulus excavated in the Rescue Season contained only two legs, with the other two buried in the fill (Edwards 1998d, 124).

2. The covering of the body, either with blanket or shroud.

3. Provision of some form of clothing. In several cases where these garments could be identified, this apparently often comprised a form of 'apron', or kilt, made of leather

4. Provision of at least a jar, or jar and a bowl.

5. Presence of personal adornments, including anklets and necklaces.

However, there is evidently a degree of variation in the form and decoration of individual items, particularly those items forming the grave furniture, which may be considered as fulfilling the same 'role' in the different burials. Amongst the pottery, for instance, only the oil bottles are common between two burials. Even here they differ visually, being plain red in Tumulus 2 but painted in Tumulus 4. The *kohl*-stick and the spatula in Tumulus 5 and 4 are generally similar in form, but differ in detail. Regarding the textile, much of this is similar in the different burials, but again there are differences in the presence of decoration.

One of the main hypotheses for the interpretation of Post-Meroitic burial practices is that elaborated by Lenoble. He considers that the objects in burials of this period can be viewed as elements of a burial rite developed initially during the late Meroitic period, but which substantially continued into the earlier Post-Meroitic period, at least into 5th century AD. Aspects of this rite, in the Meroitic period, included pouring of libations of water or milk near the tomb, using ceramic and bronze vessels, purification with perfumed oil, a funerary banquet, appropriate positioning of the body, provision of a funerary bed and animal sacrifice. Other elements, particularly military symbolism, could also be present (Lenoble 1987a, 96-99; cf. Yellin 1982). These aspects of ritual are linked to the persistence of Isiac rituals in the Shendi Reach and further south (cf. Lenoble and Sharif 1992, 630-632) where the various combinations of vessels, including cup and jar, have been found in Meroitic to Post-Meroitic burials (Edwards 1996, 42).

Lenoble's hypothesis of the continuity between Meroitic and Post-Meroitic can be exemplified by his interpretation of the artefacts in a burial at Berber, dated to the 5th century. Here, he identifies the presence of the two most essential elements of the ritual: the libation, seen in the provision of a decorated cup, and censing, indicated by two biconical vessels. These are described as 'pot-stand or incense-burner' in the grave catalogue, but are taken to be the latter in the interpretation. There is also evidence for a funerary banquet, shown by the presence of ovoid bowls and small cups, whilst iron weapons are interpreted as a rite based upon a symbolic expression of military power (Lenoble 1991, 170-174).

When the Gabati Test Excavation burials are considered, Tumuli 1, 4 and 5, although probably later than the Berber tomb, do present features that could fit with this interpretation. Tumulus 5 has no cup similar to that at Berber, but the 'table amphora' could have served the same purpose. The 'pot-stand' is very similar in form and basic style of decoration to the two 'incense-burners' in the Berber tomb, whilst there is also an ovoid bowl which could have been used in the 'funerary banquet'. In this tumulus, there is nothing clearly linked to military symbolism, but this may correlate with the burial being that of a woman. Tumulus 1 contains the decorated bowl with very upright sides, which could have been used as a cup for pouring libations. Evidence for a 'funerary banquet' may be seen in the broad black bowl, possibly containing foodstuff, covered by textile. The dom nuts present in Tumulus 4 might also be related to this aspect of ritual. Tumulus 4 includes the arrowhead and knife blade or spear point, which have military connotations. There are features that do not conform to all the specific elements of ritual mentioned by Lenoble. For example, the 'incense-burner' in Tumulus 5 does not exhibit any sign of having been used for burning material, and the burial in Tumulus 1 lacks a vessel that could be considered to be either an incense burner or a container for oils.

The oil bottles present in Tumuli 2 and 4 could be related to purification, or used for libations. In Tumulus 2, the accompanying bowl could have been used for libation, or in the 'funerary banquet', as could the double-compartmented dish recovered from the fill (Smith 1998c, 179-180). Despite these points, there do not appear to be elements that can be related to all the main aspects of the 'Isiac' ritual, as is the case with the Berber burial. This could be interpreted as indicating that the rites observed by the community buried at Gabati were not strongly prescriptive, and that they could allow the presence of material expressing other social practices. If this were the case, it may be that not all artefacts associated with the burial were determined by the same set of values. Such a concept has been suggested, albeit for a largely late Meroitic context, by Näser. In a re-investigation of material from Cemetery 214 at Abu Simbel North, she makes a distinction between ceramic and non-ceramic grave-goods, noting that the former may be related to funerary libations and/or feasting. She points out that the non-ceramic goods, including mainly jewellery, clothing, and a few other artefacts such as bronze bowls, kohl equipment and a knife, are mostly closely associated with the body, and considers these to be 'almost supplementary', not being determined by religious ritual or, in this case, social rank (Näser 1999, 21-24).

In situations where the form of burials cannot evidently be linked solely to the religious beliefs of a society, interpretations of burial practices can be related to a wide range of aspects of society. A survey of the literature by Tarekegn (1997, 8-17) showed that aspects including gender, social status at death, (which could include such factors as age attained and occupation), individual 'wealth', group ideology, and concepts of purity and impurity have been adduced in the archaeological study of burial. Although the sample of burials excavated during the Test Excavations is far too small to make any definitive statements about correlation with the above aspects of society, it is possible to put forward some observations about the contents of the Tumuli. Possible associations with gender may be seen in that the metal jewellery and the kohl items are in the burials of females, whilst the only occurrence of arrowhead and knife blade/spear point is in a male grave. The oil bottles appear to be associated with female burials, both in the Test Excavation graves and in those excavated during the Rescue Season (Edwards 1998f, 205).

There is insufficient evidence as yet to determine whether there is any definite correlation between the burials and age of the deceased. There does not seem to be a great degree of variation between the quality of the grave assemblage between Tumulus 2, containing an adult woman and an infant, and Tumulus 4, with a sub-adult female. The only striking difference is in the ceramics, with the single vessel provided in Tumulus 4. This vessel, being the painted imported oil bottle, may have been an item of 'prestige' value and indicate that the occupant was not of so low a status as the older woman in Tumulus 2. The burial of Tumulus 6, an elderly man, contains no objects other than the textiles associated with the body. A bier, presumably used in the burial, was left in the grave fill, rather than being placed in the burial chamber. In this case, it is not clear whether the lack of grave goods relates to the age of the man at death, his lack of personal 'wealth', or to changes in beliefs that were taking place during the process of adopting Christian burial rites (cf Edwards 1998f, 206, relating to the burial of pottery under the tumulus rather than in the grave itself).

The variation in the amount and type of grave goods in the five burials may be associated with social status and/or personal wealth; at the present state of research it is not possible to distinguish between these aspects. Interpretation in terms of social status involves the assumption that the people buried at Gabati were part of the same social system as the occupants of the other cemeteries in the region investigated to date. It could then be argued that since the burials at Hobagi (Lenoble et al. 1994; cf Lenoble 1999), containing the highest, or amongst the highest, status individuals in the society, contained an outstanding amount and range of grave goods, that the elaboration of burial in terms of the multiplication of objects does indicate higher social position. On this basis, it would be possible to view the Tumulus 6 burial as a 'low' status individual, with the caveats noted above. This may correlate with the state of the skeleton, exhibiting osteophytosis (Filer 1994, 28). The individuals in Tumuli 1, 2 and 4 could be regarded as of 'intermediate' status, although they included objects possibly of 'prestige value' through being imports. The individual in Tumulus 5 would clearly be of 'high' status, at least locally, the burial containing two definite imports, one probable import and exhibiting at least a moderate multiplication of items of the same class relative to the other four burials under consideration.

## Conclusions

Even the small number of burials investigated during the Test Excavation season has provided much data of interest for the study of burial practices in the Shendi Reach. The evidence from the six fully excavated tombs has features consistent with more than one hypothesis for its interpretation. It will not be possible to decide which is the most likely without considering the full range of material exhibited by all the burials excavated in the two seasons' work at the site, together with information from more cemeteries in the area. The tomb assemblages have elements that could be considered as expressions of differences in gender, social status and/or wealth and, conceivably, age at death. These elements could also be interpreted as evidence of burial rites involving libations, purification and possibly the consumption of food. The burials may give some indication of changes in burial ritual and, by implication, associated religious beliefs, even within the later Post-Meroitic period. These changes, presumably related to the adoption of Christianity, would ultimately result in the virtual cessation of the practice of depositing goods within graves.

These tentative conclusions will be subject to revision when there is sufficient evidence available from Upper Nubia to undertake a large-scale study of the rural Post-Meroitic burial practices. Material for such a study is being collected, from excavations such as those already conducted at el-Kadada, near Shendi, at Jebel Makbor and other sites (Geus and Lenoble 1985; Geus et al. 1986; Lenoble 1987a, 1987b, 1992), but more evidence needs to be gathered. It is to this body of evidence that the investigations at Gabati will contribute.

## Acknowledgments

I am indebted to Mr W. V. Davies for permission to examine the artefacts brought back to Britain in the Department of Ancient Egypt and Sudan, British Museum and to Ms J. M. Filer, Dr D. A. Welsby, Mr C. Johnson and Ms B. Wills, Mr J. Hayman, Mr B. Green, Mr A. Brandon and Ms J. Johnson. Thanks are due to Ms C. Thorne and Mr W. Schenck for advice on drawing delicate items and to Mr A. England and Ms J. Doole for inking the small find drawings. Particular thanks are due to Ms C. Honeycombe, formerly of the Fitzwilliam Museum, Cambridge, for advice on conservation in the field for the Test Excavation season. For help and discussion in the course of the project I am grateful to Dr J. A. Alexander, Dr L. Bassell, Dr R. Coombes, Dr S. W. G. Davies, Dr D. N. Edwards, Dr D. Q Fuller, Ms S. Harrop, Dr J. Phillips, Dr P. J. Rose, Dr K. Seetah, Dr K. Smith and Dr T. Whitelaw. Most work was undertaken whilst the author was a Junior Research Fellow and member of Wolfson College, an Affiliated Scholar at the Department of Archaeology, and Fellow of the McDonald Institute, University of Cambridge. I am grateful to these institutions and to the members of staff: the Heads of Department, Dr C. Shell, Ms J. Rippengal, and Ms J. Boreham, then of the Department of Archaeology. All the photographs are by the author except for the mirror (GBT T5/6) by D. N. Edwards. Find GBT T1/22 described in the preliminary report as 'Coloured beads', was recorded in error for 'Textile/Leather sample' for which the present author takes responsibility.

# 9. Bioarchaeological Report from the Excavations from Meroe to Atbara 1994

Rebecca Whiting

## Introduction

This report outlines the bioarchaeological analysis of material recovered from excavations between Meroe and Atbara which are detailed in this volume. These remains, due to the nature of the project, were not found in close proximity to one another. They also span a considerable temporal period, from possible Neolithic, through Meroitic, Post-Meroitic and Christian. Therefore the remains cannot be treated as a coherent assemblage, between which comparisons can be drawn, and each burial will be presented individually.[1]

Sex is determined through examination of the dimorphic features of the pelvis and skull as set out in Buikstra and Ubelaker (1994). Analysis of the composite arch, pubic proportions and ventral arc after Bruzek (2002) are also used. The five classifications for recording sex are as follows; Male, Probable Male, Unknown, Probable Female and Female. Age determination of the individuals is based on the morphology of the pubic symphysis (Buikstra and Ubelaker 1994) and auricular surface (Lovejoy 1985) as well as molar wear patterns after Brothwell (1981). It should be noted that dental wear can vary between populations. Brothwell did not use Sudanese collections to determine his wear dependant age categories, and this method may, therefore, be less reliable than that using the pubic symphysis and auricular surface. It has only been used when these elements are unavailable. Age estimation for sub-adults was made according to dental development as set out in Ubelaker (1989, 64). The age cohorts used in the current report are as follows:

| | |
|---|---|
| Infancy | – 0-1 year |
| Early Childhood | – 2-5 years |
| Late Childhood | – 6-10 years |
| Puberty | – 11-15 years |
| Adolescence | – 16-19 years |
| Young Adult | – 20-34 years |
| Middle Adult | – 35-49 years |
| Old Adult | – 50+ years |

Where possible an estimation of the living stature was calculated, after Raxter *et al.* (2008), as these regression equations have been calibrated using an Egyptian sample. However, in most cases the level of preservation was too poor to calculate accurate results. Pathological changes were interpreted using, when appropriate, the operational definitions set out in Waldron (2009), with any additional references shown in the text.

An inventory of the dentition and the wear scores for each tooth present was set out for each individual, as follows:

Present, but not in occlusion (in crypt)
Present, development complete, in occlusion
Missing, no alveolar bone
Missing, alveolus resorbing/ed (ante-mortem loss)
Missing, no resorption (post-mortem loss)
Missing, congenital absence
Present, damage renders measurement impossible
Present, but unobservable (in crypt)

The wear scores were based on Smith's scores (1984, 45-46) for the anterior teeth, and Scott's (1979) molar wear scoring method which assigns a number (according to the level of wear) to each quadrant of a molar. The wear score '99' denotes that wear was not observable. In the case of sub-adults, where possible the development stages for each tooth are included (after Buikstra and Ubelaker 1994, 50).

Below is a list of wear scores for incisors, canines and premolars taken from Smith (1984, 45).

Incisors and canines
0. Missing or cannot be coded
1. Unworn to polished or small facets (no dentine exposure)
2. Point or hairline of dentine exposure
3. Dentine line of distinct thickness
4. Moderate dentine exposure no longer resembling a line

---

[1] It should be noted that a preliminary draft report was complied by Joyce Filer and Francis Thornton after the 1994 excavations, some elements of which have been incorporated into this final report.

5. Large dentine area with enamel rim complete
6. Large dentine area with enamel rim lost on one side or very thin enamel only
7. Enamel rim lost on two sides or small remnants of enamel remain
8. Complete loss of crown, no enamel remaining: crown surface takes on shape of roots

Premolars
0. Missing or cannot be coded
1. Unworn to polished or small facets (no dentine exposure)
2. Moderate cusp removal (blunting) exposure
3. Full cusp removal and/or moderate dentine patches
4. At least one large dentine exposure on one cusp
5. Two large dentine areas (may be slight coalescence)
6. Dentinal areas coalesced, enamel rim still complete
7. Full dentine exposure, loss of rim on at least one side
8. Severe loss of crown height; crown surface takes on shape of roots

These dental inventories are shown with the following layout, with UR representing the Upper Right side of the dentition for example:

Permanent Dentition

| UR | M3 | M2 | M1 | P2 | P1 | C | I2 | I1 | I1 | I2 | C | P1 | P2 | M1 | M2 | M3 | UL |
|---|---|---|---|---|---|---|---|---|---|---|---|---|---|---|---|---|---|
| LR | M3 | M2 | M1 | P2 | P1 | C | I2 | I1 | I1 | I2 | C | P1 | P2 | M1 | M2 | M3 | LL |

Deciduous Dentition

| UR | DP2 | DP1 | C | I2 | I1 | I1 | I2 | C | DP1 | DP2 | UL |
|---|---|---|---|---|---|---|---|---|---|---|---|
| LR | DP2 | DP1 | C | I2 | I1 | I1 | I2 | C | DP1 | DP2 | LL |

During excavation, each skeleton was given a unique identity number, comprising of Area, Tumulus (T) and Organic Material (OR) number. The orientation of each skeleton is given so that the long axis of the burial was read from the end at which the skull is located.

## Results

The figures below give an overall view of the preservation and sex and age distributions of the skeletal remains.

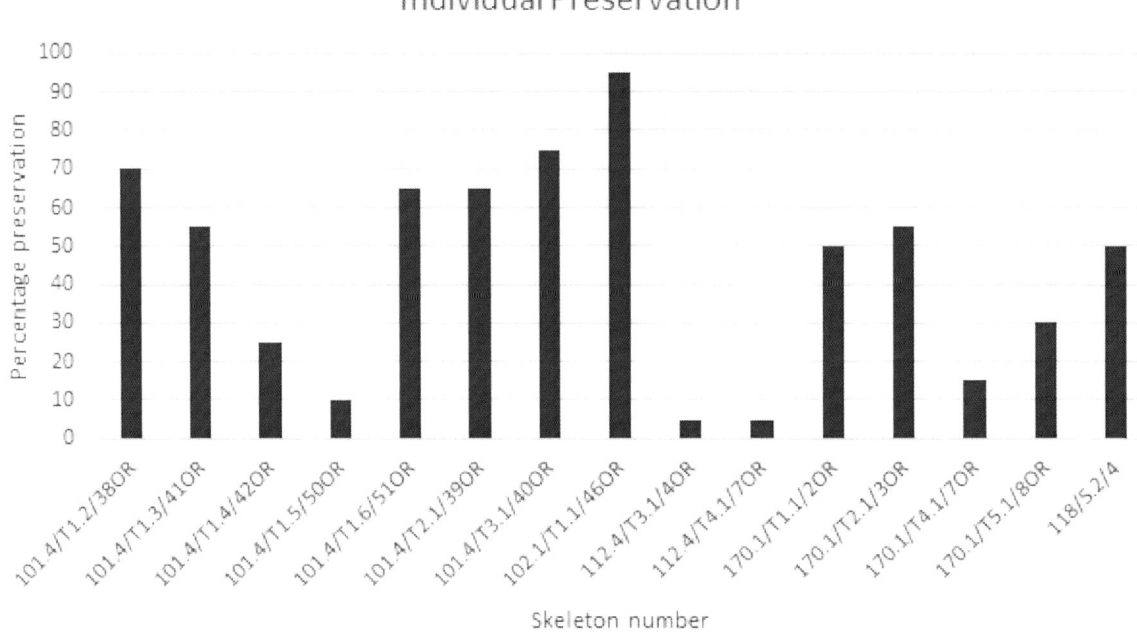

**Figure 9.1.** Approximate percentage of each individual skeleton present upon analysis, in 5% increments.

As is evident from Figure 9.2 three of the adult skeletons could not be assigned to a specific age category due to poor preservation. Table 9.1 provides scores for the pubic symphysis and auricular surface set out in their raw data form.

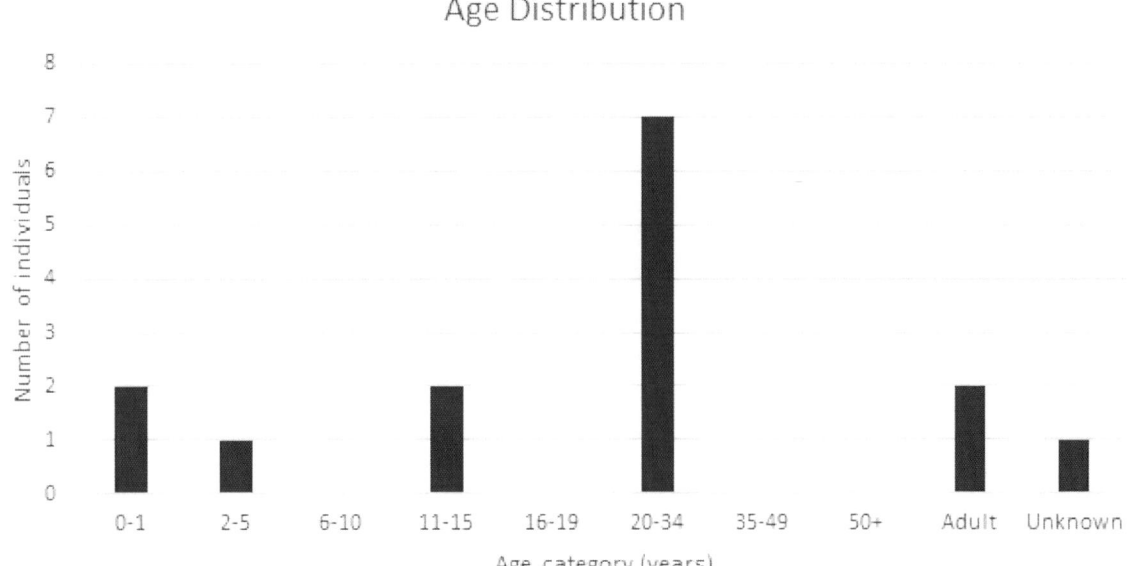

**Figure 9.2.** Bar chart showing the distribution of individuals in each age category across the collection.

**TABLE 9.1.** SCORES FOR PUBIC SYMPHYSIS ACCORDING TO TODD AND SUCHEY-BROOKS AS SET OUT IN BUIKSTRA AND UBELAKER (1994) AND SCORES FOR AURICULAR SURFACE AS SET OUT IN LOVEJOY (1985).

Note that in all columns results are displayed by left side/right side, with N.D denoting that the data is unavailable. Where data for all three is missing the dental wear according to Brothwell (1981) was used to determine age.

| ID | Sex | Age | Todd | Suchey/Brookes | Auricular Surface |
|---|---|---|---|---|---|
| 101.4/T1.2/38OR | Male | 20-34 | N.D | 3/N.D | N.D/3 |
| 101.4/T1.3/41OR | Male? | 20-34 | N.D | N.D | N.D |
| 101.4/T1.4/42OR | Male? | 20-34 | N.D | N.D | N.D |
| 101.4/T2.1/39OR | Male? | 20-34 | 6/5 | 3/4 | N.D/3 |
| 101.4/T3.1/40OR | Female | 20-34 | N.D | N.D | 2/2 |
| 102.1/T1.1/46OR | Female | 20-34 | 4/4 | 2/2 | 3/3 |
| 112.1/T4.1/7OR | Unknown | Adult | N.D | N.D | N.D |
| 118/T5.2/4 | Unknown | 20-34 | N.D | N.D | N.D |
| 170.1/T5.1/8OR | Unknown | Adult | N.D | N.D | N.D |

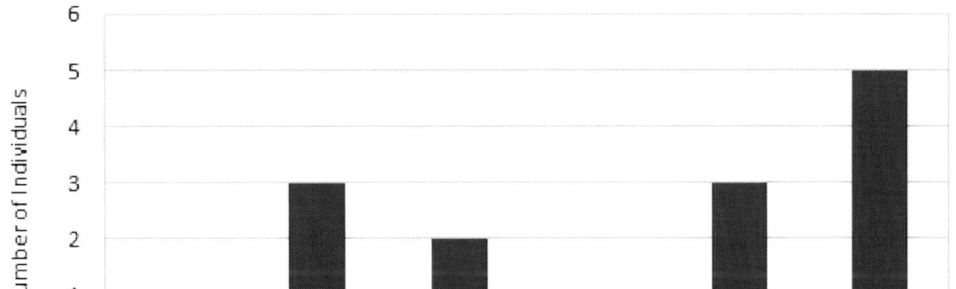

**Figure 9.3.** Distribution of individuals by sex category. Unknown refers to skeletons for which sex could not be determined due to poor preservation.

## Individuals Analysis

**Site 101.4**
A mixture of Post-Meroitic and Christian period burials.

*T1.2/38OR (Plate 9.1)*
*Burial Details* – Christian period
Skeleton in an extended supine position, with feet together and the lower arms and hands across the pelvis; buried in an east/west orientation.

*Condition and percentage present*
Approximately 70% of the skeleton is present. Bone surface is, for the most part, in good condition some epiphyses, where broken, were friable. Some areas of the feet and teeth are stained purple, a possible indication of Phosphatidate Phosphatase, an enzyme excreted by some fungi (Garrard Cole and Tony Waldron, UCL, pers. comms).

*Sex* – Male. Determined using sexually dimorphic features of the pelvis and skull.

*Age* – Adult 20-34 years. Determined only from dental attrition (Brothwell 1981), as poor preservation prevented use of any other method.

*Stature* – 169.1cm (±3.002) - Tibia Maximum Length 39.1cm

*Dental Inventory*

Inventory

| UR | 3 | 3 | 3 | 3 | 3 | 2 | 3 | 3 | 3 | 3 | 3 | 3 | 3 | 3 | 3 | 3 | UL |
|----|---|---|---|---|---|---|---|---|---|---|---|---|---|---|---|---|----|
| LR | 5 | 5 | 5 | 2 | 5 | 5 | 5 | 5 | 5 | 5 | 2 | 5 | 5 | 5 | 5 | 5 | LL |

Wear

| UR | 99 | 99 | 99 | 9 | 9 | 5 | 9 | 9 | 9 | 9 | 9 | 9 | 9 | 99 | 99 | 99 | UL |
|----|----|----|----|---|---|---|---|---|---|---|---|---|---|----|----|----|----|
| LR | 99 | 99 | 99 | 4 | 9 | 9 | 9 | 9 | 9 | 9 | 1 | 9 | 9 | 99 | 99 | 99 | LL |

*Pathology*
This individual presents some interesting pathological changes; joint disease of the left ankle in conjunction with the depression and compression fractures described below may suggest single or multiple traumatic events.
Trauma: The left tibia displays a depressed fracture (c. 22 × 14mm) on the lateral articular facet of the proximal epiphysis. Healing is present with new bone found around the edges of the fracture site.
Joint disease: Both clavicles exhibit false joints on the posterior side of the inferior surfaces, medial to the conoid tubercle. These pseudo-joints may have articulated with the coracoid process of the scapula, however neither scapula survives to corroborate this (see Plate 9.2). Rotator cuff disease is present on the right humerus at the insertion of the subscapularis, with new bone and pitting. Osteoarthritis is present in the left ankle joint on the tibia and talus, both eburnation and grooving are observable on the articular surfaces. This may be secondary to the trauma in the knee joint.
Spinal pathology: Several features are evident in the spinal column; cervical vertebrae 5 and 6 are fused possibly secondary to trauma (see Plate 9.3). Compression fractures are present in the bodies of the 8th thoracic and 1st and 2nd lumbar vertebrae. Disc herniation, or Schmorl's nodes, were observed in 1st and 2nd lumbar, perhaps as a result of the compression fractures. Marginal osteophyte is also present on the remaining lumbar vertebrae; this again may be secondary to the trauma seen elsewhere (see Plate 9.4).
Other: On the 8th left rib is an area of new bone, possibly indicating an infection of the lungs or pleura such as tuberculosis. No other new bone is observed on the ribs to assist in diagnosis.

*T1.3/41OR (Figure 2.2 and Plate 2.3)*
*Burial Details* – Post-Meroitic period
This individual was buried in a flexed position with a west/east orientation, facing south. The skeleton was found on a mat and the remains of a metal band were observed *in situ* found near to the skull (see pg. 13).

*Condition and Percentage Present*
Approximately 55% of this skeleton remains, all of which is in a poor condition, fragmentary and friable. None of the bone has a sound cortical surface effecting observations such as age and pathological changes.

*Sex* – Probable male. Determined using sexually dimorphic features of the skull only, as the pelvis is poorly preserved. Measurements of the femoral head (50mm) also suggest male according to the standards set out in Bass (1987, 218) and in comparison with other skeletons in this project (female range = 39-40; see below).

*Age* – Adult 20-34 years. Determined only from dental attrition (Brothwell 1981), as poor preservation prevented use of any other method. This may be affected by differences in diet and food processing methods compared with Brothwell's sample, and may, therefore, be less reliable than other methods of age determination.

*Dental Inventory*

Inventory

| UR | 5 | 5 | 5 | 2 | 5 | 5 | 5 | 5 | 5 | 5 | 2 | 2 | 2 | 2 | 5 | 5 | UL |
|---|---|---|---|---|---|---|---|---|---|---|---|---|---|---|---|---|---|
| LR | 2 | 2 | 2 | 2 | 5 | 5 | 5 | 5 | 3 | 3 | 3 | 3 | 5 | 2 | 2 | 2 | LL |

Wear

| UR | 99 | 99 | 99 | 4 | 9 | 9 | 9 | 9 | 9 | 9 | 9 | 4 | 4 | 22 | 99 | 99 | UL |
|---|---|---|---|---|---|---|---|---|---|---|---|---|---|---|---|---|---|
| LR | 4 | 12 | 16 | 4 | 9 | 9 | 9 | 9 | 9 | 9 | 9 | 9 | 9 | 15 | 12 | 4 | LL |

*Pathology*
Joint disease: Os acromiale was found in the left scapula; this non-metric variation is a failure of fusion of the ossification centres of the scapula. A pseudo articulation has been created between the two sections. Some pitting and osteophytic growth is observed, suggesting osteoarthritis, however OA can only be present in a true synovial joint, of which this is not an example.
Spinal pathology: Fine examples of Schmorl's nodes were observed in seven thoracic and one lumbar vertebrae. These are an expression of pressure exerted upon the vertebral bodies by small herniations of the inter-vertebral discs.
Dental pathology: Furrow form hypoplasia was found in the lower and upper molars. This may suggest a disruption during the developmental stage of life.

*T1.4/42OR* (Plate 2.4)
*Burial Details* – Post-Meroitic period
This skeleton showed signs of disturbance, even so it was possible to discern that it lay in a flexed position on its right side, in a west/east orientation with the face to the south. Arms appear to be flexed with the hands near the face.

*Condition and Percentage Present*
Approximately 20% of the skeleton remains, with most bones friable and fragmented, making identification difficult. Some areas of the bones and teeth have a mottled black discolouration.

*Sex* – Probable Male. Determined through the sexually dimorphic features of the pelvis and skull.

*Age* – Adult 20-34 years. Determined only from dental attrition (Brothwell 1981), as poor preservation prevented use of any other method.

*Dental Inventory*

Inventory

| UR | 3 | 3 | 2 | 5 | 5 | 7 | 7 | 5 | 5 | 5 | 7 | 5 | 5 | 7 | 2 | 3 | UL |
|---|---|---|---|---|---|---|---|---|---|---|---|---|---|---|---|---|---|
| LR | 2 | 2 | 2 | 5 | 5 | 5 | 5 | 5 | 5 | 5 | 5 | 5 | 5 | 2 | 2 | 2 | LL |

Wear

| UR | 99 | 99 | 29 | 9 | 9 | 9 | 9 | 9 | 9 | 9 | 9 | 9 | 9 | 99 | 16 | 99 | UL |
|---|---|---|---|---|---|---|---|---|---|---|---|---|---|---|---|---|---|
| LR | 13 | 17 | 27 | 9 | 9 | 9 | 9 | 9 | 9 | 9 | 9 | 9 | 9 | 27 | 17 | 11 | LL |

*Pathology*
Few observations could be made due to the poor condition of the remains.
Spinal pathology: Some pitting is observed on the body of the 1$^{st}$ thoracic vertebra and there is ossification of the ligamentum flavum throughout the surviving thoracic and lumbar vertebrae. The ligamentum flavum joins the neural arches of contiguous vertebrae and often ossifies in bony spurs. Positive correlation can be seen between the prevalence of ossification and activities involving excessive bending and loading of the spine (Waldron 2009, 81).

*T1.5/50OR*
*Burial Details* – Christian period
This burial was found covered by a rock, which obscured some of the burial features, however it was possible to determine that the individual was in a flexed position with an east/west orientation.

*Condition and Percentage Present*
Less than 10% of this individual is present. The bones are broken and only 50-75% have a sound cortical surface.

*Sex* – Undetermined due to age.

*Age* – Infant, birth-2 years. Age estimate is determined from dental development but due to poor preservation only the unerupted upper right canine was observed, exhibiting some root development suggesting an age of 6-9 months.

*Dental Inventory*

Inventory

| UR | 3 | 3 | 7 | 3 | 3 | 3 | 3 | 3 | 3 | 3 | UL |
|----|---|---|---|---|---|---|---|---|---|---|----|
| LR | 3 | 3 | 3 | 3 | 3 | 3 | 3 | 3 | 3 | 3 | LL |

*Pathology*
No pathological changes evident.

*T1.6/51OR* (Plate 9.5)
*Burial Details* – Christian period
This sub-adult was buried in a flexed position with an east/west orientation, laying on the left side with face to the south. The skull was near the feet of burial 101.4/T1.2/38OR and two large stones were found in association with the burial, one at the feet and one at the head of the remains.

*Condition and Percentage Present*
Approximately 65% of this individual survives; most bones are broken and fragmentary with some flaking.

*Sex* – Undetermined due to age.

*Age* – Puberty 11-15 years. Dental development suggests 15 years (± 3 years) with the 3$^{rd}$ molar present but not in occlusion. Skeletal maturation suggests 12-13 years from measurements of the clavicle. Suggesting this individual was small for this age, however measurements fall within the error range for dental development.

*Dental Inventory*

Inventory

| UR | 2 | 2 | 2 | 2 | 2 | 7 | 2 | 2 | 2 | 2 | 2 | 2 | 2 | 2 | 2 | 7 | UL |
|----|---|---|---|---|---|---|---|---|---|---|---|---|---|---|---|---|----|
| LR | 2 | 2 | 2 | 2 | 2 | 7 | 7 | 3 | 7 | 3 | 3 | 2 | 2 | 2 | 2 | 7 | LL |

Wear

| UR | 4 | 5 | 10 | 1 | 1 | 1 | 1 | 1 | 1 | 1 | 1 | 1 | 1 | 7 | 6 | 0 | UL |
|----|---|---|----|---|---|---|---|---|---|---|---|---|---|---|---|---|----|
| LR | 4 | 7 | 10 | 1 | 1 | 1 | 1 | 9 | 1 | 9 | 9 | 1 | 1 | 12 | 10 | 0 | LL |

*Pathology*
Other areas of new bone can be seen on seven of the ribs, near to the head or neck of each bone, hypervascularity is also present around these areas of new bone. New bone is also observable in the thoracic region on the vertebral bodies and on the manubrium. These observations would suggest an infection of the lungs or surrounding pleural tissue, such as tuberculosis. This may have been a chronic infection due to the frequency of sites of new bone found in the individual. Dental pathology: Carious lesions are present on the lower right 1$^{st}$ and 2$^{nd}$ molars on the occlusal surface and in the buccal groove respectively. Furrow form hypoplasia is also evident on the incisors, canines and right 3$^{rd}$ molars suggesting a disruption in development.

*T2.1/39OR* (Plate 2.5)
*Burial Details* – Christian period
This adult skeleton was buried in an extended, supine position with an east/west orientation.

*Condition and Percentage Present*
Approximately 65% of this individual is present, the majority of which has good preservation with little to no breakage and a sound cortical surface. Some purple staining to the teeth is suggestive of Phosphatidate Phosphatase, an enzyme excreted by some fungi, and would have happened post-mortem.

*Sex* – Probable male, determined through sexually dimorphic features of the skull and pelvis.

*Age* – Adult 20-34 years. Determined through the examination of dental attrition, pubic symphysis and the auricular surface. However there is some unusual enamel chipping which may make the dental attrition scores unreliable.

*Dental Inventory*

Inventory

| UR | 7 | 2 | 2 | 7 | 5 | 7 | 7 | 5 | 5 | 5 | 7 | 7 | 7 | 7 | 7 | 7 | UL |
|----|---|---|---|---|---|---|---|---|---|---|---|---|---|---|---|---|----|
| LR | 7 | 7 | 7 | 7 | 2 | 5 | 5 | 5 | 5 | 5 | 5 | 5 | 7 | 7 | 2 | 2 | LL |

Wear

| UR | 15 | 23 | 27 | 5 | 9 | 9 | 9 | 9 | 9 | 9 | 9 | 5 | 5 | 28 | 21 | 13 | UL |
|----|----|----|----|---|---|---|---|---|---|---|---|---|---|----|----|----|----|
| LR | 99 | 99 | 28 | 9 | 1 | 9 | 9 | 9 | 9 | 9 | 9 | 9 | 9 | 28 | 21 | 15 | LL |

*Pathology*
Spinal pathology: Inter-vertebral disc disease is found in the cervical and thoracic regions. This is due to the degeneration of the disc causing the space in between the vertebrae to collapse. It is considered a disease of age and is rare under 40 years, although it may be that this individual engaged in particularly strenuous activity causing early degeneration of the discs. Also a possible compression fracture is seen in the 12th thoracic vertebra. This is characterised by a marked height difference descending from the posterior to the anterior of the vertebral body (height difference of 9.6mm). A depressed lesion is found on the superior surface of the centrum, with some evidence of remodelling.
Other: Though the ribs are very fragmentary it is possible to observe new bone on three of the right ribs. This suggests possible infection of the lungs or pleura.
Biological variation: A non-metric variation known as a hypertrocanteric fossa is found on both femurs, giving the proximal shafts a flattened appearance. Dental pathology: The upper left 1st molar and lower left 2nd molar display chipping of the enamel and dentine, in the disto-lingual and mesio-lingual positions respectively. The teeth may have been used as tools, or broken during mastication; it was clear that this was not a post-mortem event as there is a build up of calculus on the chipped areas. The use of dental attrition as an age estimation may be unwise in this case, as the results could be skewed, with the unusual wear indicating an older age category. There are also large interproximal carious lesions between upper left 2nd and 3rd molar; 8 × 3mm and 9 × 4mm in size respectively. This may imply that the diet included some sticky carbohydrates which were hard to clean from in between the teeth.

*T3.1/40OR* (Plate 2.6)
*Burial Details* – Christian period
This individual was found in an extended, supine position with an east/west orientation and the face looking to the south. The forearms were crossed across the pelvis.

*Condition and Percentage Present*
Approximately 75% of this skeleton is present, with the bone surface in good condition though some elements are broken. Some purple staining on the skull and tibiae suggest possible fungal activity and rodent gnawing is also observable on the right ulna.

*Sex* – Female, determined using sexually dimorphic features of both the skull and pelvis.

*Age* – Adult 20-34 years. Determined using dental attrition and morphological changes to the auricular surface.

*Stature* – 159cm (± 2.517) – Femur Maximum Length 43.7cm.

*Dental Inventory*

Inventory

| UR | 7 | 2 | 7 | 7 | 7 | 5 | 5 | 5 | 5 | 5 | 5 | 5 | 7 | 7 | 7 | 7 | UL |
|----|---|---|---|---|---|---|---|---|---|---|---|---|---|---|---|---|----|
| LR | 5 | 7 | 7 | 7 | 7 | 7 | 5 | 5 | 5 | 5 | 5 | 7 | 7 | 7 | 7 | 7 | LL |

Wear

| UR | 6+ | 12 | 13 | 9 | 9 | 9 | 9 | 9 | 9 | 9 | 9 | 9 | 9 | 18 | 10+ | 8 | UL |
|----|----|----|----|---|---|---|---|---|---|---|---|---|---|----|-----|---|----|
| LR | 9+ | 99 | 21+ | 9 | 3 | 9 | 9 | 9 | 9 | 9 | 9 | 4 | 9 | 18 | 99 | 4+ | LL |

*Pathology*
Joint disease: Pitting on both humeral heads suggests possible early onset of impingement syndrome. This occurs when inflammation around the tendons holding the humerus in place, cause the superior part of the humeral head to contact with the inferior part of the acromion. The operational definition calls for the presence of eburnation (not present in this case) so diagnosis of this condition cannot be confirmed. Spinal pathology: Some osteoarthritis is observable in the lower thoracic region, associated with the costo-vertebral and costo-transverse joints.
Other: There is new bone present in the left and right maxillary sinus, particularly in the left where the root of the 1st molar has protruded through into the sinus. This may suggest a small infection in the sinus causing the production of new bone. An erosive lesion is present on the left clavicle at the insertion of the costo-clavicular ligament. This may suggest an injury to the ligament itself during life. Most notably there are small meningiomas present on the right parietal bone at the sagittal sinus. The meningeal vessels run over the inside of the parietal bones radiating from pterion to the sagittal suture. A meningioma is a small, usually benign, tumour at the sagittal end of the vessels. Waldron (2009) defines the meningioma as a scooped out depression that must be fed by an enlarged meningeal vessel as was the case here.
Dental pathology: Supernumerary teeth are also present in this individual in the mandible between P1 and P2, while

the sockets for these are observable on both sides only one supernumerary tooth is present. The other has most likely been lost post-mortem (see discussion, pg. 132).

**Site 102.1**
Christian period burial.

*T1.1/46OR* (Figure 2.6, Plates 2.7 and 2.19)
*Burial Details* – Christian period
This skeleton was buried in an extended, supine position with an east/west orientation and the face looking to the north.

*Condition and Percentage Present*
This skeleton was well preserved with approximately 95% of the skeleton present. The bone surface had little to no breakage or flaking though it had some post-mortem blue/grey staining on the ribs.

*Sex* – Female, determined using sexually dimorphic features of both the skull and pelvis.

*Age* – Adult 20-34 years. Determined using dental attrition, pubic symphysis and auricular surface analysis. As in skeleton 101.4/T2.1/39OR, the dental attrition is not reliable due to the pathological changes (see below).

*Dental Inventory*

Inventory

| UR | 4 | 2 | 2 | 2 | 2 | 2 | 2 | 2 | 2 | 2 | 2 | 2 | 2 | 2 | 2 | 2 | UL |
|----|---|---|---|---|---|---|---|---|---|---|---|---|---|---|---|---|----|
| LR | 2 | 2 | 2 | 2 | 2 | 2 | 2 | 2 | 2 | 2 | 2 | 2 | 2 | 2 | 2 | 2 | LL |

Wear

| UR | 99 | 26 | 28 | 7 | 7 | 4 | 4 | 4 | 4 | 5 | 4 | 7 | 6 | 28 | 26 | 16 | UL |
|----|----|----|----|---|---|---|---|---|---|---|---|---|---|----|----|----|----|
| LR | 16 | 26 | 28 | 6 | 4 | 4 | 4 | 4 | 4 | 5 | 5 | 7 | 6 | 28 | 25 | 16 | LL |

*Pathology*
Spinal pathology: Many pathological conditions can be observed in this individual. Osteoarthritis is present in the thoracic vertebrae on the inter-vertebral and costal joint facets and on the corresponding ribs. Inter-vertebral disc disease is seen on the 5$^{th}$, 6$^{th}$ and 7$^{th}$ cervical as well as the 1$^{st}$ lumbar and 1$^{st}$ sacral segment, observed as pitting and marginal osteophyte. There is also an enlarged arterial passage on the left transverse foremen of the axis. It has a scooped-out appearance and is notably more porous that the opposite foremen. This is possibly a sign of a tortuosity (a loop in the vertebral artery) or aneurism (swelling in the vertebral artery), both of which create pressure defects on the bone (Antoine and Waldron, 2002).
Dental pathology: Some florid pathological changes have taken place around the upper right 1$^{st}$ molar of this individual. The pulp chamber has become exposed with a marked loss of alveolar bone around the tooth. There is a large lytic lesion at the apex of the root on the internal surface of the maxilla – probably a drainage hole; this is accompanied by new bone formation around the lesion on the maxilla and extending onto the palette, due to an extensive infection. The drainage hole, new bone and ragged texture of the lesion suggest a periapical abscess possibly caused by non-carious pulp exposure due to a crack or chip in the tooth during life. Many of the teeth display secondary dentine laid down when the attrition had reached the pulp chamber. Possibly as a result of substantial pyogenic secretions (pus), there is a general resorption of the alveolar process at the back of the mouth, as well as reactive new bone in most areas of the upper and lower jaw. Over half of the roots of all 3$^{rd}$ molars are exposed, this is a typical sign of periodontal disease (Waldron 2009, 240). The lower left 1$^{st}$ and 2$^{nd}$ molars, 1$^{st}$ premolar and lower right 1$^{st}$ molar, display ante-mortem grooves in the enamel. In the case of the right 1$^{st}$ molar and left 1$^{st}$ premolar, the dentine is also affected. This may be due to use of the teeth as tools or an anomaly in the masticatory process. There is a small (<1mm) carious lesion which has exposed the dentine on the mesio-lingual cusp of the lower right 3$^{rd}$ molar.

**Site 112.4**

*T3.1/4OR* (Plate 2.9)
*Burial Details* – Post-Meroitic/Christian period
Poor preservation had left only hints of how this individual was buried. The angle of the surviving bones suggest an extended burial, probably supine with a west/east orientation.

*Condition and Percentage Present*
Less than 5% of the skeleton is present, with all elements splintered and friable. The bone surface is flaking and weathered with no cortical integrity. Only damaged fragments of the femurs and left tibia are present.

*Sex* – Undetermined due to low preservation.

*Age* – Adult? This can only be suggested due to the size of the bone fragments.

*Pathology*
No pathological changes evident.

T4.1/7OR
*Burial Details* – Post-Meroitic/Christian
Despite the fragmentary nature of this skeleton, some burial features could be observed. It is probable that the remains were in an extended, supine position with a south-east/north-west orientation.

*Condition and Percentage Present*
Less than 5% of this individual is present. Some fragments of long bone can be identified as the shaft of a tibia (unsided), and some are fragments of skull. All of the surviving bone is in poor condition, weathered, flaking and broken.

*Sex* – Undetermined due to poor preservation.

Age – Adult? Based on the size and morphology of the tibial fragment.

*Pathology*
No pathological changes evident.

## Site 170.1

Burial positions and pottery found in the area could suggest a Neolithic date for this group of burials, though there is some evidence for a Post-Meroitic/Christian date as well.

*T1.1/2OR* (Plate 2.41)
*Burial Details*
This juvenile was in a flexed burial position, with a north/south orientation, laying on the right side with the face to the west. The exact position of the arms and legs could not be determined as the burial had been disturbed.

*Condition and Percentage Present*
Approximately 50% of this individual is present, with more than 75% of the bone surface in a sound cortical condition. The disturbance seen in this burial may be due to animal activity as there is evidence of rodent gnawing particularly on the left femur.

*Sex* – Undetermined due to the age of the individual.

*Age* – Early childhood 2-5 years. Specifically, dental development suggests around 4 years (± 1 year), measurements of the humerus suggest 2-2.5 years, meaning that skeletal maturity does not fall within the error range for dental development. This discrepancy may indicate this individual was small for their age, possibly due to malnutrition or illness.

*Dental Inventory*

Permanent Inventory

| UR | X | X | 1 | 1 | 1 | 5 | 5 | 1 | 5 | 5 | 5 | 5 | 1 | 1 | X | X | UL |
|----|---|---|---|---|---|---|---|---|---|---|---|---|---|---|---|---|----|
| LR | X | X | 5 | 5 | 1 | 1 | 1 | 1 | 1 | 3 | 3 | 3 | 3 | 3 | X | X | LL |

Deciduous Inventory

| UR | 2 | 2 | 5 | 5 | 5 | 5 | 2 | 2 | 2 | 2 | UL |
|----|---|---|---|---|---|---|---|---|---|---|----|
| LR | 5 | 5 | 5 | 5 | 5 | 3 | 3 | 2 | 3 | 2 | LL |

*Pathology*
There is no evidence of pathological changes in this individual. The possible lesions, seen in some fragments of the skull, upon close examination are most likely to be taphonomic. They do not exhibit the features associated with a lesion, specifically undercut edges, a scooped appearance or exposed trabecular bone (Waldron 2009, 51). As well as this the presence of animal gnawing in other parts of the skeleton and the lighter colour surrounding these holes support the conclusion of taphonomic origin rather than pathological changes.

*T2.1/3OR* (Plate 2.40)
*Burial Details*
This burial contained an infant in a flexed position with an east/west orientation. Lying on its right side with the face to the north. Right arm was extended and left arm flexed.

*Condition and Percentage Present*
Approximately 55% of the skeleton is present, with only the hands, feet and pelvis having poor preservation. Bone

surface is in good condition.

*Sex* – Undetermined due to the age of the individual.

*Age* – Infant 0-1 years. Dental development suggest 6 months (± 3 months), osteometric measurements of the humerus suggest 3-6 months falling within the error range for dental development.

*Dental Inventory*

Deciduous Inventory

| UR | 1 | 1 | 1 | 1 | 1 | 1 | 1 | 1 | 1 | 3 | UL |
|---|---|---|---|---|---|---|---|---|---|---|---|
| LR | 1 | 1 | 5 | 5 | 5 | 1 | 1 | 5 | 5 | 1 | LL |

*Pathology*
No pathological changes evident.

T4.1/7OR
*Burial Details*
The skeleton of this sub-adult is in a very poor condition. Only a few skeletal elements are present, though it is believed to have been in a flexed position with an east/west orientation.

*Condition and Percentage Present*
Approximately 15% of this individual is present. Most long bones are represented though only in fragments, as well as some pieces of the skull and mandible. The lower molars, upper premolars and canines survive and can provide a small amount of information. The bone that is present is flaking and broken, with little sound cortical bone found.

*Sex* – Undetermined due to damage and age.

*Age* – Puberty 11-15 years. Determined using developmental stage of the lower right 3rd molar, for which the crown is half formed suggesting an age of 12 years (± 2.5 years).

*Dental Inventory*

Inventory

| UR | 3 | 3 | 3 | 7 | 7 | 7 | 3 | 3 | 3 | 3 | 3 | 3 | 3 | 3 | 3 | 3 | UL |
|---|---|---|---|---|---|---|---|---|---|---|---|---|---|---|---|---|---|
| LR | 1 | 2 | 5 | 5 | 5 | 5 | 5 | 5 | 5 | 5 | 7 | 5 | 5 | 2 | 2 | 3 | LL |

Wear

| UR | 99 | 99 | 99 | 1 | 1 | 1 | 9 | 9 | 9 | 9 | 9 | 9 | 9 | 99 | 99 | 99 | UL |
|---|---|---|---|---|---|---|---|---|---|---|---|---|---|---|---|---|---|
| LR | 99 | 99 | 99 | 9 | 9 | 9 | 9 | 9 | 9 | 9 | 9 | 9 | 9 | 6 | 4 | 99 | LL |

*Pathology*
Dental pathology: Lower left 1st molar displays a large carious lesion covering a third of the occlusal surface.

T5.1/8OR (Plate 2.17)
*Burial Details*
This flexed burial of an adult was lying on the left side with an east/west orientation.

*Condition and Percentage Present*
Approximately 30% of the skeleton remaining. Bones are fragmentary and friable, with flaking of the cortical surface.

*Sex* – Undetermined due to poor preservation.

*Age* – Adult, specific age undetermined. The 3rd molar is erupted and in occlusion, however, the enamel defects seen on the molars would make age estimation unreliable (see below).

*Dental Inventory*

Inventory

| UR | 2 | 2 | 7 | 3 | 3 | 3 | 3 | 3 | 3 | 3 | 3 | 5 | 5 | 7 | 2 | 2 | UL |
|---|---|---|---|---|---|---|---|---|---|---|---|---|---|---|---|---|---|
| LR | 7 | 7 | 3 | 2 | 2 | 7 | 3 | 3 | 3 | 3 | 3 | 3 | 3 | 2 | 2 | 2 | LL |

Wear

| UR | 4 | 9 | 99 | 9 | 9 | 9 | 9 | 9 | 9 | 9 | 9 | 9 | 9 | 99 | 9 | 4 | UL |
|---|---|---|---|---|---|---|---|---|---|---|---|---|---|---|---|---|---|
| LR | 2 | 6 | 99 | 3 | 1 | 1 | 9 | 9 | 9 | 9 | 9 | 9 | 9 | 99 | 8 | 4 | LL |

*Pathology*
Dental pathology: The socket in the alveolar process for the lower right 3$^{rd}$ molar is located up the side of the ramus. The tooth would have been positioned with the occlusal surface pointing mesially, a condition usually referred to as an impacted molar. Pits are visible on the cusps of all upper and lower 1$^{st}$ molars, as well as the lower left 3$^{rd}$ molar and lower right 2$^{nd}$ premolar. On the lower molars these defects are 2-3mm in diameter on the buccal cusps and smaller on the lingual (0.8mm). This evidence points to pit/plane form hypoplasia rather than caries (see discussion below). In addition, furrow form hypoplasia is present on the molars and five distinct furrows can be seen in the lower right canine. This suggests that there was some disruption in the developmental stages of this individual's life.

**Site 118**
Site 118 at Wadi Dein was examined in both the 1993 and 1994 seasons. During the former a single burial was found and in the latter no human remains were recovered. Little detail was given regarding the remains found during the 1993 season in the corresponding report, therefore they will be presented here.

FS.2/4
*Burial Details* – Post-Meroitic/Christian period
Exact burial details are unknown although the walls of the tomb are noted as being in a north/south orientation. The tomb is reported as being from the Christian period (Mallinson *et al.*, 1996, 57).

*Condition and Percentage Present*
High level of preservation with 50% of the skeleton present with more than 75% having a good cortical surface. Most elements represented except for the hands, feet, ribs and dentition. Some post-mortem mottled black staining on the lower limbs and pelvis.

*Sex* – Unknown, sex is undetermined as features of the skull and pelvis were ambiguous, with a mixture of both male and female traits.

*Age* – Adult 20-34 years. Determined from examination of the pubic symphysis and auricular surface.

*Dental Inventory*: One small fragment of a molar is the only piece of the dentition present in this individual.

*Pathology*
Joint disease: An erosive lesion can be seen on the right humerus head at the joint margin. As well as this some new bone and pitting are seen around the entire margin of the joint. Corresponding changes on the joint margin of the glenoid fossa on the right scapula are also visible, where new bone and an erosive lesion are present in the centre of the joint. These changes suggest a form of arthropathy though specific diagnosis could not be made from these changes alone. Other: Covering the occipital bone and both parietals, as well as a portion of the posterior part of the frontal bone, this individual exhibits increased porosity. On the left side of the occipital near to the lambda, is a healed lesion around 17 × 14mm in size. These two changes (lesion and porosity) are probably not connected, an increase in porosity is often found in these areas and has been occasionally referred to as ectocranial porosis (Mann and Hunt 2005, 19) and is not considered a pathological change.

## *Burials unavailable for examination*

It should be noted that 112.4/T2.1/3OR, 112.4/T6.1/12OR, 154.5/T1.1/3OR, 154.5/T1.2 and 170.1/T3 were also excavated in the course of this project, however these remains were very fragmentary and fragile and were, therefore, not transported back to the British Museum with the others. Those from site 112.4 displayed such low levels of preservation that no details could be discerned, however the burial in 170.1/T3 was described as a crouched sub-adult burial in a shallow grave with no accompanying grave goods. It had an east/west orientation and preservation was noted as very soft and fragile. Like the other burials from this site no specific period was assigned and Neolithic, Post-Meroitic and Christian dates are suggested. 154.5/T1.1/3OR was described as a prone burial with the feet close together and the hands behind the back, an unusual positioning associated with executions in some cultures, though there is no evidence for this here (Plates 2.15 and 2.16).

**Site 159** (Plates 2.24-26, 2.28-29 and 2.32)
Site 159 was found to be part of a large cemetery at Gabati which was later fully excavated, the remains found during this project and the later excavations are discussed in Judd (2012).

## *Non-metric traits*

Non-metric traits are variations in the skeleton which are not present in all individuals and have implications in phylogenetic study. They can indicate familial links and even biological distance between populations. The non-metric traits examined here are those considered most important and most commonly found (Buikstra and Ubelaker 1994, 85-94). Since the sample size examined here is small and from different locations any relationships based on non-metric traits could not be seen as significant, however the occurrence of variations found are set out in Tables 9.2, 9.3 and 9.4.

## Osteometrics

Metric analysis of skeletal elements can be used to estimate stature and make comparisons between age, sex and population. Due to the low levels of preservation, in the majority of this collection many measurements were not possible, however, those that were are set out in Tables 9.4, 9.5 and 9.6.

## Osteoarthritis

There were only four individuals with osteoarthritis, all in the age category of 20-35. Two are male (101.4/T2.1/39OR and 101.4/T1.2/38OR) and two female (101.4/T3.1/40OR and 102.1/T1.1/46OR). All osteoarthritic sites observed are presented in Table 9.7. Note should be made of the OA found in the left ankle and patellae of skeleton 101.4/T1.2/38OR, it is probably secondary to the trauma elsewhere in the body, see individual description and discussion for details.

# Discussion

There are some limiting factors to the discussion of this group of individuals. Not only are they from varying locations and time periods but it is a small sample; only 14 skeletons have been examined of which 35% have very poor preservation (20% or less). Due to the limits of the project, it is not possible to draw many broad bioarchaeological conclusions. However, we can discuss some of the pathological conditions found in these individuals and their possible causes. Some of these conditions are also discussed in the context of other sites in the area which have been more extensively excavated and have a greater sample size, such as Gabati. Percentage preservation is an estimate determined through examination of groups of skeletal elements: hands, feet, upper or lower limbs, axial skeleton, skull and dentition. These percentages are then averaged to produce an overall percentage of preservation. Though some individuals may appear in a particular state of preservation in site photographs, after excavation, lifting and transportation the percentage of preservation may be more or less than expected.

With regards to the poor condition of some of the remains, it seems prudent to mention the climatic fluctuations which may have caused this. While the areas excavated currently seem arid, the condition of some burials would suggest that water levels may have been high enough, at least once, to cause some of the observed damage. Many areas in Sudan are known to experience water table fluctuations and flash flooding during the wet season, usually experienced between May and September (Conway *et al.*, 2005). The observed taphonomic changes are unlikely to indicate major environmental changes as there is evidence that the much wetter period in Northern Sudan ended around 3,600 BC (Abell and Hoelzmann 2000), a period from which the skeletons are unlikely to originate. Furthermore, detailed analysis from archaeological sites representing the time periods covered by this collection display evidence that the climate was similar to the present day (Chaix and Grant 1993).

The burial positions and orientation of the individuals in this collection suggest varying burial practices in different time periods. Specifically the Post-Meroitic burials are flexed and have differing orientations, whereas the majority of the Christian burials are extended in an east/west orientation. These findings, again, must be tempered by the fact that this sample is small and some individuals examined have not been assigned to a specific time period. However these findings are similar to those described at other sites where, Christian burials in particular, have been found in west/east orientations, though some vary, with the head position being at the east end as seen here. Additionally Christian burials are almost always extended, while Post-Meroitic burials are usually flexed but many Kushite and New Kingdom burials are extended as are some Mesolitic ones like those found recently at el-Salha near Omdurman (Adams 1968, 202; 1999, 16; Welsby and Daniels 1991, 26; 2001, 206-224).

The pathology found in this collection is described above, however several findings warrant further discussion. Firstly the trauma apparent in skeleton 101.4/T1.2/38OR (Christian period) includes compression fractures of the vertebral bodies, a depression fracture in the proximal articular surface of the tibia and florid osteoarthritis of the ankle. It should be noted that osteoarthritis of the ankle is rare. Waldron (2009, 38) states that it occurs nine times less frequently than that of the hip or knee. This occurrence is most likely to be secondary to the trauma seen in the back and knee. It is possible that this traumatic incident damaged the cartilage in the joint or changed the way in which the joint was used to compensate for the primary trauma. Although some reports from the surrounding area suggest a high prevalence of osteoarthritis of the ankle joint, sometimes as high as 46% of all males excavated (Judd 2012, 57), it is likely that a different operational definition of osteoarthritis was used, overestimating this condition. All examples of osteoarthritis in this sample come from individuals in age category 20-34 years. This is unusual as osteoarthritis is a disease of age, Waldron notes that both prevalence and incidence increase with age (2009, 28). However, genetics, sex, trauma and movement also affect the prevalence of the disease. As mentioned above, the osteoarthritis seen in skeleton 101.4/T1.2/38OR most likely results from trauma and such additional physical strain may have led them to develop osteoarthritis at a younger age. Though more prevalent in older individuals, the age indicators used here – such as the pubic symphysis – can become more variable with increasing age. Therefore, some of these individuals may belong to a higher age cohort than assigned.

There are three individuals in the sample with new bone on the ribs, T1.6/51, T2.1/39, T2.2/38OR, all from the Christian period site 101.4. This produces a prevalence of 37.5% (3/8) when we take into account that not all of the skeletons had a level of preservation at which the ribs could be examined. Sometimes new bone on the ribs can be associated with a specific disease such as tuberculosis. However, in this case, no other signs of tuberculosis such as spinal lesions and kyphosis, were observable in any of these individuals. Therefore it can only be suggested that these were infections in the lungs or surrounding pleural tissue or possibly caused by trauma to the chest area.

There is some evidence of disruption during the developmental years. This presents in two forms; hypoplasia, and discrepancies between dental development and skeletal maturation. Hypoplasia is the appearance of defects in the enamel caused by systemic disruptions and has been linked to malnutrition and infectious disease (Hillson 2014, 184-185). Fever, for example, can affect ameloblasts secretion. In skeleton 170.1/T1.1/2OR (burial period unknown) there is also a discrepancy between the development of the dentition (4 years ± 1 year) and the skeletal maturation (2-2.5 years), suggesting that the skeleton had not reached the size regarded as standard for the age indicated by the teeth. Dental development is seen as a more reliable age indication than skeletal maturation (Hillson 2014, 1), therefore it is likely that this disruption of skeletal development was possibly brought about by stress, malnutrition or disease. However, it should be noted that these measures of skeletal and dental development are based on modern standards.

Not all developmental disturbances were limited to individuals who died while still developing. Skeleton 170.1/T5.1/8OR (burial period unknown) displays enamel defects known as pit form hypoplasia. These defects involve whole areas of enamel forming cells, ameloblasts. However, this individual died long after completion of the enamel matrix, the defects had become exposed and seen as circular pits in the teeth as dental attrition had reached the affected plane. Clearly this individual had survived a period of malnutrition or disease during development.

Skeleton 102.1/T1.1/46OR (Christian period) exhibits a large and advanced abscess in the maxilla. This particular abscess is likely to have been caused by cracking or chipping leading to pulp exposure. Grooves found on some of the teeth suggest they may have been used as tools, causing the observed cracking and chipping. The resulting infection and formation of pyogenic secretions (pus) from the abscess also caused a secondary inflammation and infection in the jaw, the resorption of the alveolar bone and the production of new bone on the surface of the alveolar process. The high level of tooth wear may be attributed to unusual masticatory activity or a highly abrasive diet. Earlier study of the nearby Gabati population (also covering the Post-Meroitic and Christian periods) suggests that dental pathology was generally low; with a diet of protein and dairy products as well as primarily C3 plants. However sorghum (a C4 plant) may also have been consumed (see Judd 2012, 65). Although it is likely that the individuals in this collection had a similar diet to those at Gabati, it is composed of too few examples to draw any firm conclusions between dental pathology and diet.

The presence of a supernumerary tooth found in skeleton 101.4/T3.1/40OR (Post-Meroitic period) is uncommon. Hillson (2005, 281) notes that a very low percentage of the population exhibit supernumerary teeth. These teeth usually present in one of two forms, either a conical single cusped crown or a complex multi-cusped crown. The latter form is found in this individual. The position between the 1$^{st}$ and 2$^{nd}$ premolar in the mandible is rare, most supernumerary teeth occur in the maxilla (Barrocal *et al.*, 2007). Although in this case only one of the supernumerary teeth is present, there is a socket for a corresponding tooth on the opposite side, and this was a bilateral occurrence (post-mortem loss is likely for this second tooth). Additionally, supernumerary teeth are more frequently found in males whereas this skeleton is that of a female (Brook 1984).

In summary, although this collection covers a wide geographical and temporal range, and is relatively small, some useful data was generated. Further analysis and excavation of these areas may one day be able to put these individuals and some of the findings into a broader context, and inform us about the inhabitants of this part of Sudan during those periods.

**TABLE 9.2.** CRANIAL NON-METRIC TRAITS (AFTER BUIKSTRA AND UBELAKER 1994, 85-94).

| Trait | Total No. L/R | 101.4/T1.2/38 Side | Cat | 101.4/T1.3/41 Side | Cat | 101.4/T1.4/42 side | Cat | 101.4/T2.1/39 Side | Cat | 101.4/T3.1/40 Side | Cat | 102.1/T1.1/46 Side | Cat | 118/S.2/4 Side | Cat | Score Category |
|---|---|---|---|---|---|---|---|---|---|---|---|---|---|---|---|---|
| Supraorbital foramen | 4/5 | | | | | | | L/R | 2/2 | | | | | R | 2 | 2 = multiple foramina |
| Mastoid foramen | 5/6 | | | R | 1 | L/R | 1/1 | L/R | 1/1 | L/R | 1/1 | L/R | 1/5 | L/R | 1/1 | 1 = temporal 5 = temporal & occipital |
| Mastoid foramen nb | 5/6 | | | R | 2 | L/R | 3/3 | L/R | 1/2 | L/R | 1/1 | L/R | 1/3 | L/R | 1/2 | 1 = 1 foramen 2 = 2 foramen 3 = > 2 foramina |
| Flexure sag. Sulcus | 5 | | | ✓ | 2 | | | ✓ | 1 | ✓ | | ✓ | 2 | ✓ | 2 | 2 = flex to the left |
| Supraorbital notch | 4/5 | | | | | | | L/R | 2/2 | L/R | 1/1 | L/R | 1/1 | R | 1 | 1 = <half occluded 2 = >half occluded |
| Mental foramen | 6/6 | L/R | 1,1 | R | 1 | L/R | 1/1 | L/R | 1/1 | L/R | 1/1 | L/R | 1/1 | L | 1 | 1 = 1 foramen |
| Maxilliary Torus | 4/4 | | | | | | | L/R | 2/2 | | | | | | | 2 = moderate |
| Mandibular torus | 7/6 | L/R | 2,2 | | | | | | | | | | | | | 2 = moderate |
| Parietal foramen | 2/4 | | | | | R | 1 | | | L/R | 1/1 | | | R | 1 | 1 = parietal |
| Lambdoid ossicle | 4/4 | | | | | | | | | R | 1 | | | | | 1 = present |
| Condylar Canal | 3/3 | | | | | | | | | L | 1 | R | 1 | | | 1 = present |
| Divided Hypoglossal canal | 4/5 | | | | | | | | | L | 1 | | | | | 1 = partial |
| Zygomatical-facial foramen | 4/4 | | | L | 6 | R | 2 | L/R | 5/5 | | | L/R | 3/1 | | | 1 = 1 large 2 = 1 large,1 small 3 = 2 large 5 = 1 small 6 = multi |
| Infraorbital suture | 3/3 | | | | | | | | | | | L/R | 2/2 | | | 2 = complete |
| Multi infraorbital foramen | 3/3 | | | | | | | | | | | L/R | 2/3 | | | 2 = 2 foramen 3 = >2 foramina |

**TABLE 9.3.** POST-CRANIAL NON-METRIC TRAITS (AFTER BUIKSTRA AND UBELAKER 1994, 85-94).

| Trait | Total No. L/R | 101.4/T1.2/38 Side | Cat | 101.4/T1.3/41 Side | Cat | 101.4/T1.4/42 side | Cat | 101.4/T2.1/39 Side | Cat | 101.4/T3.1/40 Side | Cat | 102.1/T1.1/46 Side | Cat | 118/S.2/4 Side | Cat | Score Category |
|---|---|---|---|---|---|---|---|---|---|---|---|---|---|---|---|---|
| Atlas lateral Bridging | 3/4 | - | - | - | - | - | - | L | 2 | - | - | - | - | - | - | 2 = complete |
| Atlas posterior Bridging | 4/5 | - | - | - | - | - | - | L | 2 | - | - | - | - | - | - | 2 = complete |
| Accessory transverse foramen | 2/3 | - | - | - | - | - | - | RC6 | 2 | RC6 | 2 | LC5/RC6 | 2/2 | - | - | 2 = complete |
| Calcaneus - articular surface | 4/4 | - | - | - | - | - | - | - | - | - | - | L/R | 1/1 | - | - | 1 = 3 discrete facets |
| Squatting facets Talus | 3/3 | L/R | 1/1 | - | - | - | - | - | - | - | - | - | - | - | - | 1 = present |
| Lateral squatting facets tibia | 4/5 | L/R | 1/1 | - | - | - | - | - | - | L/R | 1/1 | L | 1 | R | 1 | 1 = present |
| Vastus notch | 3/5 | L | 1 | - | - | - | - | - | - | - | - | L/R | 1/1 | - | - | 1 = present |
| Septal aperture | 5/5 | - | - | - | - | - | - | - | - | L/R | 1/1 | L/R | 2/2 | - | - | 1 = small pinhole; 2 = true perforation |
| Os acromiale | 3/5 | - | - | L | 1 | - | - | - | - | - | - | - | - | - | - | 1 = present |
| Medial squatting facets - tibia | 4/3 | - | - | - | - | - | - | - | - | - | - | L/R | 1/1 | - | - | 1 = present |
| Third trocanter - femur | 4/4 | - | - | - | - | - | - | - | - | - | - | - | - | L/R | 1/1 | 1 = present |

**TABLE 9.4.** OSTEOMETRIC MEASUREMENTS (AFTER BUIKSTRA AND UBELAKER 1994, 69-84).

| Individual Number | 101.4/1.6/51 | 170.1/1.1/2 | 170.1/2.1/3 |
|---|---|---|---|
| Age | 11-15 | 2-5 | 0-1 |
| Measurement (mm) | | | |
| Basilar occipital length | | | 14 |
| Petrous and mastoid portion of temporal length | | 41 | 40 |
| Petrous and mastoid portion of temporal width | | 29 | 20 |
| Mandible width of arc | | | 21 |
| Humerus length | | 132 | 84 |
| Humerus width | | | 19 |
| Humerus diameter | | 11 | 7 |
| Clavicle length | 109 | | |
| Clavicle diameter | 10 | | |

**TABLE 9.5.** CRANIAL OSTEOMETRIC MEASUREMENTS (AFTER BUIKSTRA AND UBELAKER 1994, 69-84).

| Individual Number | 101.4/1.2/38 | 101.4/1.3/41 | 101.3/1.4/42 | 101.4/2.1/39 | 101.4/3.1/40 | 102.1/1.1/46 | 118/S.2/4 |
|---|---|---|---|---|---|---|---|
| Sex | M | M? | M? | M? | F | F | ? |
| Measurement (mm) | | | | | | | |
| Maximum cranial length | | | | | | | 170 |
| Maxillo-alveolar breadth | | | | | | 59 | |
| Maxillo-alveolar length | | | | | | 54 | |
| Upper facial height | | | | | | 67 | |
| Upper facial breadth | | | | | | 109 | |
| Nasal height | | | | | | 48 | |
| Nasal breadth | | | | | | 25 | |
| Orbital breadth | | | | | | 40 | |
| Orbital height | | | | | | 35 | |
| Interorbital breadth | | | | | | 27 | |
| Frontal chord | | | | | | | 108 |
| Parietal chord | | | | | | | 116 |
| Occipital chord | | | | | | | 86 |
| Foramen magnum length | | | | | | 37 | |
| Foramen magnum breadth | | | | | | 26 | |
| Mastoid length | | 34 | 30 | 20 | 28 | | 25 |
| Chin height | | 36 | 32 | 29 | 33 | | |
| Height of mandibular body | | 14 | 33 | 27 | 32 | | |
| Breadth of mandibular body | 10 | 12 | | 13 | 18 | 12 | |
| Bigonial width | | | | 99 | 70 | 81 | |
| Bicondylar breadth | | | | 112 | 78 | 104 | |
| Min. ramus breadth | 30 | 32 | | 34 | 32 | 32 | |
| Max. ramus breadth | | | 41 | 41 | 43 | 40 | |

**TABLE 9.5. (CONT.).** CRANIAL OSTEOMETRIC MEASUREMENTS (AFTER BUIKSTRA AND UBELAKER 1994, 69-84).

| Individual Number | 101.4/1.2/38 | 101.4/1.3/41 | 101.3/1.4/42 | 101.4/2.1/39 | 101.4/3.1/ 40 | 102.1/1.1/46 | 118/S.2/4 |
|---|---|---|---|---|---|---|---|
| Max ramus height | | | | 50 | 50 | 48 | |
| Mandibular length | 93 | | | 73 | 91 | 84 | |
| Mandibular angle | | | | 126° | 109° | 115° | |


**TABLE 9.6.** POST-CRANIAL OSTEOMETRIC MEASUREMENTS (AFTER BUIKSTRA AND UBELAKER 1994. 69-84).

| Individual Number | 101.4/1.2/38 | 101.4/1.3/41 | 101.3/1.4/42 | 101.4/2.1/39 | 101.4/3.1/ 40 | 102.1/1.1/46 | 118/S.2/4 |
|---|---|---|---|---|---|---|---|
| Sex | M | M? | M? | M? | F | F | ? |
| Measurement (mm) | | | | | | | |
| Clavicle max length | | | | | | 145 | |
| Clavicle ant-post diameter at midshaft | | | | | | 12 | |
| Clavicle sup-inf diameter at midshaft | | | | | | 10 | |
| Humerus max length | 309 | | | 289 | | | |
| Humerus epicondylar breadth | 60 | | | 61 | 55 | 56 | 57 |
| Humerus vertical diameter of head | 46 | 54 | | 39 | | | 40 |
| Humerus max diameter at midshaft | 21 | | | 20 | | | |
| Humerus min diameter at midshaft | 17 | | | 17 | | | |
| Radius max length | 239 | | | 243 | 240 | | |
| Radius ant-post diameter at midshaft | 11 | | | 12 | 11 | 10 | |
| Radius med-lat diameter at midshaft | 14 | | | 14 | 13 | 14 | |
| Ulna max length | 270 | | | 265 | | 254 | |
| Ulna ant-post diameter | 15 | | 15 | 14 | 16 | 11 | 15 |
| Ulna med-lat diameter | 12 | | 16 | 16 | 12 | 13 | 10 |
| Ulna physiological length | 248 | | | 240 | | 130 | |
| Sacrum anterior length | | | | | | 91 | |
| Sacrum ant sup breadth | | | | | | 101 | 90 |
| Sacrum max trans diameter of base | | | | | | | 37 |
| Os coxae height | | | | | | 189 | |
| Os coxae illiac breadth | | | | | | 141 | |
| Os coxae Pubis length | | | | | | 77 | |
| Os coxae Ishium length | | | | | | 68 | |

**TABLE 9.6 (CONT).** POST-CRANIAL OSTEOMETRIC MEASUREMENTS
(AFTER BUIKSTRA AND UBELAKER 1994, 69-84).

| Individual Number | 101.4/1.2/38 | 101.4/1.3/41 | 101.3/1.4/42 | 101.4/2.1/39 | 101.4/3.1/40 | 102.1/1.1/46 | 118/S.2/4 |
|---|---|---|---|---|---|---|---|
| Sex | M | M? | M? | M? | F | F | ? |
| Measurement (mm) | | | | | | | |
| Femur max length | | | | | 437 | 435 | |
| Femur bicondylar length | | | | | 433 | 432 | |
| Femur epicondylar breadth | 80 | | | | | 70 | |
| Femur max diameter of head | | 50 | | | 39 | 40 | |
| Femur ant-post subtroc diameter | | | | 23 | 24 | 20 | |
| Femur med-lat subtroc diameter | | | | 31 | 30 | 27 | |
| Femur ant-post diameter at midshaft | | | | | 26 | 25 | |
| Femur med-lat diameter at midshaft | | | | | 23 | 25 | |
| Tibia max length | 391 | | | | | 364 | |
| Tibia physiological length | 383 | | | | | 360 | |
| Tibia max prox epiphyseal breadth | 79 | | | 69 | | 70 | |
| Tibia max distal epiphyseal breadth | 48 | | | 45 | 42 | 42 | |
| Tibia max diameter at nutrient foramen | 35 | | | 35 | 31 | 27 | |
| Tibia med-lat diameter at nutrient foramen | 18 | | | 20 | 20 | 18 | |
| Fibula max length | | | | 356 | | 35 | |
| Fibula max diameter at midshaft | | | | 13 | | 12 | |
| Calcaneus max length | 82 | | | 82 | | 75 | |
| Calcaneus middle breadth | 41 | | | 40 | | 37 | |
| 2$^{nd}$ metacapal midline length | 77 | | | 65 | | | |
| 1$^{st}$ metatarsal max length | 60 | | | | | 59 | |
| 2$^{nd}$ metatarsal max length | 80 | | | | | 68 | |
| 3$^{rd}$ metatarsal max length | 75 | | | | | 64 | |
| 4$^{th}$ metatarsal max length | 72 | | | | | 63 | |

**TABLE 9.7. OCCURRENCES OF OSTEOARTHRITIS, BY SKELETON.**

● - OA is present, 0 - site was observed but no OA present, ND - site not present for observation. Skeletons 170.1/T5.1/8, 112.4/T4.1/7 and 112.4/T3.1/4 were also examined but due to preservation had no observable joint surfaces.

| Site of Osteoarthritis | 101.4/T1.2/38 Left | 101.4/T1.2/38 Right | 101.4/T1.3/41 Left | 101.4/T1.3/41 Right | 101.4/T1.4/42 Left | 101.4/T1.4/42 Right | 101.4/T2.1/39 Left | 101.4/T2.1/39 Right | 101.4/T3.1/40 Left | 101.4/T3.1/40 Right | 102.1/T1.1/46 Left | 102.1/T1.1/46 Right | 118/S.2/4 Left | 118/S.2/4 Right | No. with OA/Total observed Left | Right | Total |
|---|---|---|---|---|---|---|---|---|---|---|---|---|---|---|---|---|---|
| TMJ - temporal | ND | ND | 0 | 0 | 0 | ND | 0 | 0 | 0 | 0 | 0 | 0 | 0 | 0 | 0/6 | 0/5 | 0/11 |
| TMJ - Mandible | 0 | 0 | 0 | ND | ND | ND | 0 | 0 | ● | 0 | 0 | 0 | ND | ND | 1/5 | 0/4 | 1/9 |
| Clavicle - distal | ND | ● | ND | ND | ND | ND | ND | ● | ND | ND | ● | ● | ND | ND | 1/1 | 3/3 | 4/4 |
| Clavicle - proximal | ● | ND | ND | ND | ND | ND | ND | ND | 0 | ND | 0 | 0 | ND | ND | 1/3 | 0/1 | 1/4 |
| Scapula - acromioclavicular | ND | ● | ND | ND | ND | ND | ● | ● | ND | 0 | ND | 0 | ND | 0 | 1/1 | 2/5 | 3/6 |
| Scapula - glenoid fossa | ND | 0 | 0 | ND | ND | 0 | 0 | 0 | 0 | 0 | 0 | 0 | 0 | 0 | 0/5 | 0/6 | 0/11 |
| Sternum - clavicular notch | ● | ● | ND | ND | ND | ND | ND | ND | ND | ND | 0 | 0 | ND | ND | 1/2 | 1/2 | 2/4 |
| Humerus - head | 0 | 0 | 0 | 0 | ND | ND | 0 | 0 | 0 | 0 | 0 | 0 | 0 | 0 | 0/6 | 0/6 | 0/12 |
| Humerus - capitulum | 0 | 0 | ND | ND | ND | ND | 0 | 0 | 0 | 0 | 0 | 0 | 0 | ND | 0/5 | 0/4 | 0/9 |
| Humerus - trochlea | 0 | 0 | 0 | 0 | ND | ND | 0 | 0 | 0 | 0 | 0 | 0 | 0 | 0 | 0/6 | 0/6 | 0/12 |
| Ulna - proximal | 0 | 0 | 0 | ND | ND | ND | 0 | 0 | 0 | 0 | 0 | 0 | 0 | 0 | 0/6 | 0/6 | 0/12 |
| Ulna - radial notch | 0 | 0 | 0 | ND | ND | 0 | 0 | 0 | 0 | 0 | 0 | 0 | 0 | 0 | 0/6 | 0/5 | 0/11 |
| Ulna - distal | 0 | 0 | ND | ND | ND | ND | 0 | ND | 0 | 0 | 0 | 0 | 0 | ND | 0/5 | 0/4 | 0/9 |
| Radius - proximal fovea | 0 | 0 | 0 | 0 | ND | ND | 0 | 0 | 0 | 0 | 0 | 0 | ND | ND | 0/5 | 0/4 | 0/9 |
| Radius - prox circumference | 0 | 0 | 0 | 0 | ND | ND | 0 | 0 | 0 | 0 | 0 | 0 | ND | ND | 0/5 | 0/4 | 0/9 |
| Radius - distal radioulnar | 0 | 0 | ND | ND | ND | ND | 0 | 0 | 0 | 0 | 0 | 0 | 0 | 0 | 0/5 | 0/5 | 0/10 |
| Radius - distal lateral | 0 | 0 | ND | ND | ND | ND | 0 | 0 | 0 | 0 | 0 | 0 | 0 | 0 | 0/5 | 0/5 | 0/10 |
| Radius - distal medial | 0 | 0 | ND | ND | ND | ND | 0 | 0 | 0 | 0 | 0 | 0 | 0 | 0 | 0/5 | 0/5 | 0/10 |
| Innominate - acetabulum | 0 | 0 | 0 | 0 | ND | ND | ND | ND | 0 | 0 | 0 | 0 | 0 | 0 | 0/5 | 0/5 | 0/10 |
| Femur - head | 0 | 0 | 0 | 0 | ND | ND | 0 | 0 | 0 | 0 | 0 | 0 | 0 | 0 | 0/6 | 0/5 | 0/11 |
| Femur - femoropatella | 0 | 0 | ND | ND | ND | ND | 0 | 0 | 0 | 0 | 0 | 0 | ND | ND | 0/5 | 0/4 | 0/9 |
| Femur - distal lateral | 0 | 0 | ND | ND | ND | ND | 0 | 0 | 0 | 0 | 0 | 0 | 0 | ND | 0/5 | 0/4 | 0/9 |
| Femur - distal medial | 0 | 0 | ND | ND | ND | ND | 0 | 0 | 0 | 0 | 0 | 0 | 0 | ND | 0/5 | 0/4 | 0/9 |
| Patella | ● | ● | 0 | 0 | ND | ND | ND | 0 | ND | ND | 0 | 0 | ND | ND | 1/3 | 1/4 | 2/7 |
| Tibia - proximal lateral | 0 | 0 | ND | ND | ND | ND | 0 | 0 | 0 | 0 | 0 | 0 | 0 | ND | 0/5 | 0/4 | 0/9 |

**TABLE 9.7 (CONT.). OCCURRENCES OF OSTEOARTHRITIS, BY SKELETON.**

● - OA is present, 0 - site was observed but no OA present, ND - site not present for observation.
Skeletons 170.1/T5.1/8, 112.4/T4.1/7 and 112.4/T3.1/4 were also examined but due to preservation had no observable joint surfaces.

| Site of Osteoarthritis | 101.4/T1.2/38 | | 101.4/T1.3/41 | | 101.4/T1.4/42 | | 101.4/T2.1/39 | | 101.4/T3.1/40 | | 102.1/T1.1/46 | | 118/S.2/4 | | No. with OA/Total observed | | |
|---|---|---|---|---|---|---|---|---|---|---|---|---|---|---|---|---|---|
| | Left | Right | Left | Right | Left | Right | Left | Right | Left | Right | Left | Right | Left | Right | Left | Right | Total |
| Tibia - proximal medial | 0 | 0 | ND | ND | ND | ND | 0 | 0 | 0 | 0 | 0 | 0 | 0 | ND | 0/5 | 0/4 | 0/9 |
| Tibia - tibiofibular | 0 | ND | ND | ND | ND | ND | 0 | 0 | 0 | 0 | 0 | 0 | ND | ND | 0/4 | 0/3 | 0/7 |
| Tibia - distal talocrural | ● | 0 | ND | ND | ND | ND | 0 | 0 | 0 | 0 | 0 | 0 | 0 | 0 | 1/5 | 0/5 | 1/10 |
| Fibula - proximal | ND | ND | ND | ND | ND | ND | 0 | 0 | 0 | ND | 0 | 0 | 0 | ND | 0/4 | 0/3 | 0/7 |
| Fibula - distal | ND | 0 | NE | ND | ND | ND | 0 | 0 | 0 | 0 | 0 | 0 | ND | 0 | 0/3 | 0/4 | 0/7 |
| Tarsus | ● | 0 | NE | ND | ND | ND | 0 | 0 | 0 | ND | 0 | 0 | ND | ND | 1/4 | 0/4 | 1/8 |
| Carpus | 0 | 0 | ND | ND | ND | ND | 0 | 0 | 0 | 0 | 0 | 0 | ND | ND | 0/4 | 0/3 | 0/7 |
| Foot - small joints | 0 | 0 | ND | ND | ND | ND | 0 | 0 | 0 | 0 | 0 | 0 | ND | ND | 0/4 | 0/4 | 0/8 |
| Hand - small joints | 0 | 0 | ND | ND | ND | ND | 0 | 0 | 0 | ND | 0 | 0 | ND | ND | 0/4 | 0/3 | 0/7 |

# The Animal Remains

*Jane Sanford Gaastra[1]*

The animal bone was collected as part of the Test Excavations, and was exported to Britain for further study. An examination of the bone available was conducted by Salima Ikram. This showed that the material was only identifiable to Class and/or size category, and so only a broad indication of the range of species could be gained. A distal fragment of a camel metatarsal was identified, otherwise long bones fragments of large and medium-sized mammals were noted. A few of these were considered to be from camels, while others were likely to have been from cattle. The medium-sized mammal bones on the basis of size and available morphological characteristics, were likely to have originated from sheep or goat. The general nature of the identifications possible on the basis of this material meant that it was not considered appropriate to include a detailed report on this part of the assemblage in the present volume. Subsequently, further bone was recovered, which could be identified more readily to species level. This latter material forms the subject of the present report.

Finds from the Begrawiya-Atbara Test Excavations were identified using the reference collections of the Grahame Clarke Zooarchaeological Laboratory and Museum of Zoology at the University of Cambridge. All material was identified to element and species, where possible. Those elements not identifiable to species were identified to order (i.e. Rodentia) or family (Bovidae) or to size class of ungulate as applicable. All fragments greater than 1mm in length which bore no diagnostic morphology were classified as unidentifiable and sorted into unidentifiable classes of shaft, cancellous, cranial, rib and other. A total of 388 unidentifiable fragments were recovered from the survey areas, with a combined weight of 125.6g and an average weight of 0.32g. All unidentifiable fragments were unburnt.

The identifiable survey material was heavily fragmented, both from post-depositional and post-excavation fragmentation. Where possible, recent breaks to elements were assessed for refit in order to better aid identification. An overview of the finds recovered from these contexts is given in Tables 10.1-10.4.

Samples have been quantified by Number of Identified Specimens (NSIP) in these tables. This form of quantification provides information as to the proportional representation of taxa in each sample, but NSIP counts are subject to inflation with increased fragmentation. Because of this, samples have also been quantified by Minimum Number of Individuals (MNI) in Table 10.2. This technique is also not without its faults, as it can under-estimate the presence of fragments from multiple individual animals in these samples, but a combination of these two quantification techniques helps to illustrate the presence and distribution of taxa in the survey areas sampled.

The majority of faunal remains from these contexts comprise fragments of domestic mammals (mainly sheep (*Ovis aries*), donkey (*Equus asinus*), goat (*Capra hircus*) and camel (*Camelus sp.*)) as well as possibly cattle or other wild bovids (*Bos sp.*) along with fragments of wild taxa such as gazelle (*Gazella sp.*) and fragments of uncertain taxonomic attribution in the Bovidae and ungulate fragments.

Approximately one half (37 out of 74 fragments) of the recovered faunal material demonstrates evidence of slight to moderate weathering in the form of fine fractures and spalling with some surface erosion. Weathering was most prominent in survey contexts 155.4 and 159.2 Tumulus 5, which demonstrated evidence of faunal remains having been left exposed to the elements for a moderate period of time prior to the development of protective sediment accumulation.

The majority of fracture surfaces to bones demonstrate breakage occurring after deposition, when bones were dry. Fresh breaks (occurring on 'fresh' bone prior to prolonged exposure or cooking) are found in these same contexts. Fresh breaks were recorded from context 155.4 on the recovered *Ovis aries* humerus, radius shaft fragment of a large ungulate, the distal metapodial and proximal radius of *Camelus sp.*, the proximal metacarpal of *Equus asinus* as well as from context 159.2 Tumulus 5 on the scapula neck of a very large ungulate. These fresh fractures are not of definitive anthropogenic origin (that is, caused by human carcass processing) as cut marks are not present and carnivore gnawing is present on the metapodial shaft fragment and ilial wing fragment of large ungulates from context 155.4. However, this does demonstrate that these contexts were used occasionally as butchery/carcass processing areas either for humans or other local carnivores. Given that the evidence in these contexts for carnivore gnawing is slight, and the taxonomic bias tends towards domestic animals, it is presumed that these butchery events were likely the result of humans, with gnawing occurring while deposited faunal material remained exposed to the elements.

**SARS Survey Reference 112.3 X1**
**Context 9** comprises one complete right bovid radial carpal, one partial rib and complete sesamoid from a small ungulate and a proximal fragment of right goat (*Capra hircus*) metacarpal.

**SARS Survey Reference 151.4 S1**
**Context 4** comprises 27 small fragments of bovid teeth.

**SARS Survey Reference 155.4 S1** comprises seven contexts containing bones.
**Context 4** contains one rib fragment, one partial proximal metacarpal fragment and a fragment of ascending ramus of the mandible from small ungulates; six fragments of mandibular alveolus and 19 tooth fragments from a medium-sized ungulate; a fragment of olecranon from the

[1] With a contribution by Salima Ikram.

**TABLE 10.1.** OSTEOLOGICAL FINDS BY CONTEXT, GIVEN IN NUMBER OF IDENTIFIED SPECIMENS (NISP).

| Taxon | 101.4/T1 | 101.4/T3 | 112.3/X1 9 | 155.1/S1 4 | 155.4/S1 4 | 155.4/S1 5 | 155.4/S1 13 | 155.4/S1 17 | 155.4/S1 22 | 155.4/S1 27 | 159.2/T5 50 | 159.2/T5 40 |
|---|---|---|---|---|---|---|---|---|---|---|---|---|
| Very Large Ungulate | | | | | 1 | | | | | | 3 | |
| Large Ungulate | | | | | 1 | | | 7 | | 1 | 2 | 1 |
| Medium Ungulate | | | | | 7 | | | | | 1 | | |
| Small Ungulate | | | 2 | | 3 | 3 | | 4 | | 2 | | |
| Bovidae | | | 1 | | | | | | | 1 | | |
| Bovidae Teeth | | | | 27 | 19 | | | | | | | |
| Rodentia | | | | | 4 | | | | | | | |
| Bos sp. | | | | | | | | | 1 | | 1 | |
| Camelus sp. | | | | | 3 | | | | | | | |
| Gazella sp. | | | | | | 1 | 1 | | | | | |
| Equus asinus | | | | | 2 | | | 5 | | | | |
| Ovis aries | | | | | | | 1 | 1 | | 1 | | |
| Capra hircus | | | 1 | | | | | | | | | |
| Human | 1 | 2 | | | | | | | | | | |
| Total | 1 | 2 | 4 | 27 | 40 | 4 | 2 | 17 | 1 | 6 | 6 | 1 |

**TABLE 10.2.** OSTEOLOGICAL FINDS BY CONTEXT, GIVEN IN MINIMUM NUMBER OF INDIVIDUALS (MNI).

| Taxon | 101.4/T1 | 101.4/T3 | 112.3/X1 9 | 155.1/S1 4 | 155.4/S1 4 | 155.4/S1 27 | 155.4/S1 13 | 155.4/S1 22 | 155.4/S1 5 | 155.4/S1 17 | 159.2/T5 50 | 159.2/T5 40 |
|---|---|---|---|---|---|---|---|---|---|---|---|---|
| Very Large Ungulate | | | | | 1 | | | | 1 | | | |
| Large Ungulate | | | | | 1 | 1 | | | 1 | 1 | 1 | |
| Medium Ungulate | | | | | 1 | | | | | | | 1 |
| Small Ungulate | | | 1 | | 1 | 1 | | 1 | 1 | | | |
| Bovidae | | | 1 | | | 1 | | | | | | |
| Bovidae Teeth | | | | 1 | 1 | | | | | | | |
| Rodent | | | | | 1 | | | | | | | |
| Bos sp. | | | | | | | | 1 | | | 1 | |
| Camelus sp. | | | | | 1 | | | | | | | |
| Gazella sp. | | | | | | | 1 | | 1 | | | |
| Equus asinus | | | | | 1 | | | | | 1 | | |
| Ovis aries | | | | | | 1 | 1 | | | 1 | | |
| Capra hircus | | | 1 | | | | | | | | | |
| Human | 1 | 1 | | | | | | | | | | |

**TABLE 10.3.** NON-OSTEOLOGICAL FINDS.

| Context | Finds | Taxon | n |
|---|---|---|---|
| 155.4 F 85 | Coprolites | Camelus sp. | 3 |

ulna of a large ungulate; a small fragment of acetabulum from the ilio-ischial junction of the innominate of a very large ungulate; the right femur and tibia and both left and right innominates of a small rodent; a right proximal metacarpal and right third carpal from a donkey (*Equus asinus)* and right distal metapodial and fragment of olecranon of the ulna from a camel (*Camelus sp.*).

**Context 5** contains a fragment of left scapula (partial glenoid fossa and blade fragment) and a left fragment of ischium from a small ungulate (c.f. Bovidae) and a proximal radius fragment from a gazelle (*Gazella sp.*).

**Context 13** contains the left proximal radius of a sheep (*Ovis aries*) and the left distal humerus of a gazelle (*Gazella sp.*).

**Context 17** contains two rib fragments, a partial thoraxic vertebra and a left tibial shaft fragment from small ungulates; a rib fragment, a partial cervical vertebrae, a fragment of axis (first cervical vertebra), fragments of right ilial wing and pubic symphysis of an innominate (comparable to *Equus asinus* although not definitively attributable), a fragment of an ulnar shaft and a small shaft fragment from a right radius; as well as fragments of the right ischium, left ilium and right posterior femur of a donkey (*Equus asinus*).

**Context 22** one posterior shaft fragment of a left *Bos sp.* femur.

**Context 27** contains two small fragments of large ungulate metapodial, one eroded right bovid (Bovidae) astragalus and one distal right adult sheep (*Ovis aries*) humerus bearing fresh fractures to the diaphysis.

**SARS Survey Reference 159.2 Tumulus 5**

**Contexts 40 and 50** comprise one fragment of large ungulate ilium (wing) in **Context 40** and in **Context 50** three fragments of scapula from a very large ungulate (two fragments of blade and spine, one of neck and a portion of the glenoid fossa, most likely from the same individual), the rib and proximal humerus from a large ungulate and a right *Bos sp.* innominate.

**TABLE 10.4.** SUMMARY OF THE ANIMAL BONE DATA (CONT.).

| Context | Element | Portion | Taxon | Recent Break | Length (mm) | Weight (g) | Articulates with | Side | %Complete | Age Criteria | Age | Sex | Weathering | Weathering Type | Breakage | Gnawing | Cut Marks | M1 | M2 | M3 | Comments |
|---|---|---|---|---|---|---|---|---|---|---|---|---|---|---|---|---|---|---|---|---|---|
| 155.4 Sl 5 | Acetabulum | Ischial and shaft | Small Ungulate | Length reduced | 85.5 | 9.2 | | L | 50 | | | | Very slight | Fine fractures | Dry | | | | | | |
| 155.4 Sl 5 | Radius | Proximal | Gazella sp. | Length unaffected | 63.9 | 5.3 | | R | 25 | Fused | Adult | | Very slight | Fine fractures | Dry | | | | | | |
| 155.4 Sl 5 | Radius | Proximal | Ovis aries | Length unaffected | 55.45 | 8.6 | | L | 25 | Fused | Adult | | Slight | Fine fractures | Dry | | | Bp 31.5 | BFp 33.2 | Dp 18.6 | Adult also by muscle attachments |
| 155.4 Sl 13 | Humerus | Posterior Shaft | Gazella sp. | Length unaffected | 32.2 | 1.6 | | L | 25 | | | | Moderate | Spalling | Dry | | | | | | |
| 155.4 Sl 13 | Rib | Body section | Large Ungulate | Length reduced | 122.1 | 6.8 | | | 15 | | | | Slight | Fine fractures | Recent | | | | | | |
| 155.4 Sl 17 | Rib | Body section | Small Ungulate | Length reduced | 81.05 | 4.1 | | L | 25 | | | | Very slight | Fine fractures | Dry | | | | | | |
| 155.4 Sl 17 | Rib | Body section | Small Ungulate | Length reduced | 43.7 | 0.7 | | | 10 | | | | | | | | | | | | |
| 155.4 Sl 17 | Cervical Vertebra | Complete | Large Ungulate | Length unaffected | 51.3 | 22.4 | | | 90 | | | | Very slight | Fine fractures | | | | | | | |
| 155.4 Sl 17 | Thoraxic Vertebra | Pedicle, Anterior Facet | Small Ungulate | Length reduced | 21.5 | 2.1 | | | 15 | | | | | | | | | | | | |
| 155.4 Sl 17 | Ischium | Symphysis and shaft | Equus asinus | Length unaffected | 61.8 | 13.8 | | R | 25 | Fused | Adult | M | Very slight | Fine fractures | Fresh | | 1 | | | | Chop through ilial shaft |
| 155.4 Sl 17 | Acetabulum | Ilial and shaft | Equus asinus | | 134.3 | 45.1 | | L | 40 | | | M | Slight | Fine fractures | Fresh | | | LA 40.8 | LAR 46.4 | | |
| 155.4 Sl 17 | Ilium | Wing | Large Ungulate | Length reduced | 85.6 | 23.2 | | R | 25 | | | | Slight | Fine fractures | Dry and recent | Slight Carnivore | | | | | Equus-asinus sized, lacks diagnostic features |
| 155.4 Sl 17 | Pubis | Symphysis and shaft | Large Ungulate | Length reduced | 48.7 | 4.9 | | R | 25 | | | M | Very slight | Fine fractures | | | | | | | |

**TABLE 10.4.** SUMMARY OF THE ANIMAL BONE DATA (CONT.).

| Context | Element | Portion | Taxon | Recent Break | Length (mm) | Weight (g) | Articulates with | Side | %Complete | Age Criteria | Age | Sex | Weathering | Weathering Type | Breakage | Gnawing | Cut Marks | M1 | M2 | M3 | Comments |
|---|---|---|---|---|---|---|---|---|---|---|---|---|---|---|---|---|---|---|---|---|---|
| 155.4 S1 17 | Axis | Anterior | Large Ungulate | | 36.3 | 3.8 | | | 10 | | | | | | | | | | | | |
| 155.4 S1 17 | Humerus | Proximal Anterior | Equus asinus | Length unaffected | 43.7 | 5.8 | | L | 5 | Fused | Adult | | Slight | Fine fractures | | | | | | | |
| 155.4 S1 17 | Radius | Proximal | Equus asinus | | 74.1 | 9.3 | | L | 5 | Fused | Adult | | Slight | Fine fractures | Dry | | | | | | |
| 155.4 S1 17 | Ulna | Olecranon | Ovis aries | Length reduced | 47.1 | 4.5 | | L | 25 | Fused | Adult | | Very slight | Fine fractures | | | | SDO 25.9 | LO 39.9 | | |
| 155.4 S1 17 | Ulna | Shaft | Large Ungulate | Length reduced | 36.3 | 3.6 | | | 10 | | | | Very slight | Fine fractures | | | | | | | |
| 155.4 S1 17 | Femur | Posterior | Equus asinus | Length reduced | 81.3 | 25.7 | | R | 10 | Fused | Adult | | Slight | Eroded | Recent | | | | | | |
| 155.4 S1 17 | Tibia | Shaft | Small Ungulate | Length unaffected | 81.6 | 4.6 | | L | 50 | | | | Very slight | Spalling | Dry | | | | | | |
| 155.4 S1 17 | Radius | Shaft | Large Ungulate | | 97.4 | 16.5 | | R | 15 | | | | Slight | Fine fractures | Fresh and dry | | | | | | |
| 155.4 S1 22 | Femur | Posterior Shaft | Bos p. | Length reduced | 72.8 | 7.9 | | L | 15 | | | | | | Dry | | | | | | |
| 155.4 S1 27 | Astragalus | | Bovidae | Length unaffected | 68.2 | 35.7 | | L | 75 | | | | Moderate | Eroded | | | | | | | |
| 155.4 S1 27 | Humerus | Distal | Ovis aries | | 65.4 | 11.8 | | R | 25 | Fused | Adult | | Very slight | Eroded | Fresh and dry | | | | | | |
| 155.4 S1 27 | Metapodial | Anterior Shaft | Large Ungulate | Length reduced | 73.1 | 6.1 | | | 15 | | | | Slight | Fine fractures | Dry | Slight carnivore | | | | | |
| 155.4 S1 27 | Metapodial | Anterior Shaft | Small Ungulate | Length reduced | 23.8 | 0.1 | | | 5 | | | | Very slight | Spalling | Recent | | | | | | |
| 155.4 S1 27 | Lumbar Vertebra | Transverse Process | Medium Ungulate | Length reduced | 27.5 | 0.3 | | | 5 | | | | | | | | | | | | |

147

**TABLE 10.4.** SUMMARY OF THE ANIMAL BONE DATA (CONT.).

| Context | Element | Portion | Taxon | Recent Break | Length (mm) | Weight (g) | Articulates with | Side | %Complete | Age Criteria | Age | Sex | Weathering | Weathering Type | Breakage | Gnawing | Cut Marks | M1 | M2 | M3 | Comments |
|---|---|---|---|---|---|---|---|---|---|---|---|---|---|---|---|---|---|---|---|---|---|
| 155.4 S1 27 | Femur | Posterior Shaft | Small Ungulate | Length unaffected | 57.7 | 2.3 | | | 10 | | | | Slight | Spalling | Dry | | | | | | |
| 155.4 F 85 | Coprolite | | c.f. Camelus sp. | | 32.7 | 2.3 | | | 100 | | | | | | | | | | | | |
| 155.4 F 85 | Coprolite | | c.f. Camelus sp. | | 23.1 | 2.4 | | | 100 | | | | | | | | | | | | |
| 155.4 F 85 | Coprolite | | c.f. Camelus sp. | | 22.0 | 1.1 | | | 90 | | | | Slight | Eroded | | | | | | | |
| 159.2 T5 50 | Rib | Head, neck and body | Large Ungulate | Length reduced | 334.5 | 55.9 | | L | 75 | | | | | | | | | | | | |
| 159.2 T5 50 | Pelvis | | Bos sp. | Length reduced | 313.5 | 175.3 | | R | 75 | | | | Slight | Fine fractures, spalling | | | | | | | |
| 159.2 T5 50 | Scapula | Blade and spine | Very Large Ungulate | Length reduced | 211.6 | 146.4 | | R | 50 | Fused | Adult | | Moderate | Fine fractures | Dry | | | | | | |
| 159.2 T5 50 | Scapula | Glenoid and neck | Very Large Ungulate | | 85.4 | 30.7 | | | 15 | | | | | | Fresh and dry | | | | | | |
| 159.2 T5 50 | Scapula | Blade and spine | Very Large Ungulate | Length reduced | 69.7 | 17.3 | | | 5 | | | | Moderate | Fine fractures | Dry | | | | | | |
| 159.2 T5 50 | Humerus | Proximal | Large Ungulate | Length reduced | 33.1 | 4.2 | | R | 5 | Unfused | Sub-adult | | Very slight | Spalling | | Fresh Carnivore | | | | | |
| 159.2 T5 40 | Ilium | Wing | Large Ungulate | Length reduced | 53.9 | 6.2 | | L | 10 | | | | | | | | | | | | |

148

# 11. Environmental Material from the Begrawiya-Atbara Survey 1994

*Chris Stevens and Dorian Q Fuller*

## Introduction

Twelve samples were examined from the excavations, mainly taken through hand-collection.

### Desiccated Plant Remains

Two finds were made of desiccated plant remains. These were a single desiccated stone of desert date (*Balanites aegyptiaca*) (Plate 11.1) from Site 112.4, a shallow tumulus (112.4/T1.1/4). Desert date is native in Sudan, growing up to 10m tall and found in both savannah and open woodland, the fruit is edible and may be used medicinally, for cooking or as cosmetic oil (Wilson 1988, 48). It is a common find on Egyptian sites, e.g. within settlement middens (Fahmy *et al.* 2011), and has also been recovered from pottery vessels in tombs where it had been left as part of votive offerings (Lilyquist 2003, 12).

A single stone of date palm fruit (*Phoenix dactylifera*) (Plate 11.2) was recovered from Site 159, a Post-Meroitic tumulus (159.2/T4.1/89Or). Date palm is cultivated within Sudan and such finds are commonplace.

### Charcoal[1]

Charcoal was collected within two samples. That from 159.2/T7.1/61Or comprised small round wood and was around 5-10ml in volume. The other sample, from 155.4/S1/14Or had stem or branch wood charcoal and was around 20ml in volume. The charcoal fragments in both samples had been subjected to very high temperatures thereby altering some key diagnostic features so it was not possible to identify them to species level (only to genus level).

The charcoal fragments from Site 155.4 come from Context 46 in pit 2, a storage pit probably dating to the Christian or Islamic period, in the area under discussion, and were identified as *Acacia* sp., acacia.

The charcoal from Site BM159.2 Tumulus 7, Gabati, was associated with a male Christian burial, possibly dating to the 6th-8th century AD. The charcoal fragments were identified as *Acacia* sp., acacia.

Most African acacias are tolerant of high heat and air dryness, although they do require permanent or sub-permanent soil moisture. Consequently they may develop deep underground root systems as well as shallow ones. Acacias can often be found growing along rivers and watercourses. Many acacias have useful gums, as well as leaves and pods that may be used for animal fodder. The hard, heavy, high-quality wood of acacias is sought after as timber and for producing excellent charcoal.

### Animal Remains

A small fragment of bone was recovered from 159.2/T4.1/89Or, Site BM159 Tumulus 4, Gabati, but this material was not identifiable. This same sample however, produced a number of fragments and possible whole specimens of bird of prey pellets (Plates 11.3-11.6). Examination of the material under a low-powered stereo-binocular microscope has demonstrated that the pellets have a large number of fine animal hair fragments within them. However, none of the small animal bones which characterise owl pellets have been found within these samples. The absence of bones, and small size of the pellets might suggest that they are more likely to be from a raptor as they digest bones more thoroughly than owls. Kestrels for example, in particular common kestrels (*Falco tinnunculus*), are present in Sudan and mainly feed on small mammals. Corvids (e.g. crows, ravens) also produce pellets that often comprise large amounts of coarse grit, so might provide a further possibility.

### Shell Remains

A number of the samples contained shells and shell fragments. Whole specimens and fragments of the land-snail *Limicolaria* sp. (probably *Limicolaria cailliaudi* syn. *L. flammata*) were recovered from Site 159.2, a Post-Meroitic tumulus which provided two whole shells (Plate 11.7, from 159.2/T1/27Or in Unit 128) and broken shells of at least three individuals (159.2/T1/2Or, Unit 99). A further possible fragment came from Site 153 (Plate 11.8, 153/1 28Or). The snail is generally associated with more disturbed habitats and periods of higher rainfall than those present today in the region (Fuller and Smith 1998). However, whether the snails are contemporary with the post-Meroitic burials cannot be ascertained with any degree of certainty.

Also relatively frequent were fragments of freshwater mussel. The shells compare most favourably with *Nitia/Pleiodon* sp. or (Unionidae) or *Aspatharia/Chawahlbergi* sp. (Iridinidae), most probably *Nitia teretiuscula/Pleiodon ovatus* or *Aspatharia marnoi/A. chaiziana /Chawahlbergi wahlbergi/C. rubens*. The shells would appear to be comparable with those in the Iridinidae. Fragments of these shells were recovered from Site 153.8/S1 (Plate 11.8 153.8/S1/28Or), lying close to Wadi Gabati, with a more complete example including a hinge from 112.3/X19Or (Plate 11.9). Two larger fragments from the house site 155.4/S1 (155.4/S1/4Or and 155.4/S1/22Or) could be identified as *Unio elongatus*.[2]

A single fragment of freshwater shell with a more

---

[1] The charcoal was identified by Dr Caroline Cartwright (Senior Scientist, British Museum).

[2] Samples 155.4/S1/4Or, 155.4/S1/22Or and 155.4/S1/26Or were not separated from the bone samples, so these were identified by the faunal analyst, Dr J. Sanford.

roughly sculptured surface (Plate 11.10) was recovered from 155.4/S1/5Or. The shell most closely resembles the freshwater oyster (*Etheria elliptica* syn. *Aetherea ellipitca*) a common find within the Nile and other African rivers. The species is indicative of rapidly flowing, clear oxygenated water-bodies, attaching to rocks within them (Kröpelin 2007, 27).

Cowrie shells (*monetaria moneta/annulus*) were recovered from Site 101, tumulus 101.4/T2/23Or, Unit 72 (Plate 11.11) and from Site 159.2/T2/26a (Plate 8.7). The former site is Post-Meroitic to Christian in date, and the latter is Meroitic to Christian in date. Such shells have been previously recovered from Sudanese contexts ranging from the Neolithic (Geus 1984, 64, no. 14) to the medieval (Fuller and Edwards 2001). Whilst cowrie shells are likely to originate from the Red Sea coast such items are widely traded and used in Sudan for personal ornamentation (Manzo 2012; or at least in later periods as money see Şaul 2004).

One fragment of eggshell, from a bird (Aves), of unidentifiable species, was recovered from the house site 155.4 (sample 155.4/S1/26Or).

# BIBLIOGRAPHY

Abell, P. and P. Hoelzmann 2000. 'Holocene Paleoclimates in Northwestern Sudan: stable isotope studies on molluscs', *Global and Planetary Change* 26, 1-12.

Adams, W. Y. 1968. 'Invasion, Diffusion, Evolution?', *Antiquity* 42, 194-215.

Adams, N. K. 1986. 'Textiles at Qasr Ibrim: an Introductory Quantitative Study', *Wissenschaftliche Zeitschrift der Humboldt-Universität zu Berlin. Gesellschaftswissenschaftliche Reihe* 35, 21-26.

Adams, W. Y. 1977. *Nubia: Corridor to Africa*. Reprinted 1984. London and Princeton.

Adams, W. Y. 1986. *Ceramic Industries of Medieval Nubia*. Memoirs of the UNESCO Archaeological Survey of Sudanese Nubia 1. Lexington.

Adams, W. Y. 1996. *Qasr Ibrim. The Late Mediaeval Period*. Egypt Exploration Society, Fifty-Ninth Excavation Memoir. London.

Adams, W. Y. 1999. *Kulubnarti 3. The Cemeteries*. Sudan Archaeological Research Society Publication 4. London.

Addison, F. A. 1930. 'A Christian Site near Khartoum', *Sudan Notes and Records* 13, 285-288.

Addison, F. 1949. *Jebel Moya*. The Wellcome Excavations in the Sudan. Vols I and II. London and New York.

Allason-Jones, L. 1991. 'Small Objects from the Western End of Mound B' in Welsby and Daniels 1991, 126-162.

Allason-Jones, L. 1998. 'The Small Objects' in Welsby 1998a, 60-81.

Arkell, A. J. 1949. *Early Khartoum*. Oxford.

Arkell, A. J. 1953. *Shaheinab*. Oxford.

Aston, D. A. 1996. *Egyptian Pottery of the Late New Kingdom and Third Intermediate Period (Twelfth-Seventh Centuries BC): Tentative Footsteps in a Forbidding Terrain*. Studien zur Archäologie und Geschichte Altägyptens 13. Heidelberg.

Aufderheide, A. C. and C. Rodriguez-Martin 1998. *The Cambridge Encyclopedia of Human Paleopathology*. Cambridge.

Bass, W. M. 1987. *Human Osteology: A Laboratory and Field Manual*. Denver.

Bates, O. and D. Dunham 1927. 'Excavations at Gammai', *Harvard African Studies* 8, 1-122.

Beck, H. C. 1928. 'Classification and Nomenclature of Beads and Pendants', *Archaeologia* 27, 1-76.

Bergman, I. 1975. *Late Nubian Textiles*. The Scandinavian Joint Expedition to Sudanese Nubia: Vol. 8, Stockholm.

Berrocal, M. I. L., J. F. M. Morales and J. M. Martínez González 2007. 'Frequency of supernumerary teeth; An observational study of the frequency of supernumerary teeth in a population of 2000 patients', *Medicina Oral, Patologia Oral y Cirugia Bucal* 12, 34-38.

Bonnet, C. 1978. 'Les Fouilles Archéologiques de Kerma (Soudan)', 1977-1978. *Genava* 26, 107-127.

Bonnet, C. 1990. *Kerma, Royaume de Nubie*. Geneva.

Bourriau, J. D. and D. A. Aston 1985. 'The Pottery', in G.T. Martin, *The Tomb-Chapels of Paser and Ra'ia at Saqqâra*, 32-55. Egypt Exploration Society, Fifty-Second Excavation Memoir. London.

Brook, A. H. 1984. 'A unifying aetiological explanation for anomalies of human tooth number and size', *Archives of Oral Biology* 29, 373-378.

Brothwell, D. R. 1981. *Digging up Bones*. Ithica.

Brothwell, D. R. and S. Browne 1994. 'Pathology', in J. M. Lilley, G. Stroud, D. R. Brothwell and M. H. Williamson (eds), *The Jewish Burial Ground at Jewbury*. The Archaeology of York Vol. 12. York, 457-494.

Bruzek, J. 2002. 'A Method for Visual Determination of Sex, Using the Human Hip Bone', *American Journal of Physical Anthropology* 117, 157-168.

Buikstra, J. E. and D. H. Ubelaker 1994. *Standards for Data Collection from Human Skeletal Remains*. Arkansas Archaeological Survey, Research Series, No. 44. Fayetteville.

Bullock, P., N. Fedoroff, A. Jongerius, G. Stoops and T. Tursina 1985. *Handbook for Soil Thin Section Description*. Wolverhampton.

Caneva. I. (ed.) 1988a. *El Geili: the History of a Middle Nile Environment, 7000 B.C.-A.D. 1500*. Cambridge Monographs in African Archaeology 29. BAR International Series 424. Oxford.

Caneva, I. 1988b. 'The Cultural Equipment of the Early Neolithic Occupants of Geili', in Caneva 1988a, 65-147.

Caneva, I. 1988c. 'Late Neolithic to Recent Graves at Geili', in Caneva 1988a, 151-225.

Caneva, I., E. A. A. Garcea, A. Gautier and W. van Neer 1993. 'Pre-Pastoral Cultures along the Central Sudanese Nile', *Quaternaria Nova* 3, 177-252.

Caneva, I. and A. Gautier 1994. 'The Desert and the Nile: Sixth Millennium Pastoral Adaptations at Wadi el Kenger (Khartoum)', *Archéologie du Nil Moyen* 6, 65-92.

Caneva, I. and A. E. Marks 1990. 'More on the Shaqadud Pottery: Evidence for Saharo-Nilotic Connections during the 6th-4th Millennium B.C.', *Archéologie du Nil Moyen* 4, 11-36.

Chaix, L. and A. Grant 1993. 'Palaeoenvironment and Economy at Kerma Northern Sudan, during the 3rd Millennium BC: Archaeological and Botanical Evidence', in L. Krzyżaniak and M. Kobusiewicz (eds), *Environmental Change and Human Culture in the Nile Basin and Northern Africa until the Second Millenium BC*. Poznań, 399-404.

Ciodnicki, M. 1984. 'Pottery from the Neolithic Settlement at Kadero (Central Sudan)', in Krzyżaniak and Kobusiewicz 1984, 337-342.

Conway, D., E. Allison, R. Felstead and M. Goulden 2005. 'Rainfall Variability in East Africa: Implications for natural resources management and livelihood', *Philosophical Transactions of the Royal Society* 363, 49-54.

Crawford, O. G. S. and F. Addison 1951. *Abu Geili and Saqadi and Dar el Mek*. The Wellcome Excavations in the Sudan. Vol. III. Oxford.

Dunham, D. 1955. *Nuri. The Royal Cemeteries of Kush* II. Boston.

Dunham, D. 1957. *Royal Tombs at Meroë and Barkal. The Royal Cemeteries of Kush* IV. Boston.

Dunham, D. 1963. *The West and South Cemeteries at Meroë. The Royal Cemeteries of Kush* V. Boston.

Dzierzykray-Rogalski, T. and E. Prominska 1978. 'Tombeaux de Deux Dignitaires Chrétiens dans l'Église Cruciforme de Dongola', in J. Leclant and J. Vercoutter (eds), *Études Nubiennes: Colloque de Chantilly 2-6 Juin 1975*. Cairo, 91-93.

Edwards, D. N. 1989. *Archaeology and Settlement in Upper Nubia in the 1st Millennium A.D.* Cambridge Monographs in African Archaeology 36. BAR International Series 537. Oxford.

Edwards, D. N. 1991. 'Three Cemetery Sites on the Blue Nile', *Archéologie du Nil Moyen* 5, 41-64.

Edwards, D. N. 1995. 'A Meriotic settlement and cemetery at Kerdurma in the Third Cataract region, Northern Sudan', *Archéologie du Nil Moyen* 7, 37-51.

Edwards, D. N. 1996. *The Archaeology of the Meroitic State: New Perspectives on its Social and Political Organisation*. Cambridge Monographs in African Archaeology 38. BAR International Series 640. Oxford.

Edwards, D. N. 1998a. *Gabati. A Meroitic, Post-Meroitic and Medieval Cemetery in Central Sudan*. Vol. 1. Sudan Archaeological Research Society Publication 3. London.

Edwards, D. N. 1998b. 'Catalogue of Meroitic Graves', in Edwards 1998a, 13-60.

Edwards, D. N. 1998c. 'Catalogue of Post-Meroitic and Medieval Graves', in Edwards 1998a, 72-111.

Edwards, D. N. 1998d. 'The Post-Meroitic and Later Finds II: Post-Meroitic and Later Finds', in Edwards 1998a, 124-137.

Edwards, D. N. 1998e. 'Fabric 21', in Edwards 1998a, 183.

Edwards, D. N. 1998f. 'Post-Meroitic and Medieval Gabati', in Edwards 1998a, 202-209.

Edwards, D. N. 1998g. 'The Beads. Beads from Post-Meroitic and Medieval Graves', in Edwards 1998a, 225-227.

Eisa, K. A. and D. A. Welsby 1996. 'A Soba Ware Vessel from the Upper Blue Nile', *Beiträge zur Sudanforschung* 6, 133-136.

Emery, W. B. and L. P. Kirwan 1938. *The Royal Tombs of Ballana and Qustul*. Cairo.

Fahmy, A. G., R. Freidman and M. A. Fadl 2011. 'Economy and ecology of Predynastic Hierakonpolis, Egypt: Archaeobotanical evidence from a trash mound at HK11C', in A. G. Fahmy, S. Kahlheber and A. C. D'Andrea (eds), *Windows on the African Past: Current Approaches to African Archaeobotany*. Frankfurt, 91-117.

Farid, S. 1963. *Excavations at Ballana 1958-1959*. Cairo.

Filer, J. M. 1994. 'The SARS Survey from Bagrawiya to Atbara: the Contents of Five Tombs from Cemetery 159.2, Wadi Gabati'. *SARS Newsletter* 6, 25-28.

Filer, J. M. 1995. 'The SARS Excavations at Gabati, Central Sudan. 1994-1995 c: The Skeletal Remains', *SARS Newsletter* 8, 23-27.

Filer, J. M. 1997 'Ancient Egypt and Nubia as a Source of Information for Violent Cranial Injuries', in J. Carman (ed.), *Material Harm. Archaeological Studies of War and Violence*. Glasgow, 47-74.

Filer, J. 1998 'The Skeletal Remains', in Welsby 1998a, 213-233.

Firth, C. M. 1912. *The Archaeological Survey of Nubia. Report for 1908-1909*. Cairo.

French, P. G. 1986. 'Late Dynastic Pottery from the Vicinity of the South Tombs', in B. J. Kemp *Amarna Reports* III, 147-188. London.

Fuller, D. Q and D. N. Edwards 2001. 'Medieval plant economy in Middle Nubia: Preliminary Archaeobotanical evidence from Nauri', *Sudan & Nubia* 5, 97-103.

Fuller, D. Q and L. Smith 2004. 'The Prehistory of the Bayuda: New Evidence from the Wadi Muqaddam', in T. Kendall (ed.), *Nubian Studies 1998. Proceedings of the Ninth Conference of the International Society of Nubian Studies, August 21-26, 1998*. Boston, 265-281.

Garstang, J., A. H. Sayce and F. Ll. Griffith 1911. *Meroë, the City of the Ethiopians*. Oxford.

Gempeler, R. D. 1992. *Elephantine X: Die Keramik römischer bis früharabischer Zeit*. Deutsches Archäologisches Institut Abteilung Kairo, Archäologische Veröffentlichungen 43. Mainz am Rhein.

Gerharz, R. 1994. 'Jebel Moya', *Meroitica* 14. Berlin.

Geus, F. 1984. *Rescuing Sudan Ancient Cultures*. Khartoum.

Geus, F. 1986. 'Des Tombes Contemporaines du Néolithique de Khartoum à el Ghaba (Taragma)', in Krause (ed.), 1986, 67-70.

Geus, F. and P. Lenoble 1985. 'Evolution du Cimetière Méroïtique d'el Kadada. La Transition vers le Postméroïtique en Milieu Rural Méridional', in F. Geus and F. Thill (eds), *Mélanges offerts à Jean Vercoutter*. Paris 67-92.

Geus, F., F. Hinkel and P. Lenoble 1986. 'Investigations Postméroïtiques dans la Région de Shendi', in Krause (ed.), 1986, 81-88.

Godlewski, W. 1991a. 'The Fortifications of Old Dongola: Report on the 1990 Season', *Archéologie du Nil Moyen* 5, 103-128.

Godlewski, W. 1991b. *Coptic and Nubian Pottery. Part II* Occasional paper, National Museum in Warsaw 2.

Griffith, F. Ll. 1924. 'Oxford Excavations in Nubia XXX-XXXIII', *Liverpool Annals of Archaeology and Anthropology* 11, 141-180.

Hayes, J. W. 1972. *Late Roman Pottery*. London.

Hayes, J. W. 1980. *A Supplement to Late Roman Pottery*. London.

Hillson, S. 1996. *Dental Anthropology*. Cambridge.

Hillson, S. 2005. *Teeth*. Cambridge.

Hillson, S. 2014. *Tooth Development in Human Evolution and Bioarchaeology*. New York.

Jacquet-Gordon, H. and C. Bonnet 1972. 'Tombs of the Tanqasi Culture at Tabo', *Journal of the American Research Center in Egypt* 9, 77-83.

Jakobielski, S. 1975. 'Old Dongola 1972-1973', *Études et Travaux* 8, 349-360.

Judd, M. A. 2012. *Gabati. A Meroitic, Post-Meroitic and Medieval Cemetery in Central Sudan*. Vol. 2.

*The Physical Anthropology.* Sudan Archaeological Research Society Publication 20. London.

Kirwan, L. P. 1939. *The Oxford University Excavations at Firka.* Oxford.

Krause, M. 1986. *Nubische Studien. Tagungsakten der 5. Internationalen Konferenz der International Society for Nubian Studies, Heidelberg, 22.-25. September 1982.* Mainz am Rhein.

Kröpelin, S. 2009. 'Wadi Howar. Climate Change and human occupation in the Sudanese Desert during the past eleven thousand years', in P. G. Hopkins (ed.), *Kenana Handbook of Sudan.* Abingdon, 19-38.

Krzyżaniak, L. and M. Kobusiewicz 1984. *Origin and Early Development of Food-Producing Cultures in North-Eastern Africa*, Poznań.

Lenoble, P. 1987a. 'Trois Tombes de la Région de Méroé. La Clôture des Fouilles Historiques d'el Kadada en 1985 et 1986', *Archéologie du Nil Moyen* 2, 89-120.

Lenoble, P. 1987b. 'Quatre Tumulus sur Mille du Djebel Makbor A.M.S. NE-36-0/3-X-1', *Archéologie du Nil Moyen* 2, 207-247.

Lenoble, P. 1989. '"A New Type of Mound-Grave" (continued): le Tumulus à Enceinte d'Umm Makharoqa, près d'El Hobagi (A.M.S. NE-36-O/7-O-3)', *Archéologie du Nil Moyen* 3, 93-120.

Lenoble, P. 1991. 'Chiens de Païens: une Tombe Postpyramidale à Double Descenderie hors de Méroé', *Archéologie du Nil Moyen* 5, 167-184.

Lenoble, P. 1992. 'Documentation Tumulaire et Céramique entre 5$^e$ et 6$^e$ Cataractes: un Exemple de "Prospection Orientée" Visant à Renseigner la "Fin de Méroé" dans la Région de Méroé', in C. Bonnet (ed.), *Études Nubiennes, Conférence de Genève. Actes du VIIe Congrès International d'Études Nubiennes 3-8 Septembre 1990 Vol. I Communications Principales.* Geneva, 79-97.

Lenoble, P. 1995. 'La Petite Bouteille Noire, un Récipient Méroéen de la Libation Funéraire', *Archéologie du Nil Moyen* 7, 143-162.

Lenoble, P. 1999. 'The Division of the Meroitic Empire and the End of Pyramid Building in the 4th Century AD: an Introduction to Further Excavations of Imperial Mounds in the Sudan', in Welsby 1999, 157-198.

Lenoble, P. and N. M. Sharif 1992. 'Barbarians at the Gates? the Royal Mounds of El Hobagi and the End of Meroë', *Antiquity* 66, 626-635.

Lenoble, P., R.-P. Disseaux, A. Ali Mohammed, B. Ronce and J. Bialais 1994. 'La Fouille du Tumulus à Enceinte el Hobagi III. A.M.S. NE-36-0/7-N-3', *Meroitic Newsletter* 25, 53-88.

Lilyquist, C. 2003. *The Tomb of Three Foreign Wives of Tuthmosis III.* New Haven.

Lovejoy, C. O., R. S. Miendl, T. R. Pryzbeck and R. P. Mensforth 1985. 'Chronological metamorphosis of the auricular surface of the ilium: A new method for the determination of adult skeletal age at death', *American Journal of Physical Anthropology* 86, 15-28.

Mallinson, M. 1994. 'The SARS Survey from Bagrawiya to Atbara: the Excavations', *SARS Newsletter* 6, 18-25.

Mallinson, M. 1998. 'SARS Survey from Omdurman to Gabolab 1997: the Survey', *Sudan & Nubia* 2, 42-45.

Mallinson, M. D. S., L. M. V. Smith, S. Ikram, C. Le Quesne and P. Sheehan 1996. *Road Archaeology in the Middle Nile Valley* Volume I: *The SARS Survey from Begrawiya-Meroe to Atbara 1993.* Sudan Archaeological Research Society Publication 1. London.

Mann, R. W. and D. R. Hunt 2005. *Photographic Regional Atlas of Bone Disease: A guide to pathologic and normal variation in the human skeleton.* Illinois.

Manzo, A. 2012. 'From the sea to the deserts and back: New research in Eastern Sudan', *British Museum Studies in Ancient Egypt and Sudan* 18, 75-106.

Marks, A. E. 1991. 'The Stone Artifacts from Shaqadud Midden', in Marks and Abbas Mohammed-Ali 1991, 95-122.

Marks, A. E. and Abbas Mohammed-Ali 1991. *The Late Prehistory of the Eastern Sahel: the Mesolithic and Neolithic of Shaqadud, Sudan.* Dallas.

Meyer, C. 1992. *Glass from Quseir al-Qadim and the Indian Ocean Trade.* Studies in Ancient Oriental Civilization 53. Chicago.

Mills, A. J. 1982. *The Cemeteries of Qasr Ibrim: a Report of the Excavations Conducted by W.B. Emery in 1961.* Egypt Exploration Society, Fifty-First Excavation Memoir. London.

Mohammed-Ali, A. S. 1982. *The Neolithic Period in the Sudan, c. 6000-2500 B.C.* Cambridge Monographs in African Archaeology 6. BAR International Series 139. Oxford.

Mould, Q. 1998. 'Leather Sample Analysis', in Edwards 1998 a, 125-126.

Myśliwiec, K. 1987. *Keramik und Kleinfunde aus der Grabung im Tempel Sethos I. in Gurna.* Archäologische Veröffentlichungen 57. Mainz am Rhein.

Näser, C. 1999. 'Cemetery 214 at Abu Simbel North. Non-elite Burial Practices in Meroitic Lower Nubia', in Welsby 1999, 19-28.

Nordström, H.-Å. 1972 *Neolithic and A-Group sites.* The Scandinavian Joint Expedition to Sudanese Nubia: Vols 3:1 and 3:2, Stockholm.

Nowakowski, J. 1984. 'The Typology of Lithic Implements from the Neolithic Settlement at Kadero (Central Sudan)', in Krzyżaniak and Kobusiewicz 1984, 343-351.

Ortner, D. and W. Putschar 1985. *Identification of Pathological Conditions in Human Skeletal Remains.* Washington.

Peacock, D. P. S. 1977. 'Ceramics in Roman and Medieval Archaeology', in D. P. S. Peacock (ed.), *Pottery and Early Commerce: Characterization and Trade in Roman and Later Ceramics*, 21-33. London.

Pellicer, M. and M. Llongueras 1965. *Las Necropolis Meroiticas del Grupo 'X' y Cristianas de Nag-el-Arab (Argin, Sudan).* Memorias de la Mision Arqueologica Española en Nubia V. Madrid.

Phenice, T. W. 1969 'A Newly Developed Visual Method of Sexing the Os Pubis', *American Journal of Physical Anthropologists* 30, 297-302.

Phillips, J. 1997. 'Punt and Aksum: Egypt and the Horn of Africa', *Journal of African History* 38, 423-457.

Phillips, J. 2001. 'Pottery', in K. Grzymski and J. Anderson, *Hambukol Excavations 1986-1989.* Canadian Expedition to Nubia 1. Mississauga, 119-146.

Plumley, J. M. 1975. *The Scrolls of Bishop Timotheos: Two Documents from Medieval Nubia*. Egypt Exploration Society, Texts from Excavations. First Memoir. London.

Pluskota, K. 1990. 'Early Christian Pottery from Old Dongola', in W. Godlewski (ed.), *Coptic Studies: Acts of the 3rd International Congress of Coptic Studies, Warsaw, 20-25 August 1984*. Warsaw, 315-333.

Pluskota, K. 1991. 'Dongola: a Pottery Production Centre from the Early Christian Period', in Godlewski 1991b, 34-56.

Rice, P. M. 1987. *Pottery Analysis: a Sourcebook*. Chicago.

Robertson, R. 1991. 'The Late Neolithic Ceramics from Shaqadud Cave', in Marks and Abbas Mohammed-Ali 1991, 123-172.

Robertson, R. and A. E. Marks 1989. 'Shaqadud Cave: the Organization of the 3rd Mil. B.C. Ceramics', *Meroitica* 10, 515-534. Berlin.

Rodziewicz, M. and E. Dinkler 1972. 'Die Keramikfunde der Deutschen Nubienunternehmungen 1968/69', *Archäologischer Anzeiger* 1972: 4, 643-713.

Rose, P. J. 1998. 'The Pottery. Part I: the Meroitic Pottery', in Edwards 1998a, 142-177.

Rose, P. J. and L. Smith 1998. 'The Pottery: Introduction', in Edwards 1998a, 138-141.

Sadr, K., A. and A. Castiglioni 1995. 'Nubian Desert Archaeology: a Preliminary View'. *Archéologie du Nil Moyen* 7, 203-236.

Şaul, M. 2004. 'Money in Colonial Transition: Cowries and Francs in West Africa', *American Anthropologist* 106(1), 71-84.

Säve-Söderbergh, T. 1989. *Middle Nubian Sites*. The Scandinavian Joint Expedition to Sudanese Nubia: Vol. 4:1. Partille.

Säve-Söderbergh, T. and L. Troy 1991. *New Kingdom Pharaonic sites. The Finds and the Sites*. The Scandinavian Joint Expedition to Sudanese Nubia: Vols 5:2 and 5:3. Uppsala.

Säve-Söderbergh, T., with G. Englund and H.-Å. Nordström 1981. *Late Nubian Cemeteries*. The Scandinavian Joint Expedition to Sudanese Nubia: Vol. 6. Solna.

Schild, R., M. Chmielewska and H. Wieckowska 1968. 'The Arkinian and Shamarkian Industries', in Wendorf 1968, 651-767.

Scott, E. C. 1979. 'Dental wear scoring technique', *American Journal of Physical Anthropology* 51(2), 213-217.

Shiner, J. L. 1968. 'The Cataract Tradition', in Wendorf 1968, 535-629.

Shinnie, P. L. 1955. *Excavations at Soba*. Sudan Antiquities Service Occasional Papers 3. Khartoum.

Shinnie, P. L. 1996. *Ancient Nubia*. London.

Shinnie, P. L. and R. J. Bradley 1980. *The Capital of Kush 1: Meroe Excavations 1965-1972*. Meroitica 4. Berlin.

Shinnie, P. L., and H. N. Chittick 1961. *Ghazali: a Monastery in the Northern Sudan*. Sudan Antiquities Service Occasional Papers 5. Khartoum.

Smith B. H. 1984. 'Patters of Molar wear in Hunter-Gatherers and Agriculturalists', *American Journal of Physical Anthropology* 63, 39-56.

Smith, L. M. V. 1994. 'The SARS Survey from Bagrawiya to Atbara: the Complete Vessels from Cemetery 159.2, Wadi Gabati', *SARS Newsletter* 6, 28-33.

Smith, L. M. V. 1996. 'Report on the Pottery Collection', in M. D. S. Mallinson, L. M. V. Smith, S. Ikram, C. Le Quesne and P. Sheehan 1996, 165-207.

Smith L. M. V. 1998a. 'The Post-Meroitic and Later Finds I: Finds from the Test Excavations', in Edwards 1998a, 112-123.

Smith, L. M. V. 1998b. 'The Beads. Beads from Post-Meroitic and Medieval Graves (Test Excavations)', in Edwards 1998a, 228-233.

Smith, L. M. V. 1998c 'The Pottery II: The Post-Meroitic and Medieval Pottery', in Edwards 1998a, 178-193.

Smith, L. M. V. 1998d. 'SARS Survey from Omdurman to Gabolab 1997: Pottery and Small Finds', *Sudan & Nubia* 2, 45-52.

Stewart, T. D. 1979. *Essentials of Forensic Anthropology*. Springfield, Illinois.

Strouhal, E. 1991. 'Further Analysis of the Fine Handmade Pottery of Egyptian Nubia in 3rd-5th Century A.D.', in Godlewski 1991b, 3-9.

Tarekegn, A. 1997. *The Mortuary Practices of Aksumite Ethiopia with Particular Reference to the Gudit Stelae Field (GSF) site*. Unpublished Ph.D. Dissertation, University of Cambridge.

Taylor, S. 1998. 'The Textiles: Sample Descriptions', in Edwards 1998a, 236-240.

Trigger, B. G. 1967. *The Late Nubian Settlement at Arminna West*. Publications of the Pennsylvania-Yale Expedition to Egypt 2. New Haven and Philadelphia.

Trotter, M. 1970. 'Estimation of Stature from Intact Limb Bones', in T. D. Stewart (ed.), *Personal Identification in Mass Disasters*. Washington, 71-83.

Troy, L. 1991a. 'Other finds', in Säve-Söderbergh and Troy 1991, 51-181.

Troy, L. 1991b. 'The Cemetery at Fadrus (No. 185)', in Säve-Söderbergh and Troy 1991, 212-293.

Ubelaker, D. H. 1989. *Human Skeletal Remains: Excavation, Analysis, Interpretation* (2nd edition). Washington, DC.

van Biek, G. C. 1983. *Dental Morphology* (2nd edition). Bristol.

Vantini, G. 1975. *Oriental Sources Concerning Nubia*. Heidelberg and Warsaw.

Vogelsang-Eastwood, G. 1998. 'The Textiles', in Welsby 1998a, 177-182.

Waldron, T. and D. Antoine, 2002. 'Tortuosity or aneurysm? The palaeopathology of some abnormalities of the vertebral artery', *International Journal of Osteoarchaeology* 12(2), 79-88.

Waldron, T. 2009. *Palaeopathology*. Cambridge.

Welsby, D. A. 1991. 'Pottery from the Western End of Mound B', in Welsby and Daniels 1991, 165-245.

Welsby, D. A. 1996. *The Kingdom of Kush. The Napatan and Merotic Empires*. London.

Welsby, D. A. 1998a. *Soba II. Renewed Excavations within the Metropolis of the Kingdom of Alwa in Central Sudan*. London.

Welsby, D. A. 1998b. 'The Pottery' in Welsby 1998a, 87-174.

Welsby, D. A. 1998c. 'Discussion', in Welsby 1998a, 269-278.

Welsby, D. A. 1998d. 'The Excavations', in Welsby 1998a, 20-59.

Welsby, D. A. 1999. *Recent Research in Kushite History and Archaeology. Proceedings of the 8$^{th}$ International Conference for Meroitic Studies*. British Museum Occasional Paper 131. London.

Welsby, D. A. 2001. *Life on the Desert Edge: Seven Thousand Years of Settlement in the Northern Dongola Reach, Sudan*. Sudan Archaeological Research Society Publication 7. London.

Welsby, D. A. and C. M. Daniels 1991. *Soba. Archaeological Research at a Medieval Capital on the Blue Nile*. London.

Wendorf, F. 1968. *The Prehistory of Nubia: Vol. II*. Fort Burgwin Research Centre Publication 5. Dallas.

Wendrich, W. Z. 1999. *The World According to Basketry: an Ethnoarchaeological Interpretation of Basketry Production in Egypt*. CNWS Publication Series 83. Leiden.

Williams, B. B. 1991. *Noubadian X-Group Remains from Royal Complexes in Cemeteries Q and 219 and from Private Cemeteries Q, R, V, W, B, J and M at Qustul and Ballana*. University of Chicago Oriental Institute Nubian Expedition IX. Chicago.

Wilson, H. 1988. *Egyptian Food and Drink*. Princes Risborough.

Woolley, C. L. and D. Randall-MacIver 1910. *Karanòg. The Romano-Nubian Cemetery*. Eckley B. Coxe Jr. Expedition of Nubia Vol. IV, Philadelphia.

Yellin, J. W. 1982. 'Abaton-style Milk Libation at Meroe', *Meroitica* 6, 151-155.

الاثرية التي وجدت ضمن هذا المشروع والحفريات اللاحقة في كتابات جود (٢٠١٢)

في الفصل العاشر ( سانفورد واكرام) وصف للهياكل الحيوانية . اوضحت الدراسات الاولية علي العظام المكتشفة ان هذه المواد يمكن تبويبها وتقسيمها حسب الفئات او الاحجام فقط ، كما يمكن التوصل لاستنتاج عام عن فصائلها. حيث تم التعرف علي بقايا من العظام العلوية لامشاط بعير، وماعدا ذلك فقد تم التعرف علي اجزاء من العظام الطويلة لثدييات كبيرة ومتوسطة ، اعتبرت انها لجمال، و ماشية . اماعظام الثدييات متوسطة الحجم فلربما ان مصدرها الضان او الماعز. كما تم العثور علي عظام اخري والتي يمكن تمييزها بكل يسر ومعرفتها وفقا للسلاله، نوع الفصيلة والاسرة. اوتميزها حسب فئات الاحجام للثدييات ذات الحافر. وجدت دلائل للتاثر بعوامل الطقس في حوالي نصف العظام التي تم اكتشافها (٣٧ من ٧٤ البقايا تم تميزه) وتتراوح ما بين طفيف الي معتدل، كما وجدت ادلة للتعرية علي السطح خلال المسح خاصة في الكوم ٥ في السياق 155.4 و 159.2 . علي ان معظم هذه الكسور قد حدثت بعد الترتيب ، عندما كانت العظام جافة، كما وجدت الكسور الحديثة ايضا في السياق 155.4 مما يشيرالي قلة وجود القوارض اكلة اللحم في هذه السياقات، بينما وجدت الحيوانات المستانسة ، ولربما ان عملية تكسير العظام الحيوانية هذه قد احدثها الانسان نتيجة لتناولها كغذاء .

احتوي السياق ٩ بالموقع 112.3 X1 علي بقايا عظمية لاحد الثديات الصغيرة، عبارة عن عظم بقري مكتمل للرسغ الايمن، جزء من عظم القفص الصدري وعظم سمساني مكتمل لقدم، بالاضافة لاجزاء من الامشاط اليمني لماعز ( كابرا هيركس ،الماعز البري/ الوعل). كما احتوي السياق ٤ بالموقع 151.4 S1 علي ٢٧ كسارة من اسنان الابقار . وعلي بقايا من عظم القفص الصدري ، اجزاء احد امشاط الاطراف السفلي، ويقايا من العظم الذي يربط الفك الاسفل بالاعلي لاحد الثديات الصغيرة ؛ سته اجزاء من الفك الاسفل و اجزاء من ١٩ سن لاحد الثديات متوسطة الحجم ؛ اجزاء من عظم الزند لمرفق لاحد الثدييات الكبيرة ؛ اجزاء صغيرة من تجويف عظمة الفخذ من منطقة التقاء الاجزاء السفلي لعظام الحوض، لاحد الثدييات الضخمة ؛ عظم الفخذ وعظم الساق الايمن والايسر لقارض صغير لم يتم تحديده ؛ الجزء الاسفل الايمن لامشاط اليد وعظم الرسخ الثالث الايمن لحمار (فصيلة ايكوس اسينوس الافريقي) ، والجزء العلوي الايمن للعظام الطويلة، وكسارة المرفق من عظم الزند من جمل . السياق٥ احتوي علي كسارة من تجويف عظم الكتف الايسر وكسارة للمخروقة اليسري لاحد البقريات الصغيرة، كسارة عظم الساعد الامامي لغزال. السياق ١٣ احتوي علي عظم الساعد الايسر الامامي لكبش (اوفيس اريس، فصيلة من الاغنام) وعلي عظم العضد االيسر الخلفي لغزال. السياق ١٧ احتوي علي كسارة اضلاع من عظم القفص الصدري، وجزء من سلسلة فقرية، و جزء من عظم الساق الايسر لاحد الثديات الصغيرة ؛كسارة لاحد اضلاع القفص الصدري ، جزء من العظام العلوية للسلسلة الفقرية جزء من الفقرة الاولي للسلسلة الفقرية، وكسارة الجناح الايمن من العظم الحوضي، و الالتحام الغضروفي الامامي للحوض لحيوان غير محدد( اشبه بالحمار) ، وكسارة من عظم الزند وجزء صغير من الجزء الداخلي للساعدالايمن، بالاضافة الي كسارة من المخروقة اليمني ، الجزء الايسرالامامي للحوض والجزء الاسفل لعظم الفخذ الا يمن لحمار. فى السياق ٢٢ تم العثور على بعض من الجزء الاسفل الايسر لعظم الفخذ من احد البقريات. السياق ٢٧ إحتوي علي اجزاء صغيرة لقدم لاحد الثدييات الكبيرة، وعظم الكاحل الايمن لاحد البقريات، تعرض للتعرية ، و الجزء الخلفي الايمن لعضد غنم بالغ و يظهر به كسرجديد. تم العثور على اجزاء لعظام احد الثديات بالموقع 159.2 الكوم ٥ السياق ٤٠ عبارة عن عظم الحوض. اما السياق ٥٠ فقد احتوى على ثلاث اجزاء من عظم الكتف لاحد الثديات الضخمة (ربما من نفس الحيوان ) والضلع والعضد الامامي والحافر من احد الثديات الضخمة، لم يتم تحديد نوعه. وبذلك، فان غالبية البقايا الحيوانية من هذه السياقات تتالف من اجزاء من عظام الثديات الاليفة ، في غالبها اغنام (اوفيس اريس)، حمار(ايكوس اسينوس) ، ماعز (كابرا هيركس) وجمال (كاميلوس) بالاضافة الي احتمالية وجود الماشية او الابقار الوحشية الاخري (مدجنة) بالاضافة الي ذلك بقايا المجموعات البرية مثل الغزالان، واخرى غير معروفة التصنيف والتي تم اسنادها الي فصائل البقريات والثديات.

الفصل ١١ ( ستيفنس ، فولر وكارتوريت) يتناول بالتفصيل دراسة١٢ عينة من المواد العضوية المكتشفة خلال الحفريات . اثنتان منها مصنوعة من بقايا النبات المجفف. وهى عبارة عن ثمرمشابه بحجرمجفف من نخيل الصحراء (هجليج) ( صورة ١١،١) من الموقع 112.4، وجدت فى كوم منخفض (112.4/T1.1/4). ثمرمشابه بحجر من فاكهة شجرة النخيل(نخلة التمر) وجد بالموقع a ,159، فى كوم يؤرخ لفترة مابعد مروي (159.2/T4.1/89Or). تم جمع عينات من الفحم احداها من الموقع 159.2/T7.1/61Or وتتالف من خشبة دائرية صغيرة، بينما العينة الاخري من 155.4/ S1/14Or لها جزع او فرع خشبي فحمي. البقايا الفحمية من الموقع 155.4 جاءت من السياق ٤٦ من الحفرة ٢، وهي عبارة عن حفرة تخزين ربما تعود الي الفترة المسيحية او الاسلامية ، في هذه المنطقة ، بالموقع BM159 الكوم ٧ بقباتي وجدت عينات الفحم مع مدفن مسيحي لذكر ، يرجع الي الفترة من القرن السادس الي الثامن الميلادي .وقد تم التعرف عى بقايا الفحم من الموقعين علي انه من شجر السنط او الطلح. بقايا وربما كل عظام احد الجوارح تم العثور عليها فى 159.2/T4.1/89Or بالموقع BM159 ،الكوم ٧ بقباتي ( لوحة 11.3-11.6). و علي الارجح ان هذه العظام لرابتور مثل طائر العوسق (صقر العوسق) او الغراب مثل (الغداف والعقعق). كل العينات وبقايا القواقع البرية مثل ليموكولاريا ( ربما قوقع ليموكولاريا المخروطى اوفلاماتا ) تم اكتشافها بالموقع 159.2/T1 وهو عبارة عن كوم يورخ الي فترة ما بعد مروي ( صورة ١٧) ، والموقع ١٥٣ ( صورة ١١.٨ ، 153/1 28Or). كما تم العثور على بقايا محار نهري، اشبه بقواقع بلح البحر . وجدت هذه القواقع بالموقع 153.8/S1 ( صورة ١١.٨ 153.8/S1/28Or ) بالقرب من وادي قباتي، كما تم العثور على نماذج مكتملة بالموقع 112.3/X19Or( صورة ١١،٩) تم العثور على كسارتين كبيرتين من الموقع السكني (155.4/S1/4Or and 155.4/S1/22Or) 155.4/S1 يمكن تعريفها علي انها محار يونو النقاتس. قطعة واحدة من الاصنداف النهرية ( لوحة ١١،١٠). اما العينة 155.4/S1/5Or من الراجح انها تمثل احد الاصنداف النهرية. كما تم العثور على ودعة بالموقع ١٠١ ، الكوم 101.4/T2/23Or السياق ٢٧ ( لوحة ١١،١١) . ويؤرخ الموقع الى الفترة من مابعد مروي الي الفترة المسيحية. ومثل هذه الاصداف تم اكتشافها في السابق فى موقع مسيحية بالسودان. بينما الودع يرجح انه اتى من ساحل البحر الاحمر ، ومثل هذه العناصر تستخد بكثرة فى التجارة وفي السودان تستخدم بكثرة وعلى نطاق واسع في الزينة الشخصية ، او نقود في الفترات اللاحقة. كما تم اكتشاف جزء من بيضة نعام بموقع المنزل 155.4( العينة .155.4/S1/26Or).

تمت ترجمة هذا النص من اللغة الانجليزية الي العربية بواسطة بشارة مرتضي بشارة محمد بشارة.

2. يتم تغطية الجسد ، اما ببطانية او كفن
3. يتم تزويد المتوفي ببعض الانواع من الملابس و دائما ماتكون من الجلد وهي عبارة عن مئزر او تنورة.
4. تزويده علي الاقل بجرة اوجرة و سلطانية
5. وجود الزينة الشخصية

تحتوي مجموعة المدافن هذه علي عناصر يمكن اعتبارها كتعبير عن الاختلاف في الجندر( النوع)، الحالة الاجتماعية او الثروة و ربما ايضا العمر عند الوفاة. كما يمكن تفسيرها كدليل علي الطقوس الجنائزية التي تتضمن السقي والتطهير وكذلك استهلاك الطعام . كما ان الاختلاف في المدافن ربما يعطي بعض الدلائل في تغيير الطقوس الجنائزية وبشكل ضمني المعتقدات الدينية المصاحبة حتي خلال الفترة المتاخرة من ما بعد مروي.

الفصل التاسع (واينينق ) تضمن الخطوط العريضة لتحاليل العظام من خلال علم الاثار الحيوي، بدراسة المواد المكتشفة من الحفريات بين مروي القديمة وعطبرة . والجدير بالذكر ان كل من جويسي فيلر وفرانسيس ثورنتون قد اعدا مسودة لتقرير اولي بعد حفريات عام ١٩٩٤، وقد تم ادراج جزء منها في هذا التقرير النهائي. اما المعلومات الرئيسة المتعلقة بوضع المتوفي ، العمر والنوع فقد تم تضمينها في وصف المواقع ( الفصل الثاني).

هنالك القليل من العوامل التي يمكن مناقشتها حول مجموعه هياكل الافراد هذه . وهي لاتنحصر فقط في التساؤل عما ان كانوا يرجعون لمناطق وفترات تاريخية مختلفة، حيث انه تم فحص عينات قليلة منها تمثل حوالي ٣٥% وهي عبارة عن ١٤ هيكل فقط، كما ان حالتها كانت سيئة من حيث الحفظ بنسبة حوالي ( ٢٠% او اقل ). ويشير وضع المتوفي واتجاه الافراد في هذه المدافن الي التنوع في طقوس الدفن في ازمان تاريخية مختلفة. و بشكل خاص فان مدافن فترة مابعد مروي تاخذ الشكل المقرفص مع اختلاف اتجاه الدفن، بينما الغالب الاعم من مدافن الفترة المسيحية ياخد الوضع الممدد في اتجاه شرق غرب. كما يجب الاشارة الي ان المدافن 112.4/T2.1/3Or, 112.4/T6.1/12Or, 154.5/T1.1/3Or, 154.5/T1. و 170.1/T3 قد تم حفرها ايضا اثناء هذا المشروع ، غير ان الهياكل العظمية وجدت متكسرة وهشه ، لذلك لم يتم تصديرها الي المتحف البريطاني للدراسة مع الهياكل الاخري .

اما فيما يختص بعلم الامراض القديمة ، فقد ظهرت الاصابات في الهيكل 101.4/T1.2/38Or (الفترة المسيحية) وتتضمن الكسور السطحية في فقرات العمود الفقري ، وكسور غائرة في سطح المفصل الاسفل لعظم الساق وهشاشة العظام عند المفصل مع الكاحل . ويجب الاشارة الي ان هشاشة عظام الكاحل قليلة . وقد قال والدرون (٣٨،٢٠٠٩) انها تحدث اقل بتسع مرات من كسور عظم الفخذ والركبة . وعلي الارجح انها اصابة ثانوية جاءت نتيجة لكسور في الظهر والركبة. كذلك اظهرت دراسة هذه العينات وجود نمو عظم جديده في الاضلاع لثلاث اشخاص من المدافن T1.6/51,T2.1/39, T2.2/38OR جميعهم من الموقع 101.4 المورخ للفترة المسيحية . وهي تمثل نسبة انتشار تقدر بحوالي ٣٧% (٨/٣) اذا ما ناخذ بعين الاعتبار كل الهياكل التي لها مستوي حفظ يمكن من فحص الاضلاع . في بعض الاحيان يمكن ان يرتبط نمو عظام جديدة في الاضلاع بامراض محددة مثل السل ، لكن في هذه الحالة ، لم توجد دلائل اخري للسل يمكن ملاحظتها في هذه الهياكل مثل تضرر النخاع والحدب. لذلك يمكن ان نشير فقط الي ان نمو هذه العظام كان نتيجة لاصابات في الرئة او الانسجة المحيطة بغشاء الرئة او ربما كسور في منطقة الصدر .كذلك تمت ملاحظة بعض الادلة للاعاقة خلال فترة النمو . والتي ظهرت في نموزجين؛ نقص التنسج والتناقضات بين نمو الاسنان وتوقف نمو الهيكل العظمي. وبالطبع لاتنحصر كل اضطرابات الاعاقة في الاشخاص الذين توفوا اثناء فترة النمو . فالهيكل 170.1/T5.1/8OR ( الفترة التاريخية غيرة معروفة ) اظهر فجوة في الفك بسبب نقص التنسج. وتصيب هذه العاهات كل مناطق اللثة متمثلة في المينا المنتجة للخلايا وخلاياء المينا. علي ان هؤلاء الاشخاص توفوا بعد انتهاء مصفوفة المينا بوقت طويل ، لذلك اصبحت هذه العلة واضحة ومكشوفة ولوحظت في شكل تجاويف دائرية في الاسنان عندما وصل احتكاك الاسنان للمنطقة المتاثرة . ويتضح جليا ان هؤلاء الاشخاص قد عانوا فترة من سوء التغذية او المرض خلال فترة نموهم . اظهر الهيكل 102.1/T1.1/46OR (الفترة المسيحية) الاصابة بخراج كبير ومتقدم في الفك الاعلي . هذا الخراج علي الارجح انه كان بسبب تكسير او تقطيع قاد الي انكشاف اللب. كما تشير الحزوز التي وجدت علي بعض الاسنان الي انها ربما استخدمت ايضا كادوات ، مما ادي الي حدوث هذا التصدع والتكسر. والذي ادي بدوره الي حدوث الالتهاب و تقيح الخراج والذي تسبب في حدوث عدوي ثانوية ادت الي التهاب الفك، وتاكل عظم اللثة، ومن ثم انتاج عظام جديدة علي سطح اللثة . ربما يعزي المعدل العالي لتاكل الاسنان الي عمليات المضغ غير الطبيعية ، او نوع الطعام الذي يحتوي علي مادة عالية الخشونة. اشارت الدراسات السابقة للمجموعات السكانية المجاور لقباتي ( والتي غطت فترة مابعد مروي والمسيحية ) الي قلة ممارسة طب الاسنان بصورة عامة ؛ مع توفر انواع من الغذاء الذي يحتوي علي البروتين ومنتجات الالبان بالاضافة الي النباتات التي تحتوي علي الكربون٣ (C3). وربما ايضا تم استهلاك نبات الذرة الذي يحتوي علي كربون ٤ ( انظر جود ٦٥،٢٠١٢ ) وعلي مايبدو ان هؤلاء الاشخاص قد استهلكوا نفس نوع الغذاء الذي استخدمه سكان قباتي،غير ان هذه المكونات القليلة للثقافة الغذائية لاتمكن من الاستنباط الدقيق للربط بين علم امراض للاسنان والغذاء . امر اخر غير معتاد وهو ظهور الاسنان الزائدة في الهياكل 101.4/T3.1/40OR فترة مابعد مروي) .حيث لاحظ هيلسون(٢٠٠٥،٢٨١) ان الاسنان الزائدة ظهرت لدي نسبه قليلة من السكان، بالاضافة لذلك، فان الاسنان الزائدة في اغلب الاحيان توجدا لدي الذكور بينما هذا الهيكل لانثي (بروك١٩٨٤). اما هشاشة العظام فقد ظهرت لدي اربعة افراد يتراوح متوسط اعمارهم جميعا ما بين ٢٠-٣٠ سنة اثنان منهم ذكور؛الهيكل101.4/T2.1/39Orوالهيكل101.4/T1.2/38Or واثنان اناث؛ الهيكل 101.4/T3.1/40Or و الهيكل 102.1/T1.1/46Or. كل المواقع التي ظهرت بها حالات الاصابة بهشاشة العظام تم عرضها في الجدول 9.7. كما يجب الاشارة الي ان حالة هشاشة العظام التي وجدت في الكاحل الايسر وعظمة الركبة للهيكل 101.4/T1.2/38Or ربما هي اصابة ثانوية بسبب كسر في مكان ما في الجسم، ولمزيد من التفاصيل اطلع علي الوصف والنقاش عن الافراد.

الموقع 159.2 (صور رقم 2.32 ,2.28-29, 2.24-26) .

تم اكتشاف الموقع ١٥٩.٢ والذي اصبح جزء من جبانة ضخمة بقباتي والتي تم تنقيبها لاحقا بصورة كاملة، تمت مناقشة المكتشفات

٢٣ مجموعة رئسية وفقا للتوسع في خطة العمل لجمع المواد من السطح، وتحتوي كل مجموعة علي مجموعات فرعية اخري وفقا للمادة المصنوع منها، معاملة السطح و الزخرفة. تم توريخ هذه المجموعات استنادا علي مقارنتها بمواد مشابهة لها من مواقع اخري.

بصورة عامة فان المجموعات j-1a و b-16a ضمت المواد الاكثر انتشارا، والتي اظهرت ان غالبية هذه المواد تؤرخ للفترة المسيحية المتاخرة او الاسلامية المبكرة. المواد التي تم ارجاعها الي الفترة الاسلامية المبكرة وهي المجموعة g-2a و 23 شكلت اكثر من 3.5% فقط من جملة المعثورات، اما اكثر المواد انتشارا فقد كانت من المجموعة g-2a والمجموعة 23 حيث تم اكتشافها من الموقع 112.3و 112.4. ولم يتم ارجاع اي مواد بصورة مباشرة الي فترة مابعد مروي، فقط المجموعة ٣ و١٣ اعتبرت انها تعود للفترة من مابعد مروي الي الفترة المسيحية المبكرة. والمجموعات b-17a18 يرجح انها تعود للفترة المروية وحتي فترة مابعد مروي. وماعدا قباتي فان المعثورات التي ترجع لهذه الفترات وجدت بشكل اساسي في المواقع 155.4 والموقع 101.4، 118FS2 والموقع 153.8 بنسب مئوية قليلة وكذلك بالموقع 112.4. المجموعات b-17a18 والتي تؤرخ من الفترة المروية الي مابعد مروي وجدت في موقع واحد فقط هو 101.4. فيماعدا قباتي، فان المجموعة 4h تحتوي ايضا علي قطعة واحدة من موقع قرية بالقرب من BM165 تم ارجاعها لفترة مابعد مروي. اما شقف الفخار في المجموعات 5c و ١١ والتي ارجعت للفترة المصرية المتاخرة من الراجح انها موازية للفترة النبتية في التوريخ، وشكلت فقط حوالي 4% من جملة المواد. الشقف الفخارية المميزة وجدت في موقع واحد فقط هو 101.4. شقف المجموعة ٢١ يحتمل ايضا ان تكون نبتية وقد شكلت حوالي 1.2% فقط من جملة الشقف المميزة. تحتوي المجموعة 6b علي شقف فخارية تؤرخ الي نهايات العصر الحجري الحديث او الالف الثالث ق م . وتشكل حوالي 2% من جملة المعثورات. معظم الشقف المميزة من المجموعه 6b جاءت من الموقع 112.3.معثورات المجموعات 15a, 15b و d-22a وجد معظمها بالمواقع 101.4و 112.3 وهي مشابهه للمجموعات 6a و 6b من حيث العجينة ، وبالرغم من ان زينيتها ليست من الخواص المميزة لفخار الفترة المتاخرة من العصر الحجري الحديث، الا انها ارجعت مبدئيا لهذه الفترة . تم التعرف علي التبادل التجاري من خلال المجموعات 5c و ١١ التي احتوت علي العجينة الطينية الكلسية والتي يحتمل انها اتت من مصر، كما احتوت المجموعة 2d علي شقف مشابهة لفخار سوبا ومن المحتمل انه تم استيراده من سوبا شرق ، وايضا المجموعة 2e بالموقع 112.3 فان طرازها مشابه لطراز ادمز NIVA لذلك يرجح انه تم استيراده من النوبة السفلي .

في الفصل الخامس (ويندريش) وصف للسلال والحبال المكتشفة بقباتي . تم العثور علي الحبال في الاكوام T1،T5 و T12 حيث تمت صناعتها باستخدام ثلاث ضفائر معزولة وملفوفة . ربما استخدمت للمحافظة علي رباط الجلد المستخدم في نسيج عنقريب الدفن في هذه المدافن، اما بقايا السلال فقد اكتشفت بالكوم T4 . وقد اظهرت حرفية متقنة ، واوضحت حيثيات صناعة السلال والتي تتضمن الوسط الحلزوني الذي يشكل مركز قاعدة السلة ، ويوضح عناية صانعي السلال في الحياكة لتكوين شكل السلال والتي كانت في الاصل مغطاة بالجلد.كما تم العثور علي بقايا صغيرة لبرش من الكوم ٤، تمت حياكته وهو مضفور ومجدول. ونمط الحياكة يتم عن طريق وضع السعف فوق ٢ و اسفل ٢ .

في الفصل السادس والسابع ( ويلس و ثيكيث ) تم تقديم النتائج لدراسة ٣٤ عينة من الجلد من الاكوام T1 و T2, T4, T5 و T6 بقباتي ، تتضمن معلومات عن انواعها، وسائل الدباغة والصناعة لهذه القطع الجلدية ومصادرها بالاضافة الي عدد اجزائها ، العينات الرئيسة التي تم التعرف عليها اتت من حقيبة(؟)، سيورجلدية ، بقايا اجزاء محاكة ،وربما غمد ( من المدفن T1 )، ربما كساء لغطاء سلة ومنزر من المدفن T6 ويمكن تقسيم هذه العينات الجلدية الي العديد من الفئات:

أ) جلد رفيع ، سمكه حتي ١ ملم ، يحتمل انه من الماعز، جدي او ضان . استخدمت هذه المواد لصناعة الحقائب ، كساء للسلال ، ملابس ، الخ.

ب) اشرطه جلدية سميكة يصل سمكها حتي ٤ ملم والراجح انها من الماشية، استخدمت هذه الخيوط او الحبال المنسوجة لتجليد العنقريب .

ج) خيوط جلدية وحبال ، وتمت كل اعمال الحياكة اما باستخدام جلود رفيعة اواشرطة من الجلد غير المدبوغ ، مع وجود تنوع في تقنية تقطيبها.

الاختبارات التي اجريت علي ٣٢ من هذه العينات لمعرفة عناصر الدباغة، مثل تلك المستخلصة من الشب والخضار اوضحت بشكل عام ان جلد الماعز الرفيع يعطي نتائج ايجابية عند دبغه بالخضار ، بينما الخيوط الجلدية او السيور السميكة ، يرجح بانها كانت خام اوانها مدبوغة بشكل جزئي، كما انها دائما ماتعطي نتائج غير محددة ،اما الشب فقد اعطي نتائج ايجابية لسبع عينات اما باضافته الي الدباغة الخضرية او بديلا عنها.

الفصل ٨ ( سميث) تناول مجموعة من الاثاث الجنائزي لمدافن فترة مابعد مروي بقباتي .الاثاث الجنائزي الرئيسي الذي اعتمد عليه في التصنيف الرئيسي للمدافن هو: المدفن T1 ويحتوي علي سلطانية سوداء مفتوحة ، سلطانية صغيرة عمودية الجوانب ، رؤس سهام حديدية ، رؤس رماح حديدية او خناجر ونصال ، اطار سرير، قطع نسيج، بقايا حبال ومشغولات جلدية ؛ المدفن T2 ويحتوي علي قنينة زيت حمراء، سلطانية سوداء، اطار سرير، قطع نسيج ،مصنوعات جلدية ' حصيرة ، خرز وودع ؛ المدفن T4 ويحتوي علي قنينة زيت حمراء ملونة ، خلخال حديد ،خاتم من خام النحاس ، سكين من خام النحاس او ملعقة كحل، اطار سرير- قماش ، مصنوعات جلدية ، خرز ، قرع ، ثمار الدوم وسلتان صغيرتان ؛ المدفن T5 ويحتوي علي حامل جرار احمر ملون ، جرة بميذاب ، امفورة حمراء ، سلطانية حمراء بها عروة ، سلطانية حمراء اللون مصقولة، ملعقة من خام النحاس معقوفة ، مسمار ، مراة من النحاس ، قارورة كحل من العاج ، مشط من الخشب سلطانية من الحجر اوالطين ، اطار سرير، قماش ، بقايا حبل، مشغولات جلدية وخرز ؛ المدفن T6 ، ويحتوي علي مشغولات جلدية قماش (كفن اوبطانية ) تابوت خشبي (؟) وجد في رديم الحفرة .و علي الرغم من اختلاف هذه المكونات الا ان هنالك بعص العناصر المشتركة بينها وهي :

١. لايتم وضع الجسد كاملا او علي الاقل جزء منه علي الارض مباشرة وانما يوضع علي سرير او بطانية او حصيرة.

اثار الطريق في النيل الاوسط ، مستخلص المجلد الثاني

هذا المجلد هو الاصدارة الثانية لنتائج المسح الاثري لطريق التحدي الذي يربط بين الجيلي وعطبرة، تم اجراء هذه الاعمال الاثرية برعاية الهيئة العامة للاثار والمتاحف، وبتمويل من جمعية ابحاث الاثار السودانية بالمملكة المتحدة. تناول الفصل الاول والثاني (مالينسون ، فتحي خضر، ماكجينز، إكرام، فيلر وثورنتون) بالشرح الحفريات الاختبارية ونتائجها. في العام ١٩٩٣ تم اجراء مسوحات اولية للمنطقة بين البجراوية وعطبرة. ومن ثم تم اجراء حفريات اختبارية في الاعوام اللاحقة للمواقع الاثرية الاكثر تهديدا نتيجة لانشاء هذا الطريق . حيث تم تنقيب ثلاث مواقع ترتيبها علي النحو التالي ابتداء من مروي وهي؛ منطقة جبل ارداب (BM101-2)، الموقع 101.4, T1 يحتوي علي دفنات رئيسية تؤرخ لاواخر الفترة المروية ودفنات ثانوية تؤرخ لفترة مابعد مروي والفترة المسيحية، بينما وجدت العديد من الدفنات المسيحية في T2 وT3 وبالموقع 102.1/T1 بالقرب من جبل البيّض (BM112) الموقع ١٠١,٤ حيث تم تنفيب خمس مدافن ركامية صغيرة، تحتوي علي دفنات سيئة الحفظ تؤرخ لفترة مابعد مروي، او ربما الفترة المسيحية. كما تم العثور علي كوخ دائري يقع اسفل الوادي ويحتوي علي فخار مسيحي/ اسلامي مبكر وموقد نار. اما الموقع 112.3/X1 فهوعبارة عن منطقة ينتشر الفخار علي سطحها وفي غالبه مسيحي واسلامي مبكر، مع جود كميات قليلة لما يعتقد انها مواد تعود للعصر الحجري الحديث والقليل من بقايا ادوات الرحي . الموقع الثالث في منطقة وادي الدان (BM118) حيث تم تنقيب موقع معسكر ربما انه في الاصل منطقة عسكرية، ويتكون من ثلاث اكواخ دائرية ذات ارضية صلبة و مواقد، من الراجح انه يعود الي فترة مابعد مروي او الفترة المسيحية. كما تم تنقيب احد المدافن التلية ووجد خاويا . في الجبل الواقع بالقرب منه (118/FS2) تم العثور علي رسم لاقدام /احد الحجاج ونقش لإسمي بنتين يؤرخ للفترة الاسلامية.

تمت دراسة خمس مواقع بمنطقة قباتي، الثلاث الاولي منها تقع الي الجنوب من خورشنقريتي وهي عبارة عن؛ كوخ دائري (153.8/S1) يقع في بداية الوادي ويحتوي علي كميات مهولة من فخار المجموعة ١، موقد ومصد للرياح . كما تم تنقيب مدفنين اسفل احد الاكوم الركامية احداهما خاوي(154.5/T2.1) لايحتوي إلا علي ثلاث من البقايا العظمية المتكسرة ، والاخر 154.5/T1.1 عبارة عن مدفن باتجاه شرق غرب لايحتوي علي اثاث جنائزي، وهو لذكر تم دفنه في وضع ممدد وجهه للاسفل، الراس نحو الغرب، ربما تم اعدامه، ويحتوي المدفن علي هيكل اخر (154.5/T1.2) غير مكتمل مستلقي علي الظهر في اتجاه شرق غرب وينظر شمال. وهذه المدافن ربما تعود للفترة المسيحية. التنقيب الاخير في المنطقة كان لمنزل كبير (155.4/S1) شهد مرحلتين من البناء، المرحلة الاولي عبارة عن جدارين من الحجارة الكبيرة تحيط بمنطقة مساحتها 15 x 10 م، بها بوابة في الناحية الجنوبية الشرقية. اما الجزء الداخلي فيحتوي علي قواعد للسواري او الاعمدة وتجاويف متاخمة للجدار الجنوبي والشرقي. في المرحلة الثانية من البناء تم استخدام حجار خفيفة الوزن لبناء غرفتين في الناحية الشرقية، تحتوي احداهما علي موقد والاخري علي حفرتين للتخزين. اما ناحية الجنوب خارج المنزل فيوجد موقد كبير محاط بالحجارة.

الموقع قبل الاخير الذي تمت دراسته عباره عن مجموعة من خمس اكوام ركامية 170.1/T1-T5 تحتوي علي مدافن اطفال صغار او صبية، اجسادهم في وضع مقرفص، الراس ناحية الشرق او الشمال، ولم يتم تحديد نوعها ان كانت لذكر او انثي. اما العظام فكانت بحالة مزرية ويرجح انها تعود للفترة المروية اوفترة مابعد مروي. تم العثور علي فخار يمكن توريخه لفترة العصر الحجري الحديث في الموقع BM164 غير انه لم يتم الكشف عن مدافن او مباني مرتبطه به. تم اكتشاف ثلاث ادوات حجرية اسفل المنطقة BM166 وهي عبارة عن نصل وشظايا اولية. اما الموقع الاخير الذي تمت دراسته فهو جبانة قباتي، حيث تم تنقيب تسعة مدافن ركامية وهي : المدفن T1 (ذكر بالغ)، المدفن T2 (انثي بالغة ومعها رضيع)، المدفن T4 (ذكر يافع)، والمدفن T5 (انثي بالغة) ويحتوي علي سرير دفن يؤرخ لفترة مابعد مروي، واثاث جنائزي عبارة عن اواني فخارية، خرز، منسوجات، جلد ومواد معدنية. المدفن T3 يحتوي علي هيكل لذكر بالغ يعود للفترة المسيحية، بينما يحتوي المدفن T6 علي ذكر بالغ يعود لفترة مابعد مروي، تمت تغطيته بقطعة من نسيج ويرتدى رداء من الجلد، كما عثر علي اجزاء صغيرة ربما لسرير او تابوت في رديم حفرة الدفن. المدفن T7 لذكر بالغ يعود للفترة المسيحية. اما المدفن T11 المؤرخ للفترة المسيحية فقد تم حفره في ركن بقايا جدارين من الطوب اللبن وتم تفسيره علي انه هرم ومدفن بمقصورة جنائزية. اسفل الهرم يوجد مدفن مروي يحنوي علي بقايا هيكلين لبالغين وعلي الاقل ١٤ انية فخارية، تسعة منها اكتشفت اثناء موسم الحفريات الاختبارية. المدفن T13 يحتوي علي طفل من الفترة المسيحية وجد معه قرط معدني واجزاء من نسيج وخرز.

في الفصل الثالث (سميث) تم وصف المعثورات الصغيرة الرئيسة المكتشفة من مواقع اخري غير قباتي، حيث تم تقسيمها وفقا للمادة المصنوعة منها. اما الادوات الحجرية فقد وجد معظمها علي السطح خلال المسوحات الاثرية، حيث تشكل الشظايا نسبة تفوق النصف ولها بعض الخصائص المميزة علي الرغم من ان بعضها شظايا اولية او ثانوية، واغلبها من حجر الصوان او الكوارتز. كما وجدت اثار قطع لتجديد الشظايا علي احد الاحجار في شكل منصة بالموقع 113.3. وجزء من مكشط من الموقع 165.1، ازميلين من الموقع رقم ٣ في كل من 150.5 و 166.2. كما تم ملاحظة التحزيز في شظيتين الاولي من الموقع 168.1 وهي شظية ثانوية تمت اعادة شحذ شفرتها، والاخري عبارة عن قاطع حاد وجد في الموقع 168.1, 99.3 وايضا في 101.4. الشظايا اليفالوازية من الموقع 181.1 تم تصنيفها وفقا لمعاملة السطح. اما الرحي فقد تم تقسيمها الي ٢٠ نوع وفقا لشكلها العام ودرجة تجويفها او تحدبها وتقعرها وتشمل كل من المساحن والمطارق والمكاشط الحجرية، القليل من هذه الانماط يمكن ارجاعها لحضارة الخرطوم الميثوليثية، وغالبها لايمكن توريخها استنادا علي شكلها العام. تم العثور على اسورة من الحجر المصقول، دائرية الشكل وحوافها في شكل مثلث اقرب في توريخها الي الفترة المروية او المسيحية. اما العناصر المعدنية فتتالف من بقايا مواد معدنية متراكمة اوعناصر متعددة وجدت بالموقع 101.4، كما وجد خرز علي السطح اثناء المسح الاثري، وبخلاف ذلك فقد وجد الخرز من القاشاني، الزجاج او مزجج ومن قشرة بيض النعام. كما عثر علي جزء من نسيج من الصوف من الموقع 101.4 وغالبا ما يكون جزء من كفن يعود لفترة مابعد مروي او الفترة المسيحية.

في الفصل الرابع (سميث) وصف للفخار الذي عثر عليه في الحفريات الاختبارية باستثناء فخار منطقة قباتي، حيث تم تقسيم الفخار الي

١

*Plate 2.1.* Tumulus 101.4 T1 with T2 and T3 behind.

*Plate 2.2.* Tomb 400 excavated by Garstang at Meroe (© Garstang Museum of Archaeology, University of Liverpool, Neg. M.503 (307 M10)).

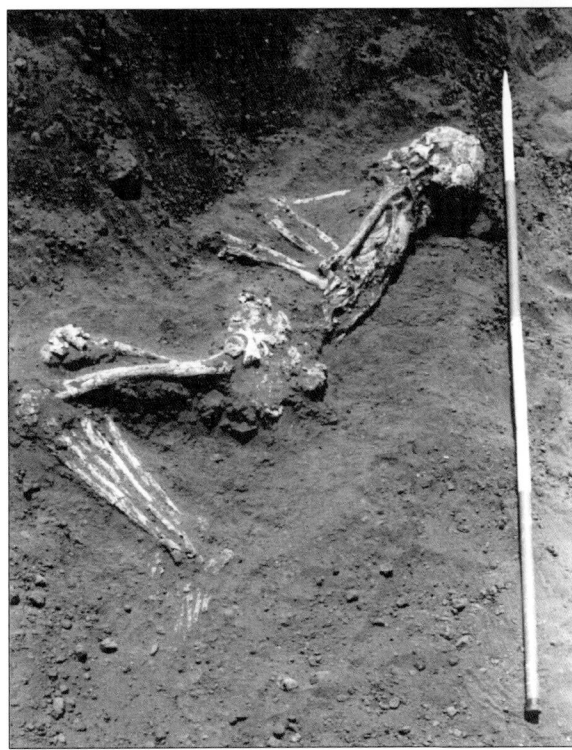

*Plate 2.3.* Burial 101.4/T1.3.

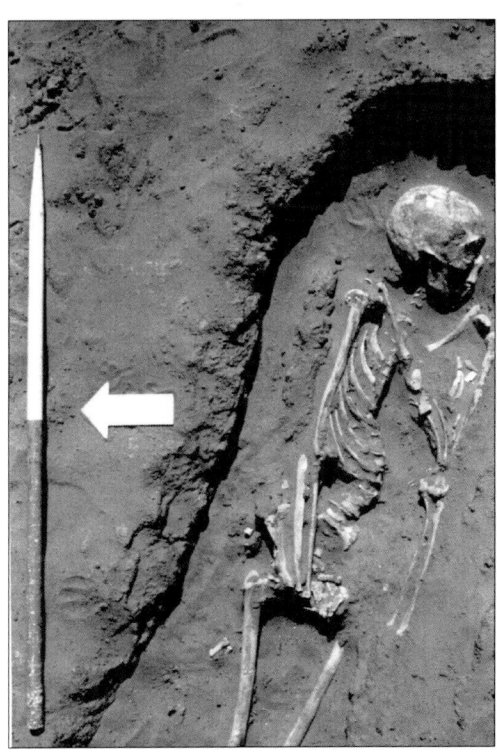

*Plate 2.4.* Burial 101.4/T1.4.

*Plate 2.5.* Burial 101.4/T.2.1.

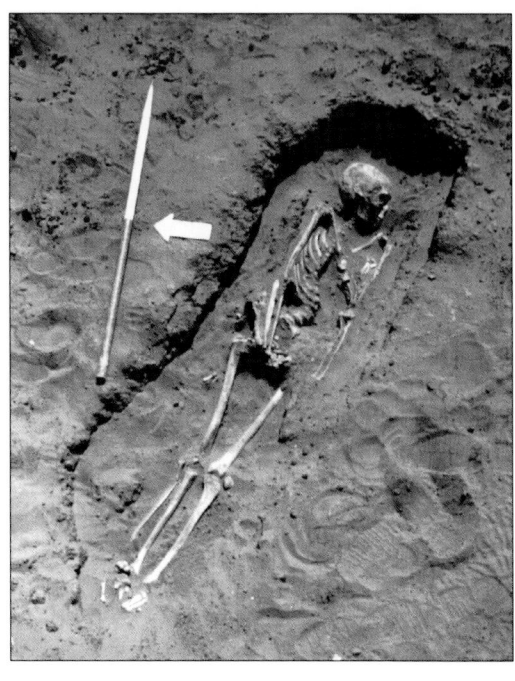

*Plate 2.6.* Burial 101.4/ T.3.

*Plate 2.7.* Burial 102.1/T1.1.

*Plate 2.8.* Tumulus 112.4/T1.

*Plate 2.9.* Burial 112.4/T3.1.

*Plate 2.10.* Site 118/FS Rock engraving.

*Plate 2.11.* Site 118/FS.3 Rock engravings.

*Plate 2.12.* Site 118/FS.2 Rock inscription.

*Plate 2.13.* Site 118/FS.1.

*Plate 2.14.* Tumulus 153.4/T1 excavation.

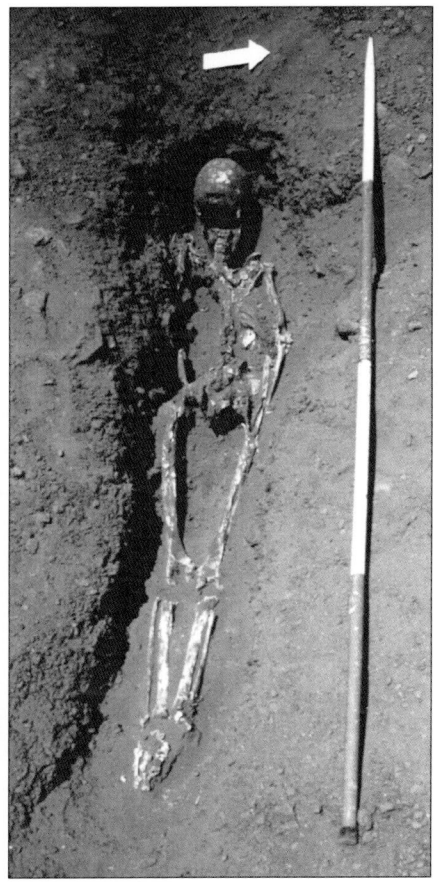

*Plate 2.15.* Burial 154.5/T1.

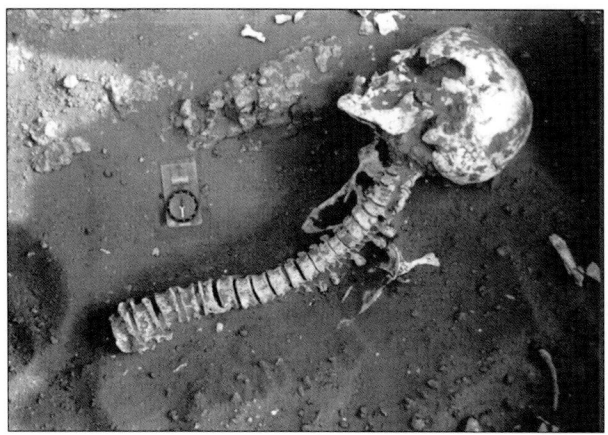

*Plate 2.16.* Burial 154.5/T1.2.

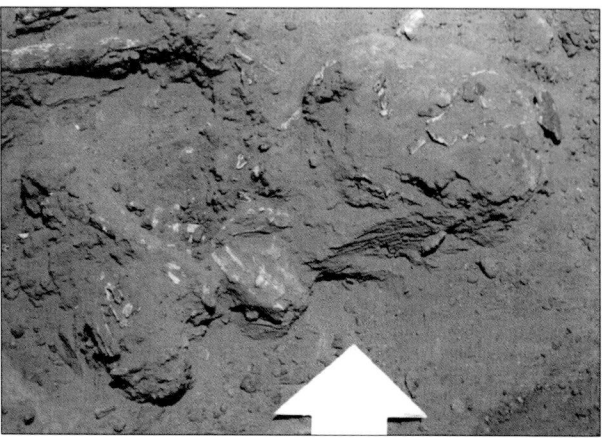

*Plate 2.17.* Burial 170.1/T5, showing fragmentary state of the skeletal material.

*Plate 2.18.* Gabati, site before excavation.

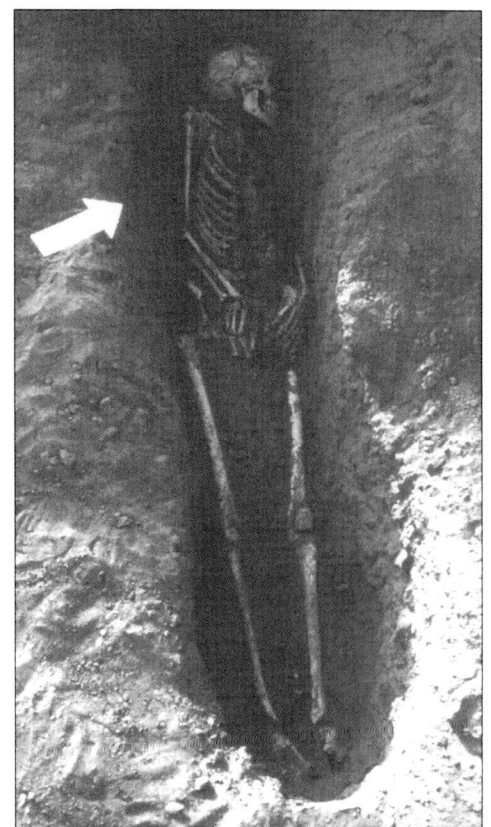

*Plate 2.19.* Burial 102.1/T1.1.

*Plate 2.20.* House 155.4 - Phase II.

*Plate 2.21.* House 155.4/S1 as cleared - Phase I.

*Plate 2.22.* House 155.4/S1 as cleared showing Phase I from east.

*Plate 2.23.* 118/FS.4 Rock engravings.

*Plate 2.24.* Burial 159.2/T4 showing detail of skeleton.

*Plate 2.25.* 159.2/T1 general view of burial.

*Plate 2.26.* Burial 159.2/T4.

**Plate 2.27.** 159.2/T4 Second layer of stones.

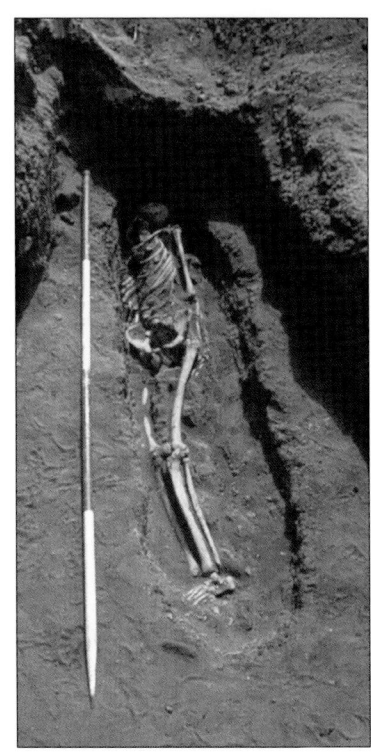

**Plate 2.28.** Burial 159.2/T7 Christian burial.

**Plate 2.29.** Burial 159.2/T5.

**Plate 2.30.** Burial 159.2/T5. Detail of ivory kohl pot as found.

**Plate 2.31.** Post-Meroitic pot-stand from Garstang's Meroe excavation (© Garstang Museum of Archaeology, University of Liverpool, Neg. M.500 (307 M10)).

**Plate 2.32.** Burial 159.2/T5. Detail of skeleton.

**Plate 2.33.** Burial 159.2/T5. Detail of pottery vessels in situ.

**Plate 2.34.** 159.2/T5. Shaft final blocking with T1 in the background.

**Plate 2.35.** 159.2/T6 Tomb chamber blocking.

**Plate 2.36.** Tumulus 159.2/T11 and T13. Jars at bottom of grave shaft.

**Plate 2.37.** 'Tumulus' 159.2/T11, general view of the superstructure.

**Plate 2.38.** Bed leg detail T1.

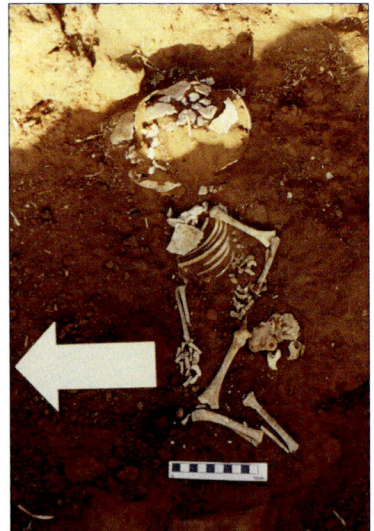

**Plate 2.40.** Burial 170.1/T2.1/3or.

**Plate 2.39.** 159.2/T11/102C. Detail of incised 'owner's mark' and Meroitic inscription.

**Plate 2.41.** Burial 170.1/T1.

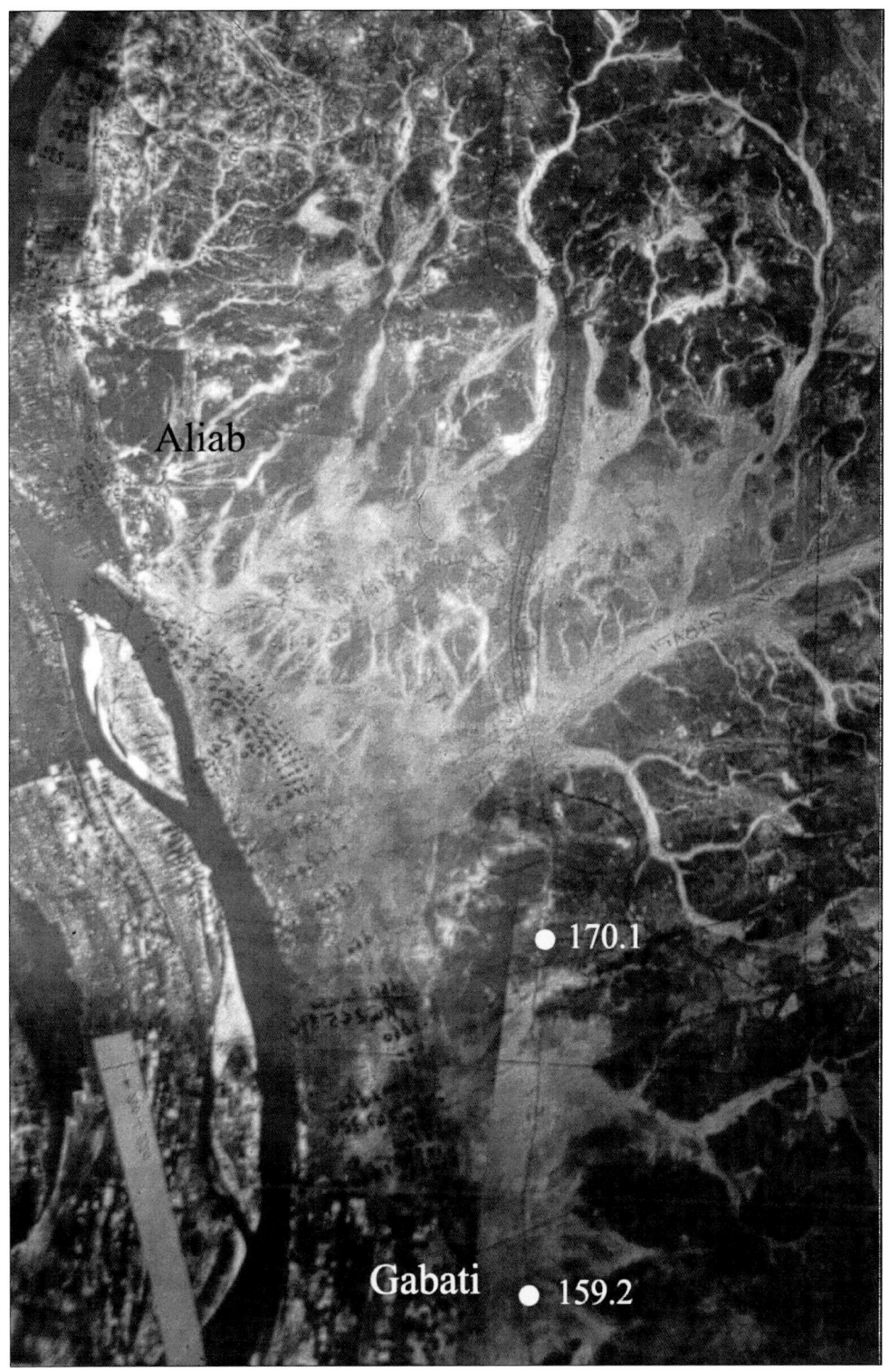

*Plate 2.42.* Aerial photograph of the Gabati/Aliab area.

a. T11/98C

b. T11/99C

**Plate 4.1.** Globular jars from 159.2/T11.

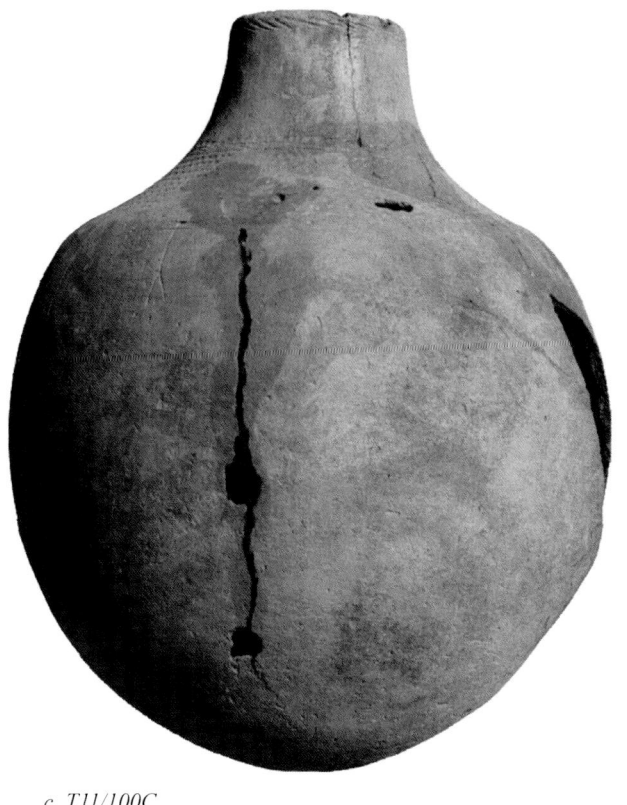

c. T11/100C

d. T11/101C

a. T11/102C

b. T11/103C

**Plate 4.2.** Globular jars from 159.2/T11.

c. T11/104C

d. T11/105C

***Plate 4.3.*** *Large bowl (Type 2:5.1) from Site 153.8. View showing lugs and incised decoration on the rim.*

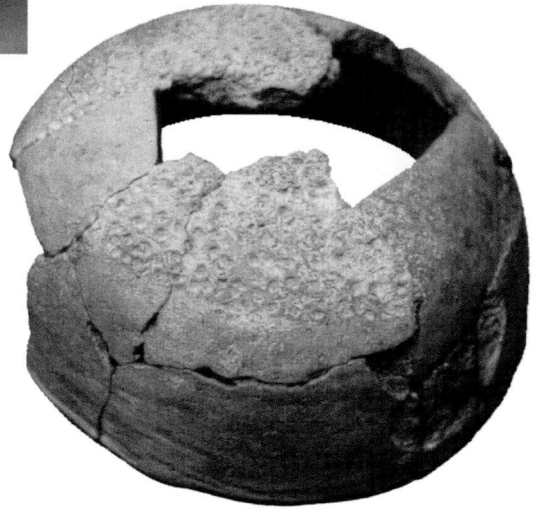

***Plate 4.4.*** *Bowl (as in Plate 4.3), showing impressed decoration on the exterior of the base.*

*Plate 8.1.* Broad open bowl 159.2/T1/41C.

*Plate 8.2.* Bowl with upright sides 159.2/T1/42C.

*Plate 8.3.* Leaf-shaped arrowhead 159.2/T1/18.

*Plate 8.4.* Knife blade or spear point 159.2/T1/24.

*Plate 8.5.* Imported oil bottle 159.2/T2/31C.

*Plate 8.6.* Examples of beads 159.2/T2/33.

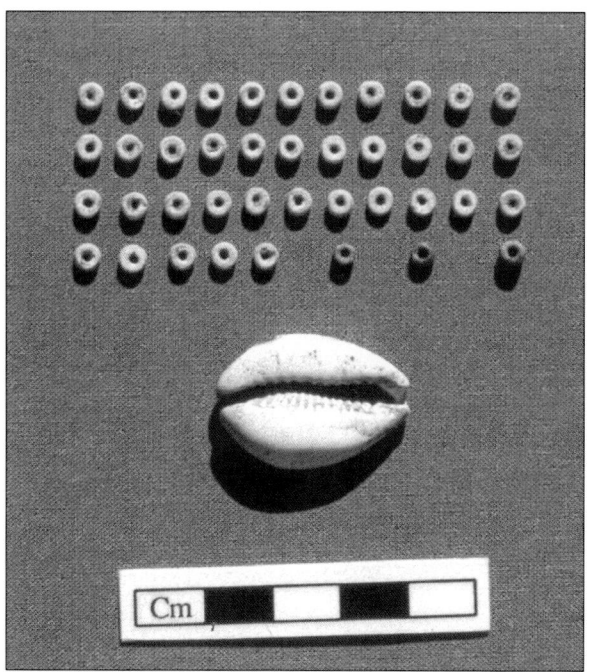

*Plate 8.7.* Worked cowrie shell and beads 159.2/T2/2b.

*Plate 8.8.* Painted imported oil bottle 159.2/T4/49C.

*Plate 8.9.* Spatula or kohl spoon 159.2/T4/12.

*Plate 8.10.* Examples of beads from 159.2/14.

*Plate 8.11.* Examples of beads from 159.2/T4.

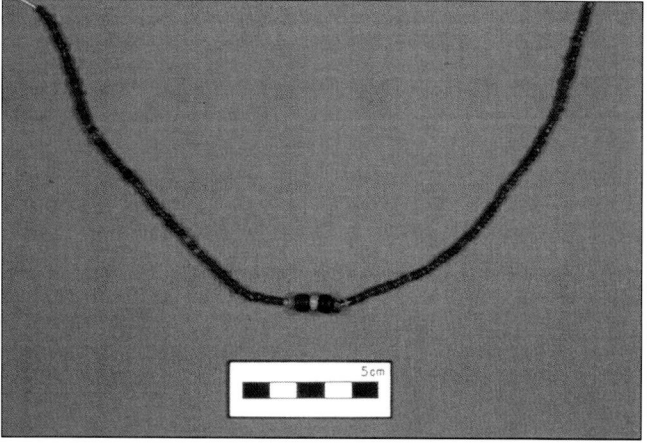

*Plate 8.12.* Examples of beads from 159.2/T4.

*Plate 8.13.* Painted pot-stand 159.2/T5/92C.

*Plate 8.14.* Painted bowl 159.2/T5/96C.

*Plate 8.15.* Ovoid bowl with lug 159.2/T5/95C.

*Plate 8.16.* Spouted jar 159.2/T5/93C.

*Plate 8.17.* Table amphora 159.2/T5/94C.

*Plate 8.18.* Circular mirror 159.2/T5/6.

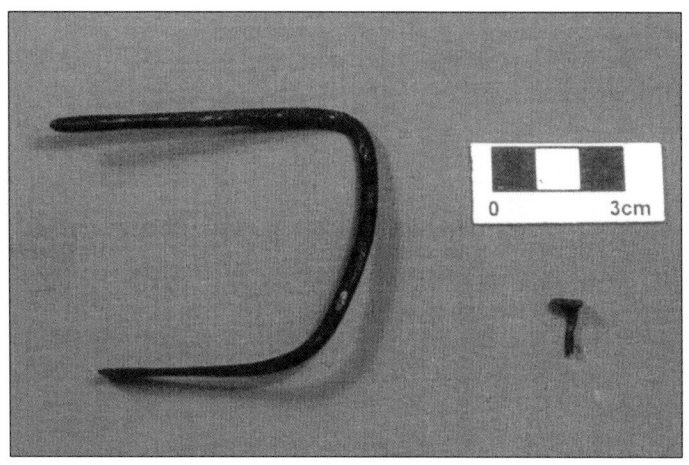

*Plate 8.19.* Spatula and nail 159.2/T5/3.

*Plate 8.20.* Stone bowl or mortar 159.2/T5/97S.

*Plate 8.21.* Ivory kohl pot 159.2/T5/5.

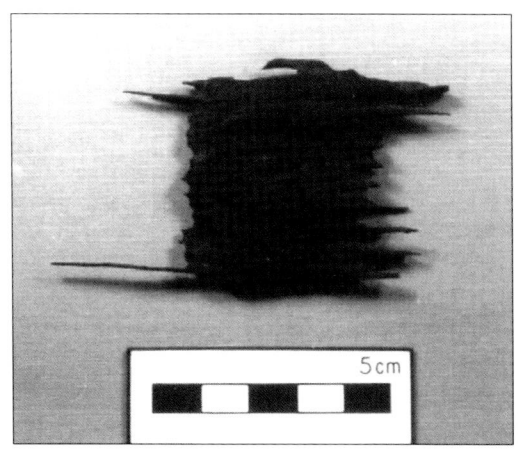

*Plate 8.22.* Wooden comb 159.2/T5/21.

*Plate 8.23.* Examples of beads from 159.2/T5.

*Plate 8.24.* Examples of beads from 159.2/T5.

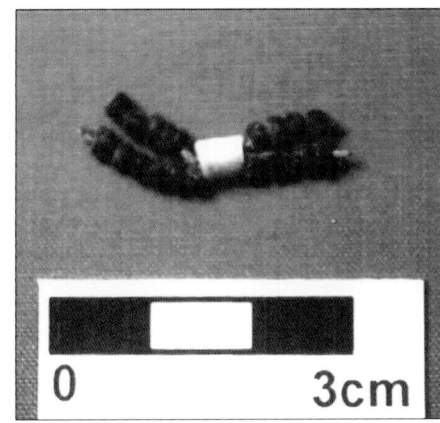

*Plate 8.25.* Part of necklace with original stringing preserved 159.2/T5/9.

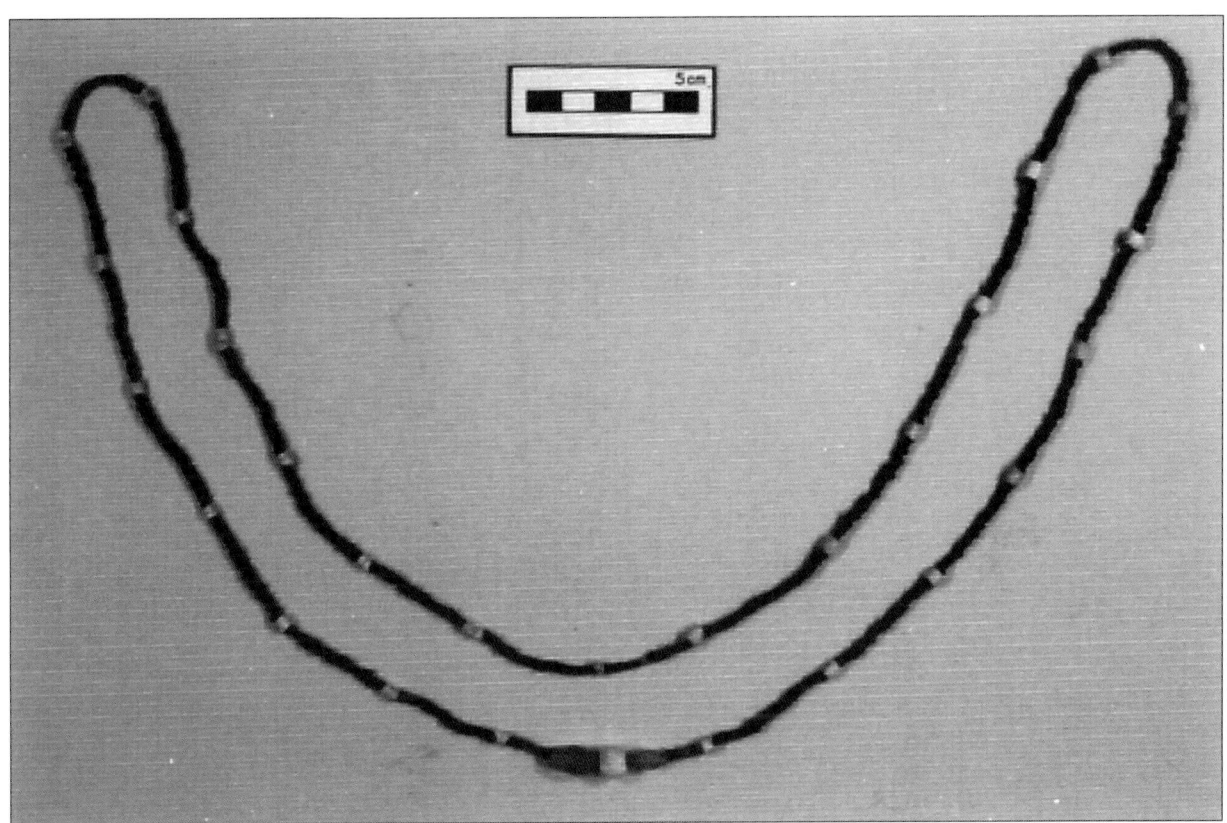

*Plate 8.26.* Necklace of beads 159.2/T5/2.

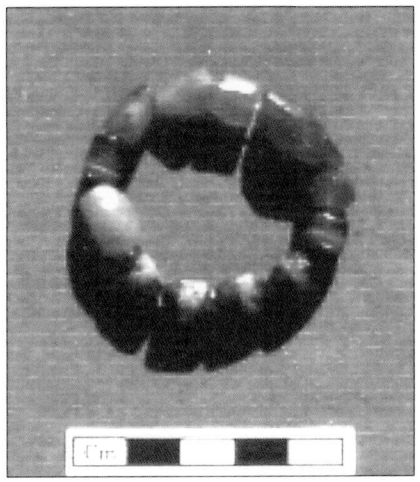

*Plate 8.27.* Beads from Christian burial, 159.2/T13.

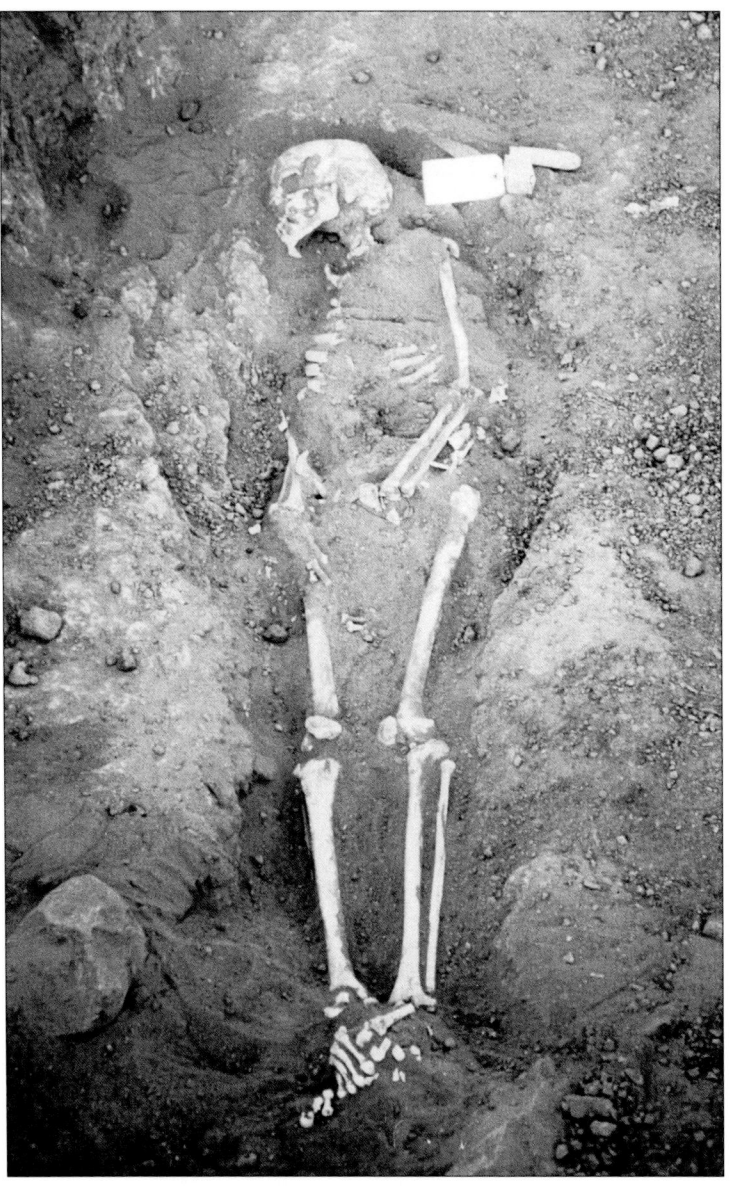

*Plate 9.1.* Jebel Ardeb, skeleton 101.4/T1.2/38Or during excavation (photo J. Filer).

*Plate 9.3.* Skeleton 101.4/T1.2/38Or, showing fusion of 5th and 6th cervical vertebrae.

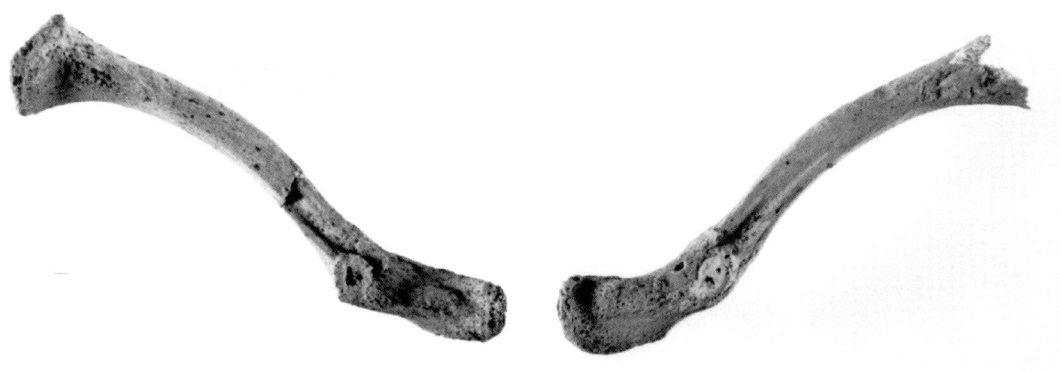

*Plate 9.2.* Skeleton 101.4/T1.2/38Or, both clavicles showing false joints.

*Plate 9.4.* Skeleton 101.4/T1.2/38Or, showing Schmorl's nodes and osteophytic growths.

*Plate 9.5.* Skeleton 101.4/T1.6/51Or, during excavation (photo M. Mallinson).

**Plate 11.1.** Desiccated stone of Balanites aegyptiaca from 112.4/T.1.1/1.

**Plate 11.2.** Date stone (Phoenix dactylifera) (dorsal and ventral views) from 159.2/T4.1/89Or.

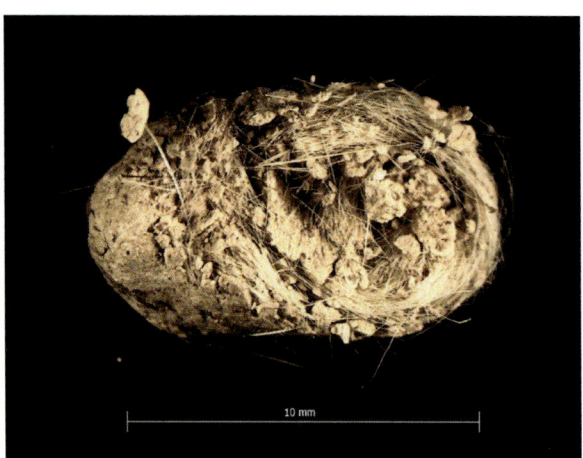

**Plate 11.3.** Possible bird of prey/carrion pellet from 159.2/T4.1/89Or.

**Plate 11.4.** Possible bird of prey/carrion pellet from 159.2/T4.1/89Or. Reverse and front view.

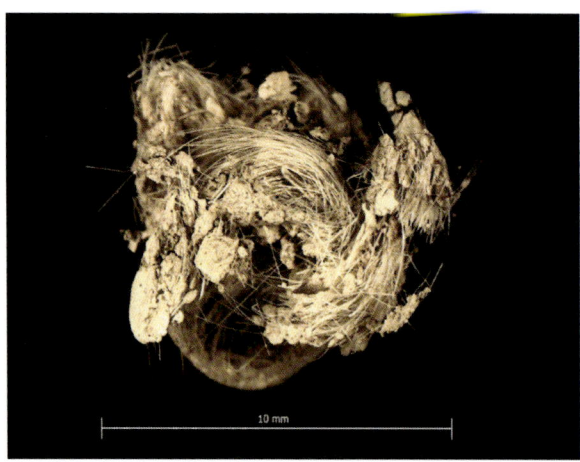

**Plate 11.5.** Fragment of possible bird of prey/carrion pellet from 159.2/T4.1/89Or.

*Plate 11.6.* Close-up of hairs within a possible bird of prey/carrion pellet from 159.2/T4.1/89Or.

*Plate 11.7.* Two shells of Limicolaria sp. from 159.2/T1/27 Or, Unit 128.

*Plate 11.8.* Two fragments of freshwater mussel and one of cf. Limicolaria sp. (left side). All from 153.8/S1/28Or.

*Plate 11.10.* Fragment of freshwater oyster shell, cf. Etheria elliptica from 155.4/51/5Or (dorsal and ventral views).

*Plate 11.9.* Fragment of river mussel shell (Unionidae/Iridinidae) from 112.3/9Or.

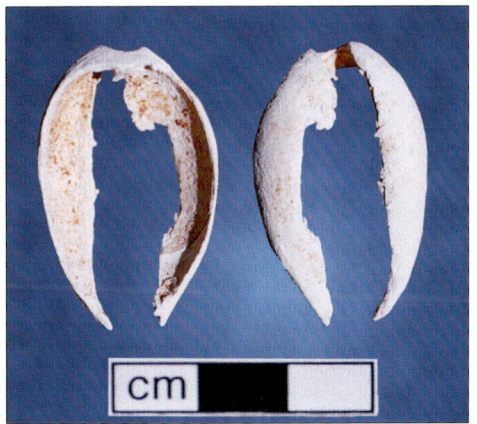

*Plate 11.11.* Cowrie shell Monetaria moneta/annulus from 101.4/T2/23 Or, Unit 72.